The GIANT Book of Amateur Radio Antennas

No. 802
$12.95

The GIANT Book of Amateur Radio Antennas

Editors of 73 Magazine

TAB BOOKS
BLUE RIDGE SUMMIT, PA. 17214

FIRST EDITION

FIRST PRINTING—APRIL 1979

Copyright © 1979 by TAB BOOKS

Printed in the United States of America

Reproduction or publication of the content in any manner, without express permission of the publisher, is prohibited. No liability is assumed with respect to the use of the information herein.

Library of Congress Cataloging in Publication Data

Main entry under title:

The Giant book of amateur radio antennas.

    Includes index.
    1. Radio—Antennas—Amateurs' manuals. 2. Amateur radio stations. I. Traister, Robert.
TK9956.G45   621.3841'35   78-21009
ISBN 0-8306-9802-7
ISBN 0-8306-8802-1 pbk.

Cover Courtesy of National Cable Television Association.

Preface

With *The GIANT Book of Amateur Radio Antennas* by your side, antennas should become the least of your worries while hamming...and possibly the most fun. Whether planning to build or buy, design or admire, or test or enjoy a ham antenna, you'll probably find what you need in these pages. With this book as a foundation, the new ham will breathe more easily, and the old-timer will awaken to novel and improved antennas.

To enlighten novices and re-educate more-experienced readers, Chapter 1 describes the basics of antenna theory and antenna testing. Antenna radiation is simplified, and reactance-versus-impedance is explained. There is an easy way to decibels, and also an easy way to design beam antennas: by computer.

Chapter 2 discusses non-directional antennas, ideal for traffic handling and round tables. Covering the hobby's mainstays—the balanced dipole and quarter-wave vertical—as well as the unorthodox—a short 160-meter antenna and an umbrella antenna—Chapter 2 has *omnidirectional* as its constant theme.

A rarity today is owning acres of land to provide for an antenna farm. College dormitories, apartments and small lots do not allow such a privilege, so Chapter 3 comes to the rescue. Abandoning the old theory that the *bigger* an antenna is, the *better* it is, In-Door and Limited Space Antennas includes a mini vee-beam, a small loop antenna, miniature antennas for 80 through 10 meters, and even an 18-inch all-band special.

Chapter 4 is particularly for those lucky few, the ones with acres of land, but any ham should feel attracted to Beams and Irrational Antennas. All sorts of directive aerials are covered, such as log periodics, take-apart beams, an 80-meter phased array, a diamond array, and a 10-meter beam for the beginning ham.

The Quad, one of the most controversial antennas in amateur radio, is the subject of Chapter 5. Quads are expensive, large, heavy, square antennas, right? Wrong! You'll find in Chapter 5 an inexpensive Quad, a mini-Quad, a lightweight version, one that has circular elements, and a 40-meter Quad. So set aside your checkbook, your barbell set and that old 40-meter half-wave dipole.

Especially when it comes to antennas, the real haven for amateur radio experimenters is vhf-uhf territory. Chapter 6 analyzes different vhf and uhf antennas from the commonplace 2-meter mobile whip to the unusual vertical J.

Chapter 7 covers antenna accessories, those inconspicuous components that permit the obvious ones to work. For without the tuners, matches, baluns, couplers, filters, rotators and watt meters that are thoroughly described in this chapter, many antennas would be virtually useless.

Imagine the possibilities! An apartment dweller with the best signal in the whole traffic net. The only member of a repeater club using a halo antenna. A double inverted vee for working DX. Your home shadowed by a large multi-band log periodic. The topic of any ragchew would be—would *have* to be—your antenna.

Or during the colder months when antenna experimentation is normally impractical, spend some time building your own antenna accessories. Homebrewing a rig just is not financially sound nowadays, but why not an accessory here and there? It isn't at all difficult to make an antenna balun, gamma match, or wide-range antenna tuner, and you can be proud that something sitting in your shack was soldered by you.

Of course, many of the concepts of antenna theory fly right over the heads of most of us. *The GIANT Book of Amateur Radio Antennas* is handy here, too, and is both a reference as well as a construction manual. So go ahead and look up your favorite antenna…or one you've heard of but know little about. Have fun!

Contents

1 Antenna Theory And Design 9
A Primer of Basic Antenna Theory 9
Antenna Radiation, Simplified 23
Measure Antenna Impedance With an SWR Bridge 32
Reactance vs. Impedance 36
The Half-Wavelength Feedline 43
Computer Design of Beam Antennas 48
Yes, I've Built 16 Log Periodic Antennas 55
An Easy Way to Decibels 86

2 Non-Directional Antennas 90
Vertical Antennas 90
A Multiband Ground Plane 99
Uses For the ⅝ths Wave Vertical 100
Simple Ground Plane Antenna 109
An Efficient 75 Meter Mobile Antenna 111
Hi-Q 80-40 Meter Vertical Antenna 121
The Umbrella Antenna 124
Multiband Vertical Antenna 129
Dipole Installation 135
The Double Inverted Vee 143
Balanced Dipole (Without a Balun) 145
Trapless Multiband Antennas 148
Ten Meter Folded Dipole Antenna 153
All Band, Rotatable Dipole 156
A Balanced Dipole Antenna 158
Old Antennas and New Baluns 162
A Simple, Short 160 Meter Antenna 168
Novel 160 Meter Antenna 171
Go-Go-Mobile 174
Whip Antenna Add-On 177
Quick Band-Change Mobile Antenna 183
Covering Additional Frequencies with Dipoles 190

3 In-Door and Limited-Space Antennas 196
Miniature Antennas for 80 Through 10 Meters 196
Apartment Dweller's Antenna System 202
Base-Tuned Center-Loaded Antenna 205
City Dweller's Multiband Antenna 212
The Mini Vee-Beam 215
An Efficient Indoor Antenna System 221
The Small Loop Antenna 227
The 18 Inch All-Band Antenna 238

4 Beams and Irrational Antennas..241
A Widespaced Beam 241
Two Element Beam Spaced a Quarter-Wavelength............247
Take-Apart Beam 251
Beginner's Beam For 10 Meters..............................254
Forty Meter Inverted Vee Beam 260
An 80 Meter Phased Array 261
Twinlead Phased Array 267
The Diamond Array..271
Multiband Log Periodics 274
Traingular Loop Beam For 7-28 MHz........................278
Long Wire Antenna 292

5 Quad Antennas..296
An Inexpensive Quad 296
The Miniquad 301
The Three Element Quad 305
A Light, Four-Element Quad 308
Long, Circular Quads 313
A Full-Size 7 MHz Quad 322
Giving the Quad a New Look 331

6 VHF-UHF Antennas..333
A 146 MHz Mobile Antenna................................333
Two Meter Mobile Halo 337
6 db on 450 MHz 340
A Tuneable Antenna For 432 MHz 343
Collinear-Gain Antenna For VHF-UHF Repeaters 347
Roof-Mounted VHF Whip Antennas 354
Two Meter Mobile Whip 364
The Two-Meter Groundplane as a Gain Antenna 368
The Two Dollar Amplifier 372
The Vertical J.. 375
Passive Reflectors For Amateurs 379

7 Accessories..391
VHF Dummy Load Wattmeter 391
A Rugged Rotator..395
Automatic Transmission Line Tuner 406
Remotely Tuned, Dual-Band Antenna Coupler 416
Transmitter Tuning of Mobile Antennas 423
Wide-Range Antenna Tuner................................429
An Easy, Wide-Band Balun.................................434
An Easy Gamma-Match Capacitor 436
Matching Stubs 439
A Durable Gamma-Match 444
A High Power Low Pass Filter 447

Appendix..451

Index..459

Chapter 1
Antenna Theory and Design

A PRIMER OF BASIC ANTENNA THEORY

Antennas are a very popular subject among radio amateurs today, as they have been almost from the beginning of radio science. In fact, of all the pieces of equipment an amateur owns the one he probably spends the most time talking about, both on and off the air, is his antenna. There are at least two good reasons for this. First, the antennas themselves are pretty simple, at least from the standpoint of circuitry. They contain no transistors or vacuum tubes or other amplifying, oscillating, modulating or detecting devices, the exact operations of which are difficult to comprehend. Antennas are usually simply pieces of wire or tubing with perhaps a transformer thrown in for impedance matching. Thus, they are at least easy to visualize.

The second reason for the great deal of attention antennas get is their performance. Every ham who has been around very long knows that there is no easier way to improve his station's capabilities than to improve his antenna. When a significant change for the better is made in the antenna, the improvement in the station's ability to communicate is immediately apparent, on both transmission and reception. Seven hundred dollars spent on an antenna installation will do worlds more good for the amateur than will a similar amount spent on a big "pair of shoes" for the exciter.

Also, antennas are popular because they are easier to homebrew than most pieces of equipment, and easier to make operate properly after their construction.

Thus, the antenna deserves its popularity. And if this is s, then the simple theory behind the antenna deserves to be understood by us all. So, the reason for this article. These few pages will discuss the very simple theory of antenna gain, efficiency, capture area and effective height. These are subjects of which very few amateurs seem to have a good grasp today. Much is heard about antenna "gain" especially, but few seem to have an exact understanding of what they are speaking about. Perhaps this article will help to clear up a little of the confusion.

Antenna Reciprocity

An antenna can basically be thought of as a device for converting energy from one form to another. When an electromagnetic (EM) wave strikes an antenna we find that electrical power is available at its terminals. On the other hand, when we apply electrical power to the terminals of the antenna, we find that an EM wave is radiated by the antenna. Thus, the antenna is capable of converting electrical energy to EM energy, and EM energy to electrical energy. This property of working both "backward and forward" is called "reciprocity" and is characteristic of antennas.

Now because the antenna is only a go-between for conversion of electrical to EM energy and vice-versa, a complete understanding of even basic antenna theory cannot be had without also obtaining some basic knowledge about EM fields, and electrical circuits as applied to antennas. Consequently, in this article we shall first discuss some things about EM fields before turning to antennas themselves. Later we will discuss basic antenna circuitry.

Electromagnetic Fields

Most of us realize that it is EM fields or "waves" that provide the invisible link between transmitting and receiving stations in a radio communications system. But the fact is that no one, even the professionals, knows just what an EM field is.

For several thousand years men have known that, after being rubbed, certain substances, amber for instance, would attract other bodies. This was a form of action at a distance, and came to be called "electric" attraction. Other substances, such as lodestone, could also attract matter at a distance, but did not require rubbing. This sort of attraction became known as "magnetic" to differentiate it from electric attraction from which it appeared to be different. These sorts of attraction remained a puzzle for thousands of years and to some degree are still puzzling. But thanks to the work of James Clerk Maxwell, engineers and physicists today have a mathematical grip, at least, on the elusive phenomenon of electromagnetism. Maxwell's theoretical work showed that there was a very definite relationship between electric and magnetic fields, and that they were really parts of the same natural phenomenon. Eventually he was able to express the relationship between the electric and magnetic fields as a set of two mathematical expressions which have come to be known as "Maxwell's Equations." These equations have been the foundation upon which most all mathematical EM theory has been built since.

The Traveling Wave

As far as radio systems are concerned the most significant feature of EM fields is their ability to move, that is, to transport energy from one place to another. These fields are called "traveling waves" and are composed of two components, an electric field component and a magnetic field component. If a person were able to stand in one place and watch an EM wave pass by he would see the energy in the wave alternately in the electric and magnetic form. Actually, the transformation of the energy from one form to the other is gradual (sinusoidal) and is complete only at an instant each cycle.

The speed with which the transformation of the energy takes place is known as the frequency of the wave. For example, if the transformation from electric to magnetic and back to electric (one complete cycle) takes place once every millionth of a second, the frequency of the wave is 1,000,000 cycles per second, or one megahertz.

Electromagnetic energy in the form of a traveling wave moves at a tremendous rate of speed, about 300,000,000 meters per second. The wavelength of a particular moving EM field is the distance the wave moves during one cycle. For our one megahertz wave it is 300,000,000 divided by 1,000,000 or 300 meters. This relationship between frequency and wavelength can be expressed algebraically:

(1) $C = f\lambda$,

where: C = speed of light
 = 300 million meters/second
 f = frequency, Hertz
 λ = wavelength, meters.

Power Density

If you drop a stone in a quiet pond, small waves will radiate away from the point where the stone struck the water. Each wave will have a circular form around the starting point, and will move away from the source. Thus we can say that the waves have a circular wavefront.

Now if we visualize a "point source" of EM energy situated in space it is easy to imagine that EM waves will radiate away from the source, and form a spherical wavefront. The surface area of this spherical wavefront will of course depend on its distance from the point source. From high school geometry we remember that:

$$S = 4\pi r^2,$$

where: S = surface area of sphere, meters (m^2)
 π = 3.14
 r = radius of sphere, meters.

Then if the point source were to radiate a certain amount of energy every second (i.e., power), the EM power would be distributed over an ever increasing surface area as a particular wave radiated away from the source. Notice that the total power is not diminished, it is just spread over a greater and greater area as we move further and further from the source.

The amount of electromagnetic power contained in a unit of surface area on the wave front is termed the "power density" of the EM wave, and depends on the distance from the source:

$$(2)\ D = \frac{P}{4\pi r^2}$$

where: D = power density, watts/m²
P = radiated power, watts.

Thus we see that as the wavefront moves away from the source, the power density decreases with the square of the distance from the source. This decrease of power density is termed "spherical divergence". It means that the EM wave is diverging or spreading out as it moves away from its origin.

As we can see from the above, a very convenient way of measuring the amplitude of an EM field is to measure its power density. Power density is today one of the most commonly used measures of EM field magnitude.

Field Strength

We mentioned above that the EM field is made up of two components—the electric field and the magnetic field. Since the EM field is capable of transmitting power from one place to another it seems reasonable that the power must be embodied in the two field components, and indeed this is the case. Thus if we were to increase the power level (and consequently the power density), there would necessarily be a corresponding increase in the strength of the electric and magnetic field components.

The mathematics which governs the relation between the power density, electric field strength and magnetic field strength is quite simple:

$$(3)\ D = EH,$$

where: E = electric field strength, volts/meter
H = magnetic field strength, amperes/meter.

This equation holds so long as the point of measurement is at least a few wavelengths from the transmitting source, which is a reasonable assumption in radio systems.

There is also a very simple relation between the electric and magnetic field strengths:

$$(4)\ E = Z_s H,$$

where: Z_s = intrinsic impedance of space
= 377 ohms.

You have probably noticed the similarity between Ohm's Law and Equation (4). Now if we combine Equations (3) and (4), we obtain two other simple expressions for EM power density:

(5) $D = \dfrac{E^2}{377}$, and

(6) $D = 377H^2$.

Summary

We have now discussed the very basic elements of EM theory necessary to an understanding of antennas. The concepts of point sources, electric and magnetic fields, frequency, wavelength, power density and field strength have been explained. Power density and electric and magnetic field strength are particularly important to understand, because they are what we measure in order to determine the amplitude of an EM field. Making a measurement of either D, E or H is equivalent to specifying all three, because they are all simply related by Equations (3) to (6). We are now ready to push on to antennas themselves.

We mentioned before that an antenna was a reciprocal device—that it worked both forward and backward. This means that the theory for both transmitting and receiving antennas must be identical. Thus it is only necessary to discuss one type, either transmitting or receiving, and the conclusions will be found to hold for both. For our purpose, we will discuss the antenna from the transmitting point of view, because it is easiest to visualize. Later we will have some comments about the receiving antenna.

Antenna Equivalent Circuit

When an antenna is attached to the output terminals of a transmitter, power flows from the transmitter output to the antenna and is radiated electromagnetically. The antenna loads the transmitter, the same as an impedance would. Thus the antenna could be represented, as far as the transmitter is concerned, by an impedance. Almost always the antenna is tuned to resonance (or nearly so) at the operating frequency,

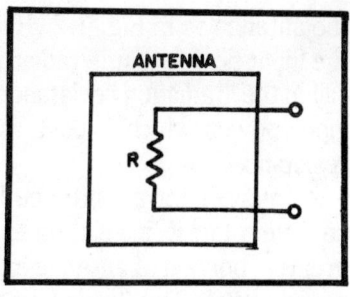

Fig. 1-1. Simplest antenna equivalent circuit.

so that the impedance it presents to the transmitter is purely resistive. Then the input impedance of the antenna is purely resistive, and the antenna could be represented by the simple equivalent circuit of Fig. 1-1.

In Fig. 1-1, R is a resistance equal in value to the antenna input impedance. The transmitter knows not whether an antenna of input impedance R, or a simple resistor of R ohms is connected to its output terminals.

Radiation Efficiency

Actually in a practical antenna not all of its input power is radiated as EM signal energy. Some is also radiated as heat, due to losses in the antenna structure. (Heat is just another form of EM energy. It is "incoherent" and at an extremely high frequency.) Losses in an antenna structure stem from several sources: dielectric losses in supporting insulators, resistive losses in the antenna system conductors, losses due to leakage currents over insulators, losses due to currents induced in nearby conductors and the ground, and corona loss. These losses can be minimized by proper design and location of the antenna.

In order to represent the splitting of the antenna input power into two parts, we split R into two parts and represent

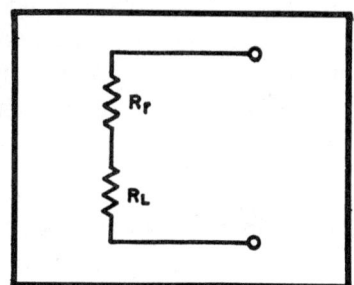

Fig. 1-2. Antenna equivalent circuit, including effect of losses. R_r is the radiation resistance, R_l the loss resistance.

the antenna as in Fig. 1-2. Here R_r represents the portion of the input power that is radiated as useful signal power, and is called the "radiation resistance." R_l represents that portion of input power which is lost as heat, and is called the "loss resistance."

In order for an antenna to be a good radiator, it should have very low R_l in relation to R_r, or in other words, by far the greatest portion of antenna input power should be radiated as useful signal. The "radiation efficiency" of an antenna is expressed mathematically like this:

$$(7) \; \eta = \frac{R_r + R_l}{R_r}$$

where: η = radiation efficiency
R_r = radiation resistance, ohms
R_l = loss resistance, ohms

If we desire to maximize this radiation efficiency, we want R_l to be small in comparison to R_r. In the extreme case of a perfect radiator, R_l would be zero and then η would be equal to 1, or 100%. When R_l increases from zero the efficiency drops, and if R_l should equal R_r, for example, then η would be only ½, or 50%.

The Isotropic Antenna

An isotropic antenna is one which radiates equally well in all directions. It is similar to the point source which we discussed before. Such an antenna is a convenient reference to use for measuring the "gain" of another antenna, although in reality there is no antenna which is truly isotropic.

Antenna Gain

When discussing the "gain" of an antenna, it should be emphasized that there are two kinds of gain—power gain and directive gain (sometimes called directivity). The two are related by a simple expression which includes radiation efficiency:

$$(8) \; G_p = \eta \, G_d,$$

where: G_p = antenna power gain
G_d = antenna directive gain.

Suppose, for example, that an antenna with a radiation efficiency of 50% had a directive gain of 4 (6 db). Using Equation (8) then, power gain would be ½ times 4, or 2 (3 db). But what do these terms power gain and directive gain mean?

Any antenna which is not isotropic has some directive gain. The directive gain is only a measure of the ability of the antenna to radiate in one direction to the exclusion of others. In other words, directive gain is a measure of the *shape* of the "radiation pattern" of an antenna, as compared with the circular shape of the radiation pattern of an isotropic antenna.

Power gain, on the other hand, includes not only the shape of the radiation pattern, but also the size of the pattern, or, in other words, it measures how effectively the antenna radiates in a particular direction.

An example should serve to make the above concepts clear. In Fig. 1-3 we have the radiation patterns (in two dimensions) of three antennas—a 100% efficient isotropic antenna (antenna A), and two directive arrays, each with a directive gain of 4. One of the directive arrays, antenna B, has a radiation efficiency of 100%, and the other, antenna C, an η of only 25%. Notice that the *shape* of the patterns of antennas B and C is the same, only the size is different. Also note that in the favored direction, antenna C is no more effective a radiator than the isotropic. Nonetheless it has directive gain, due to its ability to radiate better in the favored direction than in others.

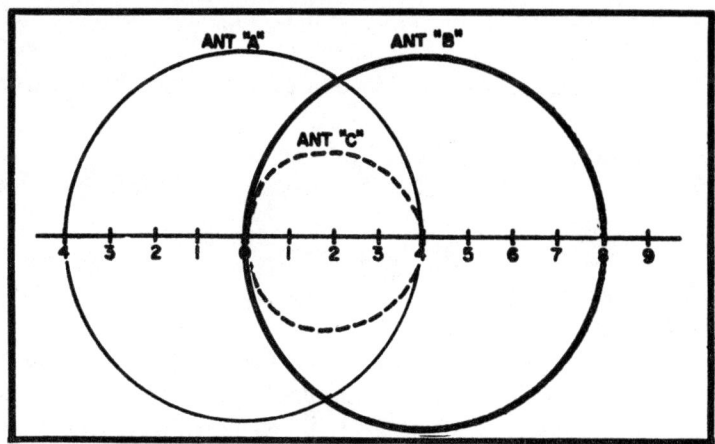

Fig. 1-3. Radiation patterns of three antennas, one isotropic.

Antenna B, on the other hand, has the same directive gain as antenna C, since the shape of its radiation pattern is the same. But its efficiency is four times greater, and thus, as shown in the diagram, is a much more effective radiator in the favored direction.

The power gain of antenna B is, from Equation (8):

G_p (antenna B) = η G_d = 1 × 4 = 4 (6 db).

For antenna C, the power gain is:

G_p (antenna C) = ¼ × 4 = 1 (0 db).

The G_p, G_d and η for the three antennas is summarized in Fig. 1-4.

Power gain can also be thought of as the power density of the EM wave radiated by a directive antenna in its favored direction, divided by the power density radiated by a 100% efficient isotropic antenna in the same direction. In Fig. 1-3 antenna B gives twice as much field strength in the desired direction than does the isotropic antenna. And since power density, from Equation (5), is proportional to field strength squared, it gives four times as much power density. Thus its power gain is 4. Antenna C, on the other hand, gives identically the same field strength, and therefore power density, as the isotropic, and thus its power gain is 1. From this it can be seen that power gain can be determined from an antenna's radiation pattern.

At this point let's look at a practical example of power gain versus directive gain. A three-element yagi, optimally designed and adjusted, has a maximum directive gain of about 11.7 (10.7 db). However, if loading coils or traps are added to the elements to decrease their size, or provide for automatic bandswitching, the efficiency is decreased due to losses in the wire from which the coils are wound. If the efficiency is decreased to as little as 80% (not an unrealistic figure in the case of tri-band beams), the power gain will suffer by about 1 db. Of course this small price has been paid for added flexibility in the antenna system.

Normally power gain and directive gain are measured with respect to an isotropic antenna, as we have done in the examples above. Historically this has not always been the case, however. In the earlier days of radio, gain was most

Fig. 1-4. A summarization of the power gain, directive gain and radiation efficiency for the antennas of Fig. 1-3.

ANTENNA	A	B	C
POWER GAIN, G_p	1	4	1
DIRECTIVE GAIN, G_d	1	4	4
RADIATION EFFICIENCY, n	1	1	1/4

often measured with respect to a half-wave dipole antenna, and this is still done today in the amateur radio community. The gain (either power or directive) is less when measured with respect to a dipole than with respect to an isotropic, because a dipole itself has gain with respect to an isotropic antenna. A 100% efficient dipole antenna has a power gain of 1.64 (2.15 db) with respect to an isotropic. Thus to convert a gain figure measured with respect to a dipole to that with respect to an isotropic, the figure should be multiplied by 1.64.

Which Gain is Important?

The question now arises, Which gain, directive or power, is important in a radio system? In the transmitting system, power gain is the most meaningful criterion of antenna effectiveness, since the ultimate aim is to radiate in the direction of the receiving station an EM field with the greatest power density. Power gain gives a good measure here because it includes both directivity and radiation efficiency.

In a receiving system, the important type of antenna gain depends on noise, and its origin. If the largest portion of the receiver output noise originates in the receiver itself, power gain is necessary in the antenna. This is normally the case in the frequency range above about 30 MHz. But if the output noise originates external to the receiving system, such as is the case with atmospheric noise below 30 MHz, then only directive gain is necessary in the antenna.

Normally below 30 MHz we amateurs use an antenna with high power gain for both receiving and transmitting, since it allows us to get by with only one antenna. When transmitting the power gain is necessary to system effectiveness, but for receiving it is not, although it certainly does no harm. By sacrificing efficiency (which we don't need anyway) we could use a physically much smaller receiving antenna, though, and get equal results. At least one manufacturer is taking advantage of this principle in small receiving antennas being built for the commercial services.

Summary

Some important conclusions can now be stated from our study of antenna gain. First, when speaking of antenna "gain" two things must be stated for clarity—the type of gain referred to, power or directive, and the reference, isotropic or dipole. Second, high radiation efficiency, while always desirable in a transmitting antenna is not necessary for receiving antennas below 30 MHz where the large majority of the receiver output noise originates outside the antenna. With low efficiency both the signal and noise are reduced proportionately in the antenna, and therefore the signal-to-noise ratio is largely unaffected.

Receiving Antennas

So far as transmitting antennas are concerned, we have now covered the basics, but for receiving antennas we have more work to do. While the antenna parameters we have specified and described above for transmitting antennas are also adequate for receiving antennas, two other unique terms have come into great usage to describe receiving antennas, capture area and effective height.

Capture Area

When using an antenna for receiving purposes, it is usually desirable to know the amount of signal power available from the antenna output, to be supplied to the receiver input. From the power density of the EM field of the signal we know the amount of power per unit area in the field. If we knew the effective capture area of the receiving antenna, then, we could find the power available from the antenna simply by multiplying the power density and capture area together. Thus, the capture area is the ratio of the power available at the antenna terminals to the power density of the intercepted EM field. Capture area is related to the power gain of the antenna, and the wavelength of the field by:

$$(9) \ A = G_p \times \frac{\lambda^2}{4\pi},$$

where: A = antenna capture area.

Let's take a simple example. Say the wavelength is 7.1 meters and G_p of the antenna is 8 (9 db). The capture area then is

$$8 \times \frac{7.1 \times 7.1}{4 \times 3.14} = 32 \text{ square meters.}$$

Then if the power density of the EM field striking the antenna were 2 nanowatts per square meter, the power available at the antenna terminals would be $2 \times 32 = 64$ nanowatts.

Large capture area is essential if a VHF antenna is to be highly effective for receiving purposes. But Equation (9) shows that capture area decreases with the square of wavelength. Therefore, as we go to higher frequencies, and consequent shorter wavelengths, power gain must be rapidly increased if we are to maintain a respectable capture area. The result of all this is that highly effective VHF antennas are just as physically large as those for the lower frequencies, despite the shorter wavelengths. They must be in order to develop the proportionately higher power gains necessary to maintain a high capture area.

Capture area is coming to be used more and more today as a measure of VHF and UHF receiving antenna effectiveness. Historically, however, antenna "effective height" came first, and we will explain that next.

Effective Height

Back in the days when regular AM broadcasting was getting its start, the amplitude of an EM field was most often specified by its electric field strength, E. Power density was very seldom used. Consequently, an antenna "transfer function" was needed which was based on field strength rather than power density. The term settled on was "effective height," and it was defined in terms of the voltage measured at the antenna terminals with no load connected across those terminals (the open-circuit voltage):

(10) $$L = \frac{\text{antenna open-circuit terminal voltage}}{\text{electric field strength, E}}$$

= antenna effective height.

This choice has turned out to be a bit ambiguous for two reasons. First, for a given field amplitude, it gives only the voltage available from the antenna. This is ambiguous for a given antenna because it depends on where the antenna is fed. If the feed terminals of a dipole antenna, for example, are located at its center the open-circuit voltage is much lower than when the terminals are located a good deal off center. And besides, it takes power to drive a receiver anyway.

Second, the term "height" has proved to be unfortunate because it implies how high the antenna is above ground, which connotation is purely incidental. A much better word here would have been "length," and indeed it is now coming into wide usage.

At amateur frequencies and higher, antenna capture area is gradually replacing the usage of effective height, especially among professionals. Its usage among amateurs should also be encouraged, but the old term is still hanging on with real tenacity.

Conclusion

You should have gained a basic insight into antenna theory. The knowledge you have gained will enable you to interpret antenna literature more wisely, and this in turn will mean better antennas at your station for the dollars you have to spend.

ANTENNA RADIATION, SIMPLIFIED

The current theory of radiation depends on the electromagnetic wave theory: When an alternating current of a specific frequency is placed on an antenna of a predetermined length (calculated by formulas familiar to all amateurs), an electrostatic field and a magnetic field are formed, one being geometrically at right angles to the other. Both of these fields are components of the thing we call the *electromagnetic wave*.

These radio waves display certain characteristics, such as the speed at which they travel being equal to the speed of light; critical angles—the angle at which they approach; the direction—horizontal or vertical—in which they peel off the antenna.

All of these things have been studied, tested, tried, and proven to some degree. Predictions can be made and results that have been achieved by this popular concept of wave formation have been for the most part satisfactory. But don't believe a word of it. Many years ago, the earth was considered to be the center of the solar system, or even further, the center of the universe. It certainly appeared to be, for it was obvious that all things in the heavens revolved about it. Mathematical formulas, predictions of eclipses, tides, moon quarters, etc., were all done with remarkable accuracy. Then Galileo looked through his telescope and realized that the sun is the center of our solar system, and we are revolving around it! A whole new system had to be devised, and of course it was.

So it is with the electromagnetic wave. If you were to hold a pebble above a still pond and let the small stone drop, a series of small concentric waves would spread out from the point of impact in all directions; as the wave distance from the point of impact increases, the size of the wave decreases. The energy of the falling stone has been transformed into a wave upon the water.

The important thing here is that the transformation of energy from the falling stone to the wave required a medium for this to happen. The medium in this case is the water;

without the water, there would be no wave of energy to move upon.

Another example is the energy of vibrating vocal cords pushing against the air, setting up an air wave that would vibrate an eardrum, allowing you to hear a voice. Take away the medium (air) and there would be no means of transporting the energy from throat to ear.

Back in the days when the electromagnetic wave theory came into its own, a small problem arose. If light (an electromagnetic wave, in theory) were to reach the earth from the sun sitting in outer space, how would it get here? If space were a vacuum, there would be no medium and hence no sunlight. Obviously, there must be a medium present. Classical physics was so sure of this that the invisible medium was called the *ether*. The explanation of why we can't see it or detect it was that we were like fish immersed in water: it's all around us, thus not readily apparent.

It bothered some people that the ether was there but was yet to be detected and confirmed. Two scientists decided to do something about this situation.

The Michelson-Morley Experiment

Albert A. Michelson conceived the brilliant idea of splitting a single ray of light into two parts:

Optical instruments would cause one portion of the ray to hit an optical screen; the other half would be deflected by a series of mirrors. The spin of the earth would cause an ether current (like wind on your face as you run forward).

The apparatus was set so that one half of the ray went downstream in the ether; the other half deflected upstream and then downstream. The two portions of the wave were now joined to form one again. Since one fraction had to travel further than the other, and to buck upstream ether flow, it seemed logical that one fraction of the ray would be out of phase with the other. This would mean heterodyning or beating of the waves to form new waves. The logic of this thought was sound and the experiment was carried out carefully.

Unfortunately for Michelson, when the rays were rejoined, they were *in* phase. This was a very disturbing result. Fortunately, Michelson was an honest man and reported his results as he found them. He did not believe that there was no ether, but that his instruments were too crude to detect the difference in phase.

Undaunted, he tried again. This time he improved the optics and had Edward Morley move farther away to get more distance between source and screen. This he hoped would show the rays to be out of phase.

To his dismay, along with that of the rest of the scientific world, the rays came back *in* phase. Over and over again, the experiment was carried out. Each time the optical instruments were perfected, but alas, the results were the same. To his dying day, Michelson believed in the presence of the ether, but simply could not prove its existence.

The experiment had a very disturbing effect in the scientific world. To some, it was almost easier to believe that the earth stood still rather than scrap the wave-ether theory.

A few years later, a scientific "bomb" was dropped, followed by a whole salvo of "bombs" by two giants in the world of physics.

Among those who scratched their heads over the experiment was Max Planck who took the bold view that light was not a wave, but a "packet of energy"—a quantum. The quantum of energy that referred to light became known as a *photon*.

Now, this was a very different situation. If light from the sun came to the earth through empty space as a particle, no medium is necessary as in the case of a wave; a new step in the understanding of the mechanics of the universe evolved.

The other giant in the scientific world was Albert Einstein. He grasped a hitherto unknown phenomenon from the Michelson-Morley experiment. The fact that the two light beams arrived at the same spot at the same time, even though one of them had to travel twice as far, brought forth the remarkable conclusion that the speed of light is *constant*, and if that is so, then distances must shrink and time must change to accommodate light!

This not an easy thing to swallow; in the beginning, there were very few people who could digest the concept. More likely the man was stark raving mad. Imagine rulers shrinking and clocks slowing down as they move fast enough to approach the speed of light! If you were never exposed to Einstein's theories before, you are probably having a difficult time trying to follow these weird statements. One might ask, "Do you really mean that my watch would slow down and that my twelve inch ruler would shrink if I were on a rocketship that went fast enough to approach the speed of light?" That is exactly what I mean!

Other predictions were made. If light were a particle, then it would be affected by gravitation. Einstein predicted that a light ray would bend if it passed near a large enough mass. He then proposed that the heavens be photographed before an eclipse and the locations of various stars be charted. Then the heavens would be photographed during a solar eclipse, when day is turned to night. The stars nearest the sun should appear to shift because the rays would be bent as they passed near the sun.

On May 29, 1919, men of science from various countries of the world, even those countries that were at war with each other, gathered in the equatorial regions in Africa to confirm Einstein's predictions. The stars in the photograph taken during the eclipse did shift, to the amount of Einstein's calculations.

EINSTEIN AND RELATIVITY

Still more was to come. One step led to another until the realization that energy and mass are one and the same, that is to say that mass is coalesced energy and that one can transform mass into energy and energy into mass. The most famous equation in history was brought forth by Einstein:

$$E = MC^2$$
$$C = \text{Speed of light.}$$

where e is energy, m is mass, and c is the velocity of light.

The dramatic proof of the validity of this equation was given on a dark night on July 14, 1945, in the Los Alamos desert in New Mexico. The United States exploded the first atomic bomb.

The world stepped into the Atomic Age.

What does all this history of quantum mechanics have to do with a radiating antenna? To answer this, we have to walk a few steps further along the path of discovery of Einstein and Planck.

Max Planck found a relationship between energy content and wavelength λ of a quantum that is linked with the second fundamental constant of the universe (now known as Planck's constant). He formed the basic equation of the quantum theory:

E = hV where V = frequency, h = Planck's constant, and E = energy.

With this equation before him, Einstein made another prediction, that of the photo-electric effect. Certain metals contain outer electrons in their atoms, which are loosely attached to their nuclei due to the relatively great distances they are from their nuclei.

Einstein reasoned that if an electron were struck by a photon of light, it would be possible to knock it out of its orbit, provided the photon had enough energy to overcome the attractive forces binding the electron to the nucleus. Once the electron was ejected from the atom, the metal atom would become charged positively.

As you can see from the equation, the higher the frequency, the greater the energy content of the photon. Therefore, he predicted that it is not the *amount* of light striking the metal that causes the photo-electric effect, but the *frequency* of the ray involved.

If the frequency of a ray of light is increased to the point that the photoelectric effect occurs, it is this frequency or any *higher* frequency that will produce this result, even though the *amount* of light is cut down to a feeble quantity.

To illustrate why it takes a certain amount of energy to knock out the electron, even though all light hits the metal at the same speed, let us imagine that you are sitting on a small dock, dangling your feet in the water. If a small canoe came toward the dock at 5 mph, you could stick out your foot and easily stop the canoe; however, if a 100,000 ton ocean liner came gliding up to the dock at 5 mph, you had better get your leg out of the way or it would be crushed, and for that matter, the dock would be smashed and a hole dug out before the big ship came to a halt.

RADIO

This effect of energy content of the quantum can be related directly with radiation of an antenna and the reflection off the ionosphere.

An amateur handbook describes *critical angle of radiation* by stating that, as the frequency of an electromagnetic wave is increased, there is less bending from the ionosphere; the *skip distance* increases until finally the angle is too great for any bending or reflecting from the ionosphere. The wave then strikes the ionosphere at an angle that allows it to pass through and not be reflected.

The quantum theory explains this from another point of view. If the particle of energy does not have a high energy level, it strikes the lower portion of the ionosphere and bounces off like a ball against a wall. The higher the frequency, the greater the energy content of the particle; it goes deeper into the ionosphere before it is reflected. The lower-frequency particle may bounce off the ionosphere at 60 miles up, whereas the higher-energy particle may bounce off the 150-mile level. (As the particle passes through the ionosphere, constant impacts with electrons drain its energy, until eventually the particle smacks into an electron of equal energy level and then rebounds back to earth.)

By increasing the frequency (and energy), the radio beam finally reaches a point where the energy level is too high for the ionosphere to stop or reflect particles, and it passes on through into outer space. It is for this reason that the higher frequencies are normally useful only for distances tangent to the earth's surface.

A word or two about the ionosphere might be useful here. In the outer reaches of the earth's atmosphere, the rarefied air is subjected to radiation from the sun. The ultraviolet or *higher* energy photons strike the air molecules and cause the ejection of electrons from the outer shells. The loss of an electron ionizes (or charges) the gases positively and causes free electrons to float around up there. These free electrons eventually find their way back to a positive ion and restore a neutral state to the gas.

However, as long as the radiation continues, the sea of electrons remains, since new electrons are continually being knocked out of their orbits. Any increase in solar activity such as solar storms (sunspots) causes the sea of electrons to deepen.

In quiet times the sea depth shrinks, and at night the ionosphere normally is less thick than in daytime.

HOW IS A PHOTON OR QUANTUM FORMED?

To best understand this, we must picture an atom (of any substance) to be a miniature solar system, the nucleus acting as the sun with the orbiting electrons as its planets. These electrons orbit the nucleus in definite levels or orbits. Each level has a maximum number of electrons that it can hold. For example, the first orbit will hold two electrons. Any more electrons have to go into the next orbit up that can hold up to a total of eight electrons. The next orbit has a definite limit of electrons it can hold, and so on.

It is in the nature of things that atoms tend to be in a state of balance or *neutral*. That is, each atom tries to be electrically neutral, with the number of electrons in orbit equal to the number of protons in the nucleus. Each electron furthermore tends to stay in an orbit at its *normal* distance from the nucleus. Any movement out of this orbit creates an unstable situation. If energy of one form or another is applied to an atom and forces an electron to move out of its normal shell or orbit level, it (the energy) is converted to a photon by the following means: the displaced electron tends to resume its normal orbital level. In bouncing back to its regular orbit, it has to give up the energy that forced it out of the orbit. The energy thus liberated is a quantum or photon. In the case of light frequencies or higher (X-ray, etc.) the electrons closest to the nucleus emit the photons. The outer electrons do not require as much energy to be knocked out of orbit and I suspect these may be responsible for the lower frequency radiations such as radio frequencies.

From this point on, the waters muddy a bit. Some of the following statements represent the latest scientific thought and some pure conjecture on my part. However, let us plunge a little deeper into the wonders of the universe.

The wave theory of light is not an easy theory to bury, for some of the characteristics of light can be explained by particles, but others seemed to be answerable only by a picture of wave mechanics. The phenomenon of diffraction gave the wave theory its greatest support.

If a beam of light were passed through a small hole in a plate and allowed to project on a white screen, it would show up as a white spot. If the hole is reduced in size to a minute opening, the spot on the screen has a different appearance. It no longer looks like a bright spot, but rather like a conventional target, with progressive concentric dark rings, going away from the center. Placing two small holes very close to each other has the effect of heterodyning the two sets of waves so that alternating dark and bright lines now show on the screen. This can be easily shown with waves of water meeting with each other causing alternative cancellation and addition of the joining waves.

To further complicate matters, the famous French physicist Louis de Broglie in 1925 threw a body blow to the quantum theory by stating that electrons (which are particles) were not hard spheres as previously thought, but also showed wave characteristics. Put to the diffraction test, lo and behold, they too showed concentric rings!

Extension of this thought led to the amazing discovery that whole atoms and even small molecules showed wave characteristics! Is anything solid?

The interference patterns (heterodyning lines) of the diffraction of light are now looked upon in a different manner. It is now thought that the photon does not travel in a straight line, but moves in a wiggling or spiral motion. That is to say, if a photon were a bullet fired from a smooth-bore gun into a pipe or long tube, it would travel straight down the center of the tube. However, if the gun bore were rifled, the bullet would travel off in a spiraling motion limited by the diameter of the pipe.

I suspect the higher the frequency, the narrower the pipe is, so that the spiral motion is tighter.

Exactly where the photon is in relation to the center of the imaginary tube at any given moment is a matter of probability.

Evidently, the photons that follow each other through the small aperture display themselves on the screen in a wavelike pattern.

"We might say then, that photons are the components of a light beam, whereas the wave is a description of it." (Scientific American, September 1968)

What makes some antennas good radiators and others mediocre or poor?

The formula for calculating antenna length works very well but does not give us insight as to why an antenna radiates.

Any conductor will radiate, but how well it radiates depends on several factors.

First, it is a matter of matching impedances from transmitter to radiator to allow a maximum transfer of energy from one to the other. Unless the capacitive and inductive reactances are properly balanced, this cannot occur.

If you remember the method of quantum or photon production and apply it to a conductor, it is only the outermost layer of atoms that do the actual radiating.

In the case of the light bulb experiment it is interesting to note that a light bulb gave a fairly good impedance match to a transmitter, allowing a transfer of energy to the filament of the bulb. It is obvious that the filament of the bulb is not 132 feet long, so that the surface area of the conductor is not as large as the conventional antenna. The amount of RF leaving the bulb is not as great.

The energy backed up in the filament has placed enough force on the deeper electrons (closed orbits) in the atoms of the conductor to form higher energy photons. The filament of a light bulb does have a resistance to AC so that it will glow (radiate light), but the impedance is a good match for the transmitter.

The quantums or photons thus formed are larger (higher energy content) so that a host of new and various energy photons are formed. Now we have RF, light, and all frequencies in between being formed. If the filament had more surface area, more RF and less of the higher energy photons would be radiated.

On the receiver end, the photons (RF) striking the antenna set up a displacement of electrons in the surface atoms, causing an electrical potential to be set up in the conductor.

The longer the antenna, the greater the *capture* area; the more quanta striking the antenna, the greater the voltage. It is as simple as that.

MEASURE ANTENNA IMPEDANCE WITH AN SWR BRIDGE

So far, the great majority of antenna impedance bridges found in construction articles are devices that function only with a low power rf source. The run-of-the-mill impedance bridge is designed to operate with a grid dipper as the source of rf excitation. Operation with tube type dippers is generally intended as the transistor dippers produce an rf level that is too low for excitation of this bridge type.

The conventional antenna bridge cannot be left in the transmission line continually as excessive rf energy would soon destroy the device. This means that each time measurement of antenna impedance is desired, the transmission line must be opened and the bridge inserted and grid dip excitation applied. Grid dippers are not necessarily the most accurate rf source for a specific frequency in an amateur band...therefore the station receiver must monitor the dipper output for any bridge accuracy. A low power bridge will not often present the true operating impedance of the antenna...especially antennas with parasitic elements. A bridge that operates under full transmitter power will present a much more accurate picture of your antenna system at a specific frequency.

Inspiration for the "In Line" full power bridge came from information concerning the standard swr bridge. Just about every amateur has in his possession some sort of swr bridge and the great majority are of the type illustrated in Fig. 1-5. This bridge consists of a section of transmission line near which are placed two inductors. These inductors are actually two bridges along with their associated diodes and resistors. One of the bridges reads forward power and the other reflected power. The resistors (Rx) at the end of the inductors L1 and L2 are critical for accurate bridge null (balance) and therefore must be the proper value for the specific transmission line used. For the average swr bridge the value for Rx is 100Ω for 75Ω line and 150Ω for 50Ω transmission line. Considering that resistor Rx is critical for the impedance of the line in use, varying the value of Rx and devising a system of

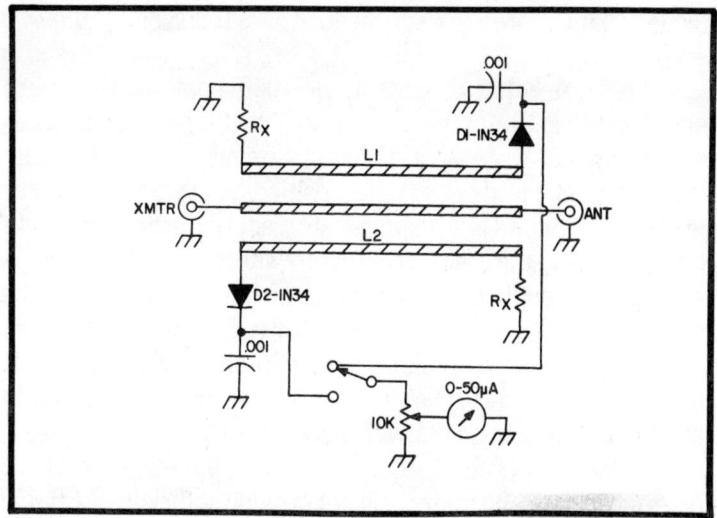

Fig. 1-5. Conventional swr bridge.

calibration for Rx would enable determination of the impedance of a line when a null is achieved on the bridge meter.

The "reflected" inductor which is L1 in Fig. 1-5 is the portion of the bridge circuit used for impedance measurements. The value of Rx and the transmission line must balance the bridge for a null to be realized. Any variation from the above parameters will mean changing the value of Rx so that the bridge again balances at a new impedance value.

By experimenting with various values of resistance at Rx, it was determined that a 1000Ω potentiometer represents a fair value. The 1000Ω potentiometer is inserted in place of Rx on inductor L1 (see Fig. 1-6). This is the inductor with the

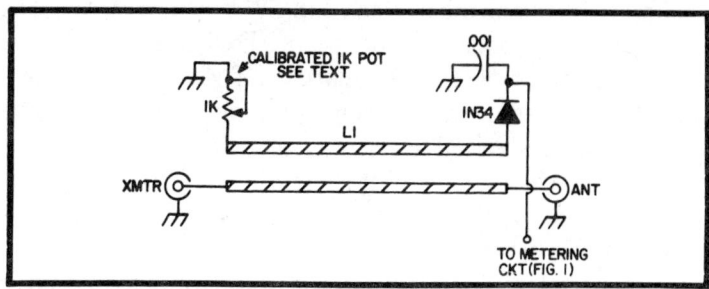

Fig. 1-6. The modified bridge leads to 1K pot should be as short as possible and shell (case) of pot grounded.

33

diode pickup located toward the load or antenna end of the swr bridge.

Make sure that all leads to the 1000Ω potentiometer are short and that the metal case (shell) of the potentiometer is well grounded. Excessive lead length or inductance will create inaccuracy of the device. The position of the potentiometer will be determined by the physical layout of the particular swr bridge. It must be set at a point where the shaft can be extended through the front panel of the bridge. Allowance must also be made for a dial or other indicating device which can be calibrated in ohms (impedance) on the front panel. It might even be desirable to mount the present bridge in another larger case so that all functions can be accommodated.

Calibration of this in-line bridge was the major problem. An ordinary grid dip meter will not provide sufficient excitation for readings. With full power applied, especially a kilowatt, it becomes difficult to find resistive dummy loads of various values to calibrate the bridge. Even with 100W of rf, proper resistive load values are not common.

The solution to the calibration problem came in the form of a CB transmitter. A CB transmitter is fortunate if it is able to put out 4W of rf and at the same time is well within the frequency range of an swr bridge. The most important fact is that a CB transmitter will provide adequate excitation for calibration of the bridge with ordinary 5W 5% carbon (garden variety) resistors. For calibration, a good assortment of these resistors is necessary. Use values such as 5, 27, 47, 75, 100, 150, 220 and 470Ω. Intermediate values can be then interpolated on the scale. The calibration procedure is simple...first obtain a CB radio, then attach the 5W resistors across the antenna coax connector of the bridge. Excitation is applied to the remaining connector on the bridge. The sensitivity should be set for a middle scale reading of the meter and the 1000Ω potentiometer is varied to reach a null on the meter. Mark the value of the calibration resistors on the potentiometer scale (dial). Do this for all of the available resistors and the bridge will be in fair calibration.

This system does not measure reactive components in the antenna system. If an antenna is reactive, either inductive or capacitive, the meter will present a shallow, poorly defined null at the operating frequency. A sharp, well defined null will indicate a purely resistive impedance.

When using the bridge in its former function as an swr bridge, set the resistance dial to the value of the transmission line. When measuring impedance, vary the dial for maximum dip on the meter and read the resistance (impedance) directly.

As a final point, it is wise to insert the swr/impedance bridge at a half-wave or an even multiple of a half-wavelength from the antenna. At half-wave points, the antenna impedance is repeated. This will enable measurements to be much more accurate. When determining half-wavelength points, take into consideration the velocity factor of the particular coax in use.

REACTANCE VS IMPEDANCE?

A problem encountered by technical people for many years involves the two fleeting quantities known as impedance and reactance. One never really knows when reading an article or text which one is being discussed due to the ambiguity traditionally associated with these terms. Many publications presently in print speak freely of reactance as having a "phase angle," or impedance as being a simple number quantity. Both of these statements are totally incorrect.

Reactance is a term applied to a quantity having magnitude only with no regard to direction. Impedance, on the other hand, not only implies magnitude, but dictates a particular direction as well. The reactance of an inductor of 1 Henry being operated at a frequency of 60 Hz would be:

$$X_L = 2\pi fL = 2\pi (60)(1) = 377 \text{ ohms}$$

This quantity is called reactance and has a value of 377 ohms, with no consideration given as to direction. If at this point we say that our inductor has a value of 377 ohms at angle (direction) of 90 degrees, we have immediately bridged the gap and developed a new quantity called *impedance*.

Consider the following analogy: A bullet is fired from a rifle at a speed of 600 miles per hour. Only one correct deduction concerning the bullet may be made with the information given; namely, that it is moving fast enough to do physical harm. It would behoove those concerned to also know the direction of travel to avoid an early demise. In other words, the information conveyed by knowing both the magnitude and the direction is most beneficial. The same is true with reactance and impedance. Reactance conveys magnitude information only; impedance denotes magnitude and direction.

In order to manipulate these quantities from a mathematical standpoint, the concept of Complex Notation must be introduced. It should be immediately pointed out that the only thing complex about complex numbers is the name. This is mentioned to overcome mental blocks, which usually arise during the initial stages of development.

A complex number is represented by the sum of two numbers, one called the real part, and the other called an imaginary part. Since both real numbers and imaginary numbers are simply numbers which we use daily (1, 2, 3.9, 2.7, 19, 140.2, etc.), we must somehow distinguish between the two. In order to do this, we introduce the imaginary operator j, which acts as an indicator much the same as a flagman would in traffic. As an example, let us assume we have the complex number: $A = 3 + j4$. In this case, the real part of the complex number A is 3; while the imaginary part is 4. Note that the function of j is only to indicate that there is something "different" about the number 4.

Relating this concept to the realm of impedance, we note that, in general, impedance is also a complex quantity. The real part of impedance is called resistance denoted by the letter R. The imaginary part of impedance is called reactance and is symbolized by the letter X.

Let us now consider a circuit containing only an ideal inductor. It is the object now to determine the total circuit impedance looking into terminals AB. In the circuit shown there is no resistance (R), therefore, there is no real part in our complex number. The imaginary part of the impedance is the inductive reactance X_L. X_L may be found knowing the inductance (L), and the frequency (f) as was shown in a previous example. The total impedance is: $Z = 0 + j X_L$ ohms, where j indicates the direction.

Since this quantity possesses both magnitude and direction, it is often more easily understood when illustrated graphically on a standard Cartesian Coordinate System. Our real axis is along the horizontal, while our imaginary is plotted vertically. By convention, always plot the real part first and then plot the imaginary in a tip to tail fashion, utilizing the

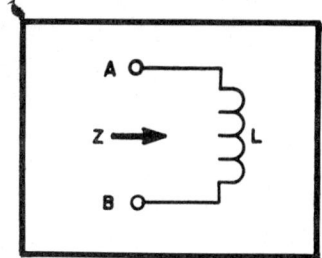

Fig. 1-7. A very simple inductive circuit. If the resistance is zero the voltage leads the current by 90 degrees and we say $Z = 0 + jX_L$ ohms.

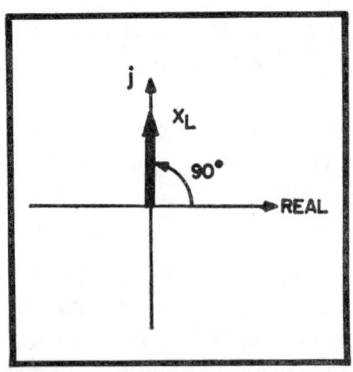

Fig. 1-8. Here we are saying $Z = 0 + jX_L$ ohms, graphically. The heavy arrow represents impedance. Its length indicates the value of the impedance, which we call reactance.

previous example, plotting resistance first $R = 0$. Now add the imaginary portion to the real obtaining the value X_L plotted vertically on the imaginary axis.

We will pick the horizontal axis to the right as being positive. All positive angles are measured with respect to this axis when moving in a counterclockwise direction. Returning to our example and measuring this angle with a protractor, we see it to be a positive 90°. The significance of this angle will be discussed later.

As a second example, consider a capacitor only, the impedance of which is to be determined. The reactance of the capacitor is given by:

$$X_C = \frac{1}{\omega C}$$

Now by definition we choose the capacitive impedance to have a $-j$ associated with it. Therefore, the capacitive impedance becomes $-jX_C$, and it is plotted on the graph previously mentioned, vertically downward as shown below:

It can be seen by these brief examples that the j designation for the impedances saves confusion when writing the values.

Let us now consider a combination of an R, L, C connected in series. It will be our objective to calculate the total impedance in both mathematical forms. Given the circuit in Fig. 1-10.

The reactances of each of the components may be found by the usual manner, X_L and X_C. We will now determine the values of the individual impedances. The impedance of the

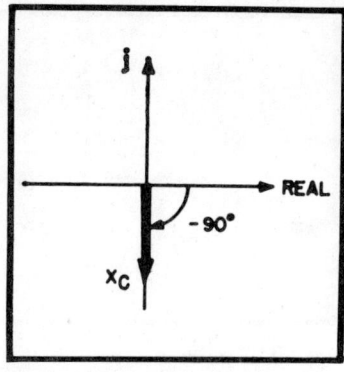

Fig. 1-9. If we replace the inductor of Fig. 1-7 with a capacitor, still assuming there is no resistance, the impedance arrow now points downward.

resistance is the easiest, for it consists of a real part only, that is, $Z_R = R + j0$ ohms. The impedance of the capacitor is $Z_C = 0 - jX_C$ ohms, and for the inductor it is equal to $Z_L = 0 + jX_L$ ohms. Impedance in series add and, therefore, the total impedance is $Z_R + Z_C + Z_L = Z_T$ or $Z_T = R + j0 + 0 + jX_L + 0 - jX_C$ ohms. We can find the sum by adding the real parts and the imaginary parts separately. $Z_T = R + jX_L - jX_C$, but since the j term is common in this case, Z_T becomes $Z_T = R + j(X_L - X_C)$. This can be shown graphically by the following:

Note that R was plotted first, then jX_L and $-jX_C$ in the tip to tail fashion described previously. jX_L and $-jX_C$ lie in the same plane, however, in the opposite direction. Therefore, the result of $j(X_L - X_C)$ can be found by algebraically subtracting X_C from X_L. The result X_T remains unchanged but X_T is now in the positive direction, this is because X_L is larger than X_C in this example.

If the other case were true (X_L smaller than X_C), the result X_T would point in the negative direction. X_T and R form a

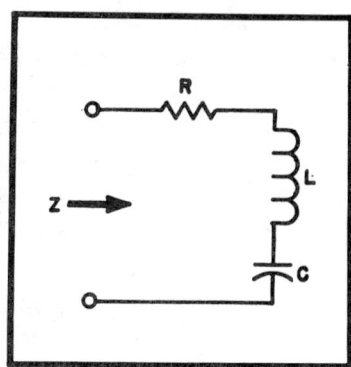

Fig. 1-10. Here is a more complex circuit. We might find this in a bandwidth-limiting application. What is its impedance?

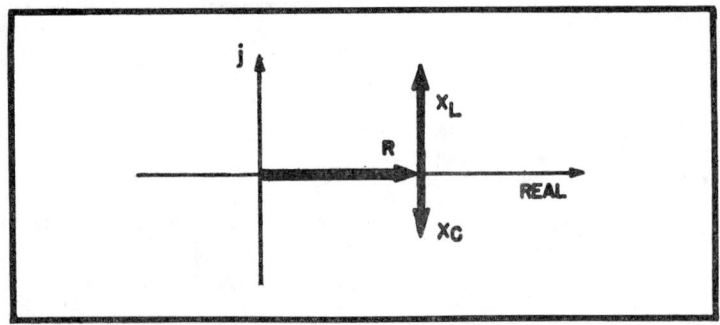

Fig. 1-11. First we must know the frequency. Then we determine impedance by adding resistance and reactance values graphically, or by math having the same meaning. Remember reactance will vary with frequency, changing the impedance.

right triangle and the line Z_T represents the hypotenuse of the right triangle. Using the theorem developed by Pythagorus which says that the hypotenuse of a right triangle is equal to the square root of the sum of the other two sides squared, the magnitude of Z_T becomes:

$$|Z_T| = \sqrt{R^2 + X^2} \text{ ohms}$$

which is the formula given in most handbooks. However, this is only half the picture; we still must have a direction. The angle θ on Fig. 1-12 can be found by the formula:

$$\theta = \arctan \frac{X_T}{R}$$

Read, theta is the angle whose tangent is X_T/R. Therefore, the impedance may be expressed in two ways:

$$Z = R + j(Z_L - X_C) \text{ or } Z_T < \theta$$

To summarize, let us now turn our attention to a numerical example. Consider an R, L, C series circuit being used at 60 Hz. The value of the individual components are as follows: R = 300 ohms, L = 0.5 Henrys and C = 10 microfarads.

Calculating the reactances:

$$X_L = 2\pi fL = 2\pi (60)(0.5)$$
$$X_L = 188.5 \text{ ohms}$$

$$X_C = \frac{1}{2\pi fC} = \frac{1}{2\pi(60)(1 \times 10^{-6})}$$

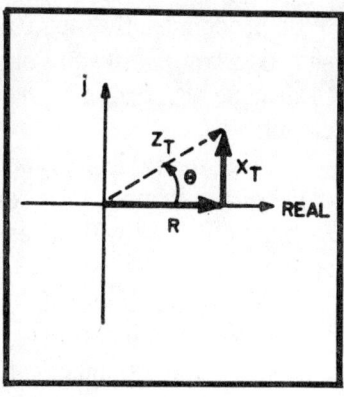

Fig. 1-12. Taking the difference between inductive and capacitive reactance, the inductive reactance wins this time. At some lower frequency they would cancel, leaving resistance only. And at a still lower frequency the capacitive reactance would predominate.

X_C = 265.0 ohms
Z = 300. + j (188.5 − 265.) ohms
or Z = 300. − j 76.5 ohms
See Fig. 1-13 for graphic illustration.

$$Z_T = \sqrt{(300)^2 + (76.5)^2} = 309 \text{ ohms}$$

$$\theta = \arctan \frac{-76.5}{300} = -14.3 \text{ degrees}$$

The total impedance for this circuit is 309 ohms at an angle of −14.3 degrees.

The angle associated with the impedance in actuality represents an angular (phase) difference between the voltage applied to, and the current in the circuit considered. By definition, if the angle associated with the impedance is positive the voltage leads the current, and the circuit appears basically inductive. Similarly, if the angle is negative, as in the previous sample, the current leads the voltage, and the circuit appears

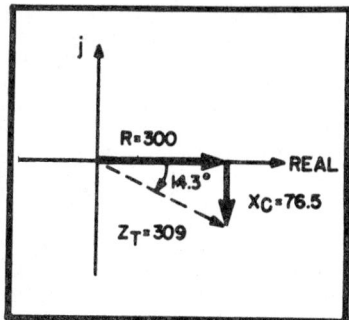

Fig. 1-13. Working out the circuit of Fig. 1-10. If we apply a 60 Hz current to this circuit we will find the voltage across its terminals lagging 14.3 degrees. Or if we trigger out scope from the voltage signal the current will appear to lead by the same 14.3 degrees.

to be predominantly capacitive. The special case of no phase shift occurring between voltage and current, (corresponding to an angle of zero degrees) simply indicates a purely resistive circuit.

One can at this point begin to appreciate the significance of the quantity impedance, and the enormous amount of information conveyed with it, as opposed to the simple quantity reactance.

Whether engineer, technician, serviceman or home experimenter, the blossoming age of electrical technology demands an understanding of the subtle distinction between these two very basic circuit concepts.

THE HALF-WAVELENGTH FEEDLINE

The antenna, one of the most important elements in a communications system, is often the least understood portion of the radio amateur's station. It has been demonstrated that many antennas, when correctly tuned, produce an increased efficiency of 10 dB or more in receive and transmit signal strength, compared to the amateur's normal tuning method of using a VSWR Bridge. The purpose of this discussion is to briefly describe the essential features of an antenna system and their optimization for maximum performance.

An antenna is basically a resonant circuit. For maximum performance it must be tuned to resonance for the same reason that the transmitter output circuit must be tuned to resonance. When the transmitter frequency is changed, it is standard practice to "dip the final," and the antenna resonant frequency should also be changed if maximum performance is desired.

The antenna is basically a series resonant circuit with a resistive component, as shown in Fig. 1-14. The resistive component is referred to as the radiation resistance. Maximum current flow will occur in the resistance only at the resonant frequency. The value of the resistive component is a function of antenna height, ratio of physical to electrical length (loaded antennas) and other factors. The resonant frequency is a function of the physical characteristics of the antenna and proximity to other objects. The major problem, when looking at the overall system, is that the antenna must be physically separated from the transmitter, hence the need for a feedline. This article is concerned with coax feedline. If an antenna

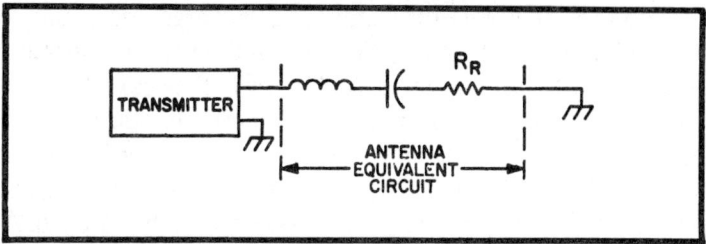

Fig. 1-14. Series resonant circuit.

tuner is used, this line is short and the antenna is virtually moved into the shack. The normal procedure for amateur antennas is to use a coax line length just long enough to go from the transmitter to the antenna, and herein lies the problem. If a random length of feedline is used and the antenna does not have the same radiation resistance as the coax characteristic impedance (nominally 50 ohms), the power doesn't get to the antenna due to losses in the mismatch and to radiation from the coax. For this case, the coax becomes part of the antenna system. On the other hand, if a half wavelength of coax (or multiple) is used, the effect of the coax may be disregarded. The coax itself is the equivalent of a series resonant circuit. When the coax and the antenna are connected, if the line length is proper, the antenna feedpoint is virtually moved to the transmitter end of the coax.

With reference to Fig. 1-15 note that the entire antenna system can be simplified to the resonant circuit of the coax and the resonant circuit of the antenna, with the radiation resistance (feedpoint impedance) as the desired "load" for the transmitter. If the coax is a half wavelength long (or multiple), it effectively will be series resonant and thus a short circuit. If the antenna is also resonant at the same frequency, maximum current will flow in the radiation resistance and hence maximum antenna efficiency.

If the coax is not a half wavelength, changing the length of the antenna can cause the entire circuit to be resonant, but the effect is to cause the coax to be part of the resonant circuit and hence radiation from the coax occurs. In addition, the antenna itself will not be a resonant circuit; hence, the resultant high impedance of the tuned circuit will prevent maximum current flow in the radiation resistance. This is not true if the antenna radiation resistance is the same as the characteristic impedance of the coax. Another effect is that the value of radiation resistance measured at the transmit end of the coax is not the same as the value at the antenna if the coax is other than ½ wavelength and the radiation resistance is other than 50 ohms.

The problem is aggravated by the fact that the antenna is resistive only at the true resonant frequency of the antenna. A few kHz from the antenna resonant frequency, the value of

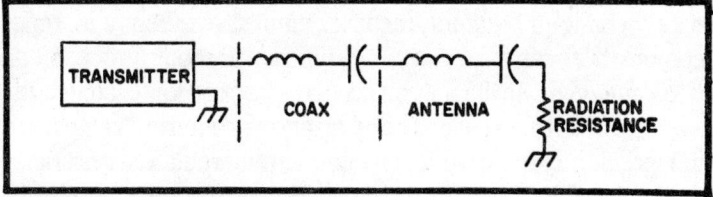

Fig. 1-15. Simplified antenna system.

capacitive or inductive reactance reaches a high value for the antenna itself. Hence, the coax is no longer properly terminated, even if it were 50 ohms at the resonant frequency. The effective bandwidth of the antenna will be greatest if a matching network is used.

Most amateur antennas, particularly 80 or 40 meter dipoles and mobile antennas, have a low value of radiation resistance. These antennas will have low efficiencies unless a matching network is used to overcome losses in the antenna conductor. Most beams use a matching network and are extremely difficult to adjust properly using a SWR Bridge. The SWR Bridge has the basic limitation that it cannot differentiate between resistive and reactive components. Most amateurs, not realizing this limitation, operate their antenna at the frequency of lowest SWR, *which is not necessarily the same as the most efficient frequency*.

If a system is tuned using the following steps, maximum efficiency will occur:

(1) Use half wavelength coax, or multiple thereof, or locate the Bridge at the antenna feedpoint.
(2) Tune the antenna to the desired frequency.
(3) Adjust the matching network for 50 ohms.

If this procedure is used, minimum VSWR will occur at the frequency of optimum efficiency. If the coax is other than ½ wavelength (during tune up), the antenna is not properly tuned, or if the radiation resistance is not the same value as the characteristic impedance of the coax, minimum SWR will not occur at the frequency of highest efficiency.

If the system is operating properly, changing the length of coax will not affect the SWR reading. (Use ¼ wavelength coax for worse case measurement.) Thus, when adjusting an antenna using a SWR Bridge, if the coax line length is alternately changed between measurements, eventually a proper combination can be reached. This is a tedious process and does not give an indication as to what to do to adjust the system. For this reason, an *rf* bridge is required to measure independently the resonant frequency and radiation resistance to allow tuning the antenna to the desired frequency first then to allow adjusting the matching network.

A mobile antenna, particularly on 40 or 80 meters, must utilize a matching network. The nominal radiation resistance is typically very low and when the system is operated at the frequency of minimum SWR without a matching network, the resonant frequency will be out of band and the antenna will not give maximum efficiency.

The same comments hold for beam antennas. Here again, minimum SWR does not necessarily mean that the antenna is resonant and the matching network is adjusted properly. Also, if a coax line of other than ½ wavelength is used, the entire system may be "down in performance" by many db.

When using a multiband antenna, the coax length should be chosen for a multiple of ½ wavelength for all bands. Example: For a 10-15-20 triband beam a ½ wavelength coax on 40 meters (approximately 44 ft. RG-8U) gives 1 wavelength on 20, 1½ wavelength on 15, and 2 wavelengths on 10. If this is too short, the next length would be 1 wavelength on 40 (approximately 88 ft. of RG-8U) which gives 2 λ on 20, 3 λ on 15, 4λ on 10 meters. If this is too long, it is better to coil up the excess in the corner rather than cut it off and not have the proper electrical length of feedline.

To achieve maximum antenna performance these considerations should be applied to all amateur antennas in addition to the normal choice of antenna type, height above ground, etc. These considerations do not rule out the case of using ¼ λ lines for matching networks.

As an example of the relatively critical tuning of antennas, the antenna conductor forms the inductive component of the tuned circuit. The capacitive component uses air as the dielectric, and the capacitive component is distributed along the conductor. If the dielectric constant changes, the resonant frequency of the antenna will change. Among other factors, the relative humidity of the air is influential. A 40 meter dipole has been observed and changes of 200 kHz in the resonant frequency are not uncommon from dry air to measurements made while rain is falling.

COMPUTER DESIGN OF BEAM ANTENNAS

The purpose of this discussion is to acquaint the reader with the technique of using a machine to help make a decision, or even to make the decision for him. Computers are used in the business world to make marketing decisions, and in the Pentagon, cost analysis is employed to justify major decisions on the defense of the country.

Many articles have presented the problems arising in the computer age, with forecasts of dire results to the average man. We hope to persuade readers who have doubts about the value of computers, how such machines can be used for helping us reach decisions, even on the merits of antenna systems, a favorite for discussion at the amateur radio club.

The capability of the computer to do arithmetic at fantastic speeds is used in various "languages" where the user commands the machine to do his bidding. One common language is called "Fortran," and a set of instructions makes a program. We can instruct the computer with a Fortran program to do fast calculations on the merits of various systems, provided we realize that initially it knows nothing about the subject and we must prime it with knowledge. If the final result is wrong, it is not the fault of the machine, but of the programmer.

A relatively new development in the use of computers is in the principle of sharing an expensive machine among many customers. The new techniques in computer design have allowed the calculations to be made in less time than a teleprinter will type out the answer. So it is possible for a large computer at a central location to be asked questions by a person many miles away, using ordinary telephone lines to carry the conversation between the computer and an inexpensive teletype terminal. This can be located in an office or even in a home a considerable distance away. This leads us to your home in the not-so-distant future, where your wife will use a distant computer to help her do her shopping! Your children will have been taught how to carry on a conversation with a machine. In fact, programmers today are working towards the ultimate language, which will approximate normal conversation. The future computer will be instructed and will answer in speech, but today we type our instructions.

Most of us who like to dream of antennas on high towers have limited funds, so there is a height of the tower, in conjunction with the number of elements in the beam, which will give best results for our money. The question we want answered is, for a variable height and variable beam size (number of elements), what is the best value in terms of gain per dollar, and is there an optimum height and number of elements we could use for each band. For our purpose, we shall consider 7 MHz, 14 MHz, 21 MHz, and 28 MHz amateur bands, heights from 20 feet to 100 feet, and beams of the close-spaced Yagi type from one (a rotary dipole) to six elements. Individuals have different ideas on construction techniques, but for amateur home-built antennas we have an idea of the approximate cost of an antenna with a tower. Based on our knowledge, we then derive a formula which seems to be reasonable. Undoubtedly, the formula used will not agree with the experience of many amateurs. However, the principles used can be exploited to suit individual ideas, so your own formulae for gains and costs can be substituted. The cost formula is a little complex, as the tower cost increases with height and also with beam size, requiring greater strength for a heavier load.

First consider power gain relative to a 20 M dipole at 33 feet, or a half wavelength high. As the number of elements increases, the gain increases. As the height increases, the gain increases. As the frequency increases at that height, the gain increases. To simplify the formula for gain we shall assume a linear relationship. So we have Gain = Height × Elements × Frequency divided by a factor A. A 20 M dipole at 33 feet is our reference. Therefore, $1 \times 14 \div A$, $A = 470$. To check A, assume 11 dB gain for a 5 element beam at 66 feet high on 14 MHz.

11 dB is a power gain of 12. Therefore, $1 = 33 \times 12 = 66 \times 5 \times 14 \div A$, $A = 390$. Select a value of 400 as reasonable. Now we have $G = (F \times H \times N)/400$ as one formula for our program. The equation for the cost of the tower plus the beam comes from personal experience and can be varied to suit the indi-

vidual. In this example, the final result was derived, Cost = 60 × Height × Square root of elements plus 600 × Elements all divided by frequency.

We shall break down the system cost into beam, tower and rotator. A reasonable formula for the beam cost is 500 N/F. For example, a 3 element beam at 14 Mc comes to $108, and at 28 Mc would be $54. Some ingenuity will give a 7 Mc 3 element beam for $216. The tower must be stronger as the beam size increases at a given height, so a factor involving N is used in the tower formula, with an estimated cost of 50 × H × $\sqrt{N}$/F. For example, a 50 foot tower to hold a 14 Mc 3 element beam would cost $310. Adding 20% to the total estimated cost for a rotator gives the final formula.

$$\text{Cost} = \frac{(500\,N + 50\,H \times \sqrt{N}) \times 1.2}{F}$$

Therefore, for the computer,

$$C = \frac{60 \times H \times \sqrt{N} + 600N}{F}$$

Value can be defined as performance versus cost, or in this case, decibels of gain for our dollars. Transfer the gain into decibels by a logarithmic function to suit our formula, and we get, V = Log G./C.

We are interested in the *maximum* value obtained from 7 MHz to 28 MHz for all combinations of heights from 20 feet to 100 feet in steps of 2 feet, with antennas from one to six elements.

Each new calculation of V is compared with the previous, and the larger of the two selected. First, at F = 7 MHz for a range of heights from 20 feet to 100 feet, the best value in dB per dollar is found for a dipole (one element). Then the computer goes through the next loop calculating the best value for a two element beam at all heights, comparing with the previous best for a dipole, and storing the maximum in its memory. The process is repeated up to six elements. The best value for the first frequency of interest, 7 MHz, is then printed. Similarly, the maximum dB per dollar is printed for 14 MHz, 21 MHz and 28 MHz. We shall see lines of print informing us of frequency, number of elements, height, gain in dB, and cost for the best value at that frequency.

Note that the figures shown apply to single beams on towers, so it should be obvious that we shall get better value for our money by using a tri-band beam on one tower.

Don Gordon, W4VTT, pointed out that an extension of the program would be to simulate the conditions when we have limited funds available, a common occurrence! He re-wrote the program to find out the best beam size versus height for fixed costs, selecting $300, $400, $500 and $600. The comparison in value now is limited at a fixed cost in each case, and the print has an additional column, allowable cost. The print-out is now frequency, number of elements, height, gain, true cost, and maximum allowable cost. The performance in gain is compared to a half-wave rotary dipole at a half wavelength high, so the dipole on 7 MHz at 24 feet has a loss of 3 dB with reference to a height of 66 feet. If you have $600 available for a 28 MHz beam, using amateur construction techniques, the computer shows a 6 element beam almost 3 wavelengths high with a gain of 15 dB over a dipole at 16 feet.

We have seen how the computer can give answers rapidly when primed with knowledge. Knowledge is gained by learning, and in the average human being is a long, slow and sometimes painful process. But the computer can be rapidly educated, or re-educated. Suppose in our case we do not agree with our educated machine. We can erase the knowledge we primed it with, in other words, alter the formulas we had.

Among DX antenna enthusiasts there is always the old argument of height against beam size. Some believe in height, others believe in large beams. Our program so far can be seen to favor large beams at medium heights for best value for the dollar. For those who disagree, we can re-educate the computer by emphasizing the value of height. Instead of a 3 dB increase on doubling the height, we can assume a 6 dB increase. We now incorporate a new formula for gain. A 6 dB increase is 4 times the power, so power increases as the square of height. The gain formula is now re-written, $G = (F \times H)^2 \times N/A$. Again we know $G = 1$ for a dipole at a half wavelength high. Recalculating for A gives the value of 220,000. We prime the computer with this new formula and run the program again. The results are shown to suit the height-oriented DX chaser.

Note that a new formula for gain does not allow us to compare the actual dB shown in the gain column from one program to another. Gain comparisons should be made only within the formula used. For example, in the list of 28 Mhz antennas at $300, $400, $500 and $600, we see that the increase in gain is 6 dB from $300 outlay to $600 , while at 7 MHz the increase is from—8 dB to 0 or plus 8 dB for a 7 MHz rotary dipole increasing in height from 24 feet to 60 feet.

We can now draw some conclusions from the four programs. The first gain formula, which emphasizes the large beams, shows that at 7 MHz we must be prepared to invest a large sum to get the best value from the antenna system, $1386 for a large 5 element beam at 50 feet high. The second formula for gain, which is oriented to suit the height-conscious amateur asks for even more money for a 3 element beam on 7 MHz at 100 feet high, quite an antenna!

The favorite DX band is 20 Meters, so our two best values there are of interest. The first formula gives a 3 element beam at 40 feet for a cost of $425. The second formula, which emphasizes height, shows almost the same cost for a rotary dipole 100 feet high. It is a debatable point which of the two would give the strongest signal at a remote location.

Now let us examine the systems available for the fixed costs of $300 to $600. Our new gain formula should not make any difference in the inexpensive 7 MHz antenna, but the higher frequencies will be height oriented again. The 14 MHz antenna for $600 in the first instance is a 4 element beam at 50 feet, and in the second case is a 2 element beam at 84 feet. For $300, the choice is between 2 elements at 34 feet and a rotary dipole at 60 feet. At the highest frequency we have considered, 28 MHz, at the maximum allowable cost of $600, we can choose between a six element beam at 88 feet, or five elements at 100 feet.

Consider the time required to calculate all the values by conventional means, then compare our first print-out of maximum values taking 11½ seconds and the second print-out taking less than half a minute of computer time!

The whole conversation with the computer for the two programs is shown, but no attempt is made to explain details.

For those interested, books on Fortran are easily obtainable. You can substitute your own formulae to determine values for quad antennas, and the computer can then answer the old question, yagi versus quad! The result will vary, of course, depending on the individual amateur's formulae, and the total result of this article will probably be to have more heated discussions in the club room.

Fortran for Antenna Cost vs. Performance

I	Freq	Elem	Hght	Gain	Cost
	7	5	50	6	1386
	14	3	40	6	425
	21	3	30	6	234
	28	2	30	6	133

At line No. 310: Stop End, Ran 69/6 Sec.

System—Fortran
New or Old—Old
Old Problem Name—ANTCST
Wait.

Ready.

List

ANTCST   13:22   W1 MON 12/11/67
100 Print "Best Antenna Gain for Fixed Cost"
110 Print 10
120 10 Format (1H14X4HFREQ4X4HELEM
    4X4HHGHT4X4HGAIN4X4HCOST4X4HMAX$)
130 DO 8 F = 7, 28, 7
140 R = Q
150 DO 7 M = 300, 600, 100
160 DO 3 N = 1, 6, 1
170 DO 1 H = 20, 100, 2
180 G = F*H*N/400
190 IF (G-R)1, 1, 2
200 2 C = (60*H*SQRT(N) + 600*N)/F
210 IF (M-C)1, 4, 4
220 4 R = G; P = N; Q = H; S = C
230 1 Continue
240 3 Continue
260 D = 4.3*LOG(R)
270 Print 6, F, P, Q, D, S, M
280 6 Format (618)
290 7 Continue
300 8 Continue
310 End

Run

ANTCST   13:25   W1 MON 12/11/67
Best Antenna Gain for Fixed Cost.

I	Freq	Elem	Hght	Gain	Cost	Max$
	7	1	24	−3	291	300
	7	1	36	−1	394	400
	7	2	26	0	486	500
	7	2	34	0	583	600
	14	2	34	3	291	300
	14	3	36	5	395	400
	14	4	38	7	497	500
	14	4	50	8	600	600
	21	4	32	8	297	300
	21	4	50	10	400	400
	21	6	46	11	493	500
	21	6	60	12	591	600
	28	4	50	11	300	300
	28	5	60	13	394	400
	28	6	70	14	495	500
	28	6	88	15	590	600

At Line No. 310: Stop End, Ran 169/6 Sec.
System—Fortran
New or Old—New
New Problem Name—Antval
Ready.

Tape
Ready.

```
100 Print "Fortran for antenna cost vs. performance"
110 Print 10
120 10 Format (1H14X4HFREQ4X4HELEM4X4
    HHGHT4X4HGAIN4X4HCOST)
130 DO 4 F = 7, 28, 7
140 Z = O
150 DO 1 N = 1, 6, 1
160 DO 2 H = 20, 100, 2
170 G = F*H*N/400
180 C = (60*H*SQRT(N) + 600*N)/F
190 V = (LOGF(G))/C
200 IF (Z — V)3, 3, 2
210 3 Z = V
220 P = N
230 Q = H
240 R = 4.3*(LOGF(G))
250 S = C
260 2 Continue
270 1 Continue
280 Print! 5, F, P, Q, R, S
290 5 Format (5I8)
300 4 Continue
310 End
```

Key
Ready.

Run
Wait

ANTVAL   13:32   W1 MON 12/11/67
170 G = (F*H) ↑ 2*N/220000
Run

Anyval   13:47   W1 Wed 12/27/67

Fortran for Antenna Cost vs. Performance

I	Freq	Elem	Hght	Gain	Cost
	7	3	100	8	1741
	14	1	100	9	471
	21	1	70	9	228
	28	1	54	10	137

At line No. 310: Stop End, Ran 72/6 Sec.

180 G = (F*H) ↑ 2*N/220000
Run

ANTCST   13:41   W1 WED 12/27/67

Best Antenna Gain for Fixed Cost

I	Freq	Elem	Hght	Gain	Cost	Max$
	7	1	24	−8	291	300
	7	1	36	−5	394	400
	7	1	48	−2	497	500
	7	1	60	0	600	600
	14	1	60	5	300	300
	14	1	82	7	394	400
	14	1	100	9	471	500
	14	2	84	10	594	600
	21	1	94	12	297	300
	21	2	84	14	396	400
	21	3	82	15	491	500
	21	3	100	17	580	600
	28	2	84	16	297	300
	28	3	90	19	398	400
	28	4	96	20	497	500
	28	5	100	22	586	600

At Line No. 310: Stop End, Ran 182/6 Sec.

YES, I'VE BUILT SIXTEEN LOG PERIODIC ANTENNAS

W4AEO from Camden, South Carolina has built sixteen log periodic antennas for experimental purposes. The broadband, uni-directional HF Log Periodic beam antenna was originally developed about 1957. Although these very excellent beams are used extensively by Commercial, Military and Government agencies for both medium and long haul circuits, their use has been rather neglected by amateurs.

It is believed the amateur fraternity may have overlooked or shied away from these antennas due to:

1) Very little information has been published on HF Log-Periodics in ham publications although there have been several articles covering these for VHF and UHF.

2) These antennas are quite complex and are highly mathematical. Several pages of formulas, reference to log tables and four or five graphs or monographs are required for optimum design.

The antenna manufacturers producing L-Ps for Commercial and Military use, program this data on a computer. By supplying the frequency range desired, gain required, etc., the computer prints out the element lengths, optimum element spacing, boom length, etc., to provide for maximum forward gain, front-to-back ratio, minimum beam width etc.

Although these formulas can be computed manually, several days may be required to design (on paper) an L-P having optimum performance in a given space.

3) Most amateurs feel that Log-Periodics are extremely expensive, which they are if purchased. The least expensive rotatable types by one commercial manufacturer are in the $1500 to $3000 range for a rotary covering 6 to 30 MHz, capable of 40, 20, 15 and 10m operation. Some of these are used by MARS stations. Rotatable L-P ham antennas have recently been announced in the $300 to $1000 class.

The larger fixed types for the 2-30MHz range having higher gain are generally in the 10-30 "kilo-buck" range. However, by assembling smaller, less complicated wire L-Ps for the 14-30MHz range on a "do-it-yourself" basis, one having an

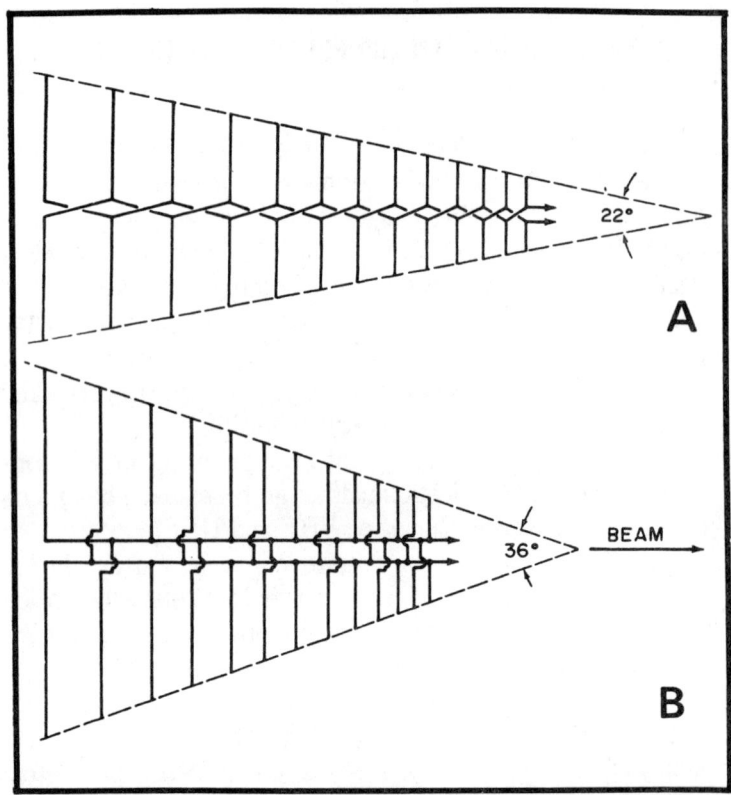

Fig. 1-16. Doublet log periodic configuration. This will cover a 2:1 bandwidth, say 7-14 MHz or 14-28 MHz. (a) has a 22° aperture angle and gives about 10 dB gain. Note the criss-cross method of transposition of the feeder. (b) is shorter, with a 36° aperture and about 8 dB gain. Note alternate method of transposition of the feeder.

8-10dB forward gain (over a doublet at the same height) can be assembled for a material cost of $15 to $25, not including masts or coax which will vary, depending on the particular site.

4) Many amateurs believe a fixed L-P requires a great deal of "acreage." This is true of the large commercial types having a 10:1 band width or a single beam covering 3 to 30 MHz. These are 63.5—127 meters (250'—500') in length, some even 203 meters (800'). However a 14—30 MHz L-P for 20—15 and 10m having an 8dB gain can be erected in a space 10.16m (40') wide by 12.7m (50') long. If the length can be extended to 17.78cm (70') the gain can be increased to

10dB compared with a doublet at the same height. By extending to 25.4m (100′), 12-13dB can be realized.

Log-Periodic Types

Log Periodic Antennas can be classified under three general types:

1) The doublet Log-Periodic (DLP) Configuration. Figure 1-16 illustrates this type covering a 2:1 (plus) bandwidth suited for a ham beam for 7-14.35 or 14-28MHz.

2) The vertical monopole Log Periodic working against ground or a ground plane counterpoise. Figure 1-17 illustrates this type, also covering a 2:1 band width.

3) The trapezoidal zig-zag or saw tooth configuration, Figure 1-18. This type being more complicated and not too suited for HF ham applications, will not be covered by this article which will deal only with the first two types.

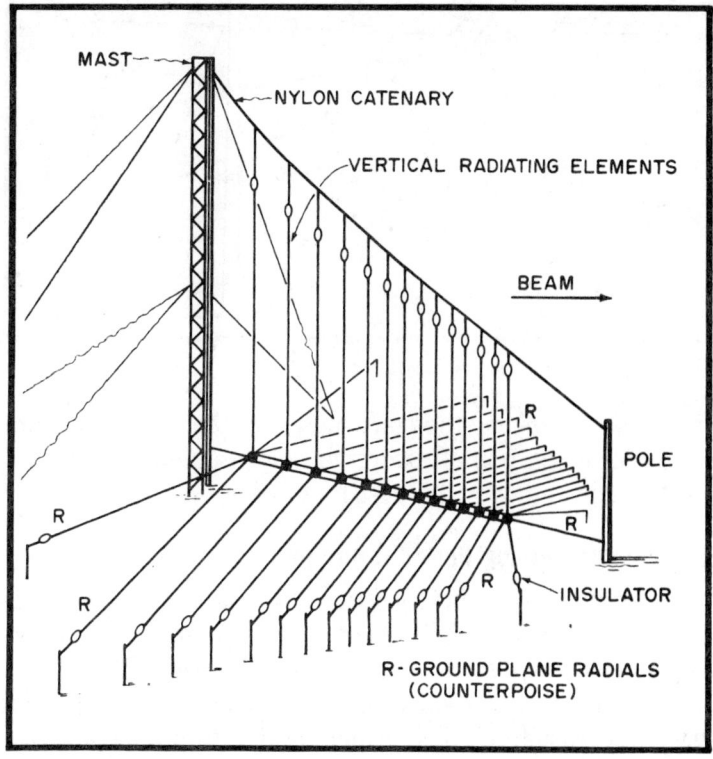

Fig. 1-17. Vertical monopole log periodic-2:1 bandwidth.

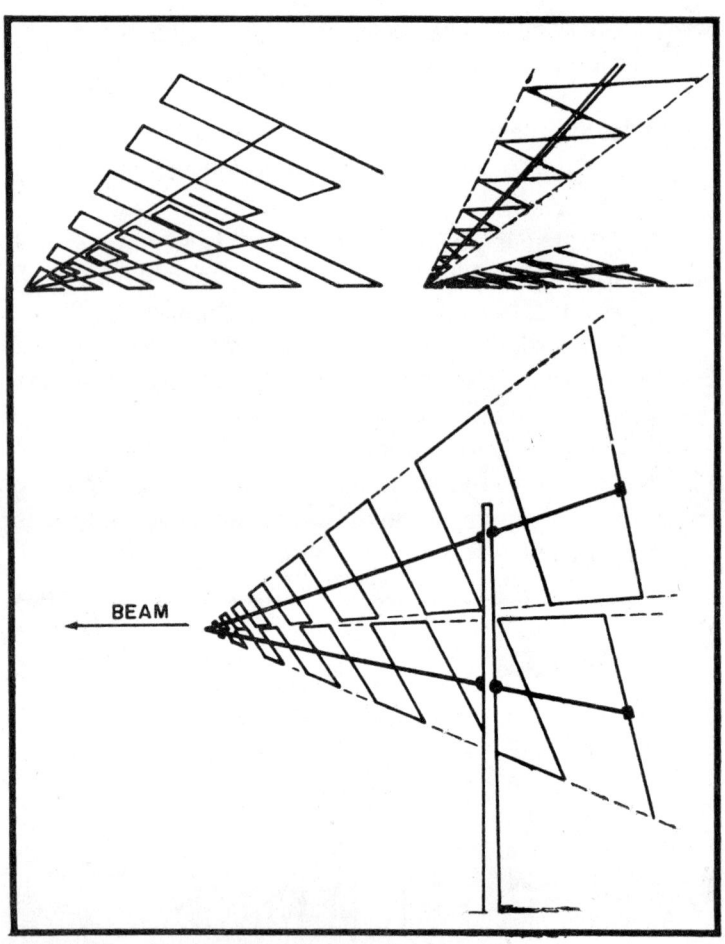

Fig. 1-18. Trapezoidal log periodics.

Before outlining the construction of the doublet and the monopole types, a brief report will be presented covering the tests conducted over the past four years.

Test Results on Log Periodic Antennas

During 1970 the first Log Periodic was put up experimentally for 20m and 15m only, to be compared with doublets and also a well known "store bought" trap vertical for 40-20-15 and 10m (using separate radials for each band). The vertical had given fair results for DX, evidently due to its low angle of radiation and its 8.9m (35′) height (at the base) above ground.

The first L-P was quite simple, using only 7-elements for 20 and 15m and only 9.7m (38′) in length. It is supported at the rear end by the peak of the roof, 10.2m (40′) above ground, and the forward end by two cedar trees about 11.4m (45′) high. It is beamed South where other hams in South America were capable of making good comparisons with the non-gain antennas previously used.

The results of these first tests were amazing as were the stations being worked. Reports on the non-gain antennas (at the same height as the L-P) normally gave reports of S8-9 on 20m from these stations. Switching to the L-P, these stations would generally report an increase of two S-units, or at least a 10 dB increase over the doublet. Usually, when the doublet was giving S-9, they would give "20 over" on the L-P. Al-

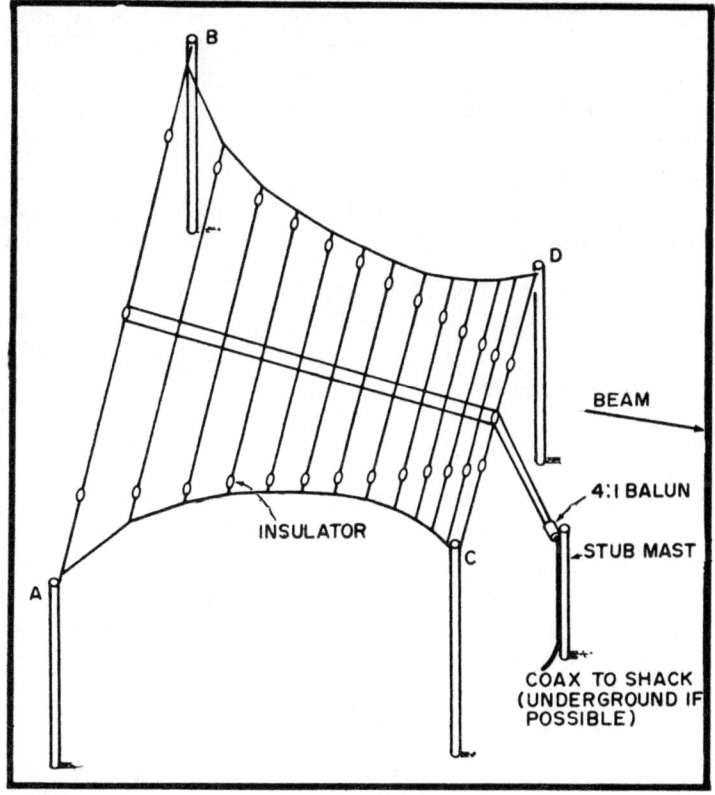

Fig. 1-19. For method of transposing the center feeder see Fig. 1-16B and Fig. 1-21. Illustrates the four masts used to support the antennas.

though a 20dB gain would seem exaggerated, the "S" meter at this end would generally confirm this increase on their signal when switching to the L-P.

It is realized that many "S" meters exaggerate but most are fairly linear and can be used for *relative* comparisons at the lower levels. Further, the "S" meter here correlated very closely with the gain figures reported when switching to the experimental L-P.

Although the original L-P, Fig. 1-19 would only have a theoretical gain of 8-10dB, L-P gain figures are often based on VHF or UHF models tested over a line-of-sight path. It is noted that one of the large manufacturers of Commercial and Military HF Log Periodics, rate their 10-12dB gains *"over average soil conditions."* It is therefore believed that this first experimental L-P gives an honest 8-10dB gain by averaging the many reports received from various stations to the South over the past 4 years. Generally, if a station reports a two S-unit or 12dB increase when switching from the doublet to the L-P, the "S" meter here normally shows the same increase in his signal.

Since the original, simple 7-element (L-P #1) for 20 and 15m was put up in 1970 it has continued to give excellent results and is still being used as of this writing. Several others having more elements and greater length, providing greater gain, have been put up and thoroughly tested. Briefly, these are (in the order tested):

L-P #2. 12-element, 17.8m (70′) in length for 20-15-10m.
L-P #3. 12-element 6.35m (25′) length for 15-10-6m.
L-P #4. 12-element, 10.16m (40′) length for 20-15-10m.
L-P #5. (#2 tested on edge in the vertical plane or vertically polarized for about two weeks).
L-P #6. 13-element, 22.86m (90′) length for 40-20-15m. This was a "skip band" type with a portion between the 40 and 20m bands omitted. Two of these are now being assembled for permanent North and South beams.
L-P #7, 5-element, 12.7 m(50′) length for 40 m only.
L-P #8. Two 5-element (same as #7) for 40 only; back-to-back in an inverted V configuration suspended by a single center support line. One beamed North, one South—exactly

180° difference. Put up to obtain additional and more accurate forward gain and better front-to-back data on 40m.

L-P #9. Improved 5-element, 40m only at increased height for additional forward gain data. Aimed South. Gave consistent 10dB gain over doublet "standard" at same height.

L-P #10. 5-element 10m monoband L-P. (See reference 18.)

L-P #11. 17-element, 25.4m (100′) length for 20-15-10, 15.24m (60′) above ground. This is the permanent West beam which has a measured 12-13dB forward gain to the West. By far the best and highest gain L-P installed to date. Side attenuation is down 25-30dB.

L-P #12. 6-element, 12.7m (50′) length. Experimental for 20m only. 10dB gain. Four additional forward parasitic directors (non-driven) were added later but little if any increase in gain could be noted.

L-P #13. 5-element vertical monopole Log-P for 40m only, using ground plane radials or counterpoise. Although this L-P gave a 10dB gain, it had an extremely low angle of radiation. Was good for DX but horizontal doublet type. L-P #7 or #9 was better for normal operation.

L-P #14. Same as #13 except inverted as an "up-side-down" inverted ground plane. Strictly an experimental antenna to try for an even lower angle of radiation.

L-P #15. 5-element vertical monopole Log-P for 80m only. Results similar to 40m monopole, L-P #13. Good for DX but poor for close stations. Gave 10dB gain (over 80m doublet at 11.43m, 45′) from stations greater than 1500 miles.

L-P #16. Trapezoidal L-P for 20 and 15m only, both the zig-zag and the saw-tooth types tested.

In addition to the above L-Ps designed and tested here, several other directional antennas were erected for comparison with the L-Ps. Some of these were:

1) A 6-element 15m "Long John" yagi mentioned later.

2) A 20m phased beam consisting of two $1/2\lambda$'s in phase, collinear with two collinear reflectors and two collinear directors beamed toward Europe. Although this showed approximately a 10dB gain, the lobe was much more narrow than the NE L-P and the band width quite narrow. At ±50kHz, the SWR exceeded 1.5:1.

3) A 5-element Bruce array on 20m beamed for Caracas. The gain was lower than any of the L-Ps tested in that direction or possibly, being vertical, the angle of radiation may have been too low for this distance. It was only tested a few weeks.

In addition to the ham L-Ps assembled here, several other L-Ps have been designed "on paper" for friends and others, one covering 12-24MHz for several MARS frequencies as well as 20 and 15m. Several commercial L-Ps for 3-30MHz, 2-4, 4-8, 6-12, 8-16MHz and several VHF and UHF for 30-50, 140-145, 150-470 MHz, including two for TV: 174-215 and 475-750MHz. Several have been completely assembled for others on "custom built" orders.

YV5DLT—W4AEO Tests

The most accurate 20 and 15m tests have been made with YV5DLT of Caracas, Venezuela. We have been constantly testing the L-Ps for several years. He is able to give very accurate readings on any changes made here.

During the original testing of the first three L-Ps, schedules were kept daily between 1200 and 1400 local time here as these hours gave the worst case conditions on 20m. Other schedules were kept on 15m.

It was during this period that the 17.78m (70′) L-P #2 and the 15 and 10m L-P #3 were put up for comparison with the original L-P #1 which had performed so well on both 20 and 15m. L-P #3 was especially good during the 15m tests, generally showing 5 dB over L-P #1 and even slightly better than L-P #2; however #3 was aimed at approximately 165°. Caracas is 149° true, 1854 miles Statute. The other two L-Ps were approximately 180°. All three were about the same height above ground.

After several months of 15m tests on #3, we wished to make a direct comparison with a good yagi aimed in the same direction. I assembled a 6-element "Long John" Yagi per. This was erected to the side of L-P #3, exactly parallel and aimed in the same direction; both 11.43m (45′), or about a full wave above ground.

Several weeks were spent comparing these two beams. Invariably YV5DLT would report L-P #3, 3-5 dB better than the yagi. The "S" meter readings here confirmed this.

40m L-P Tests

Most of the 40m tests were conducted over a period of several months with W4QS and K4FBU in Florida at the same time daily. During this period four different 40m L-Ps were beamed South for Florida at various times for comparison with a good 40m horizontal doublet at 11.43m (45'). One 40m L-P #8 was also beamed North for comparisons in that direction. All of these L-Ps produced 8-10dB gain in these directions over the dipole; however, many of the tests indicated as much as a 20dB improvement which was confirmed by the "S" meter at this end and a number of other stations in various parts of Florida.

Since the usual 2-element 40m yagi or two extended ½λ's in phase collinear do not normally exceed 3-4.8dB gain, the 10dB average gain of the L-Ps tested is worth considering; especially because of their low cost and ease of construction.

75 or 80m Vertical Monopole L-P Tests

A 5-element vertical monopole L-P, #15, was assembled for 75m. Since the mast height limited the longest rear element (the reflector) to 16.51m (65', ¼λ + 5%) this L-P was limited to 3.8—4.0 MHz, and all tests were within this range.

It was soon evident that this vertical beam was strictly for longer range communications, due to its lower angle of radiation. The ½λ 80m dipole up 45° (not an inverted V) used as the "standard" was better for distances from 400-500 miles. Beyond this range the vertical L-P was better in the forward direction. At night the doublet was better to about 1000 miles; beyond, the monopole L-P would show its increase, giving a good gain over its beam width.

For ranges greater than 1000 to 1500 miles, the 75m monopole, L-P #15, showed at least a 10dB gain over the dipole. However, for the normal working range on 80m or 75m, the doublet was better for the shorter distances.

A similar test using a 5-element 40m vertical monopole, L-P #13, was conducted with similar results as the 75m test. The horizontal doublet type 5-element 40m L-Ps #7, 8 or 9, being better for normal operations and the vertical monopole for DX. This beam was aimed NW.

During a pre-dawn 40m test with L-P #13, a W7 (working a VK on phone) in the NW about, 2000 miles from here, was monitored. On repeated "S" meter readings taken, the monopole was consistently 2 "S" units or 12dB better than on the 40m dipole when receiving the W7 in line with the monopole beam.

Receiving Advantages of the Log—Periodic

In addition to the excellent forward gain of the L-P which is quite apparent to those being worked, the received gain is also quite noticeable. Another plus factor of the L-P is its excellent diversity or "capture" effect during reception.

When QSB is bad on the dipole used as the "standard," switching to the L-P reduces fading considerably, since the "readability" on the L-P is much better.

Evidently the number of elements and its "boom length" produces the diversity effect due to its size and length compared with the doublet or even a smaller 3 or 4 element beam. The greater the number of elements and the greater its length, the better it performs for reception in addition to the increased gain apparent on both transmission and reception.

For those more acquainted with the yagi, the L-P can be considered as a multi-element, uni-directional end-fire array having a driven (rear) reflector, a ½λ driven "active" radiator and a number of forward driven directors.

L-P theory implies that for a given discrete frequency within its bandwidth, 5-elements are generally excited or driven as an "active cell." However, while testing the 17.78m (70'), 12-element, L-P #2, it was excited with low power on 20m. Rf voltage could be detected (using a neon bulb) on *all* elements except the long rear (reflector) element. The second or ½λ driven element (on 20m) was quite "hot" at the ends as would be expected. The rf voltage on the following driven director elements 3, 4—11 and 12, decreased gradually toward the forward end. Some rf could still be detected on the short forward element #12.

Evidently these multi-element, driven directors add gain and also possibly help lower the angle of radiation in the E plane and concentrate the forward lobe in the H plane. This may be the reason the apparent gain generally exceeded the theoretical during tests.

Front-to-Back Ratio

The front-to-back of the L-P is generally less than a well designed mono-band yagi. The L-P seems to be 14-15dB maximum with 10 to 13dB as typical. From the tests made here, the front-to-back improves as the L-P is raised to at least a ½λ above ground (at its lowest cut-off frequency).

The front-to-back of the 40m dipole L-Ps (DLP) tested appeared to be better for the horizontal than the inverted V configuration, as would be expected and the forward gain also better.

The Forward Lobe

The forward lobe of the L-P is generally wider (about 90-100° beam width) than that of a well designed yagi; however, for a large fixed beam this is good as it can be aimed to cover a certain part of the country or a particular DX continent. For example the NE (L-P #2) covers Europe quite well and the 30.48m long, 17-element West beam (L-P #11) seems to cover all of Australia. The side attenuation of this long L-P is down 25-30 dB.

A W1, -2 or -3 could use one or two L-Ps to cover most of the states. A W6 with an L-P beamed East would cover most of the East Coast. At this QTH 4L-Ps will cover most continents of interest: NE, Europe; East, Africa (and Australia long path); SE or South, South America; West, Australia; and NW—Alaska, Japan, etc. One for SW may be tried later for long path to Europe.

Fixed Beam Antennas vs Rotaries

An advantage in using several fixed beams over a single rotary is that they can be switched instantly from one to the other (and to the doublet used as a "standard"), whereas, it takes some time for the rotary to swing, making quantitative readings difficult, especially when QSB is bad.

The following comments are comparisons of the L-P with several other beams.

Compared with the Yagi

As more hams no doubt use yagis than other beams, these will be compared first. A well designed and properly adjusted 3 or 4-element mono-band yagi should give about the same gain as a moderate size 20-15-10m L-P when both are at the same height above ground. The L-P will, of course, cover all frequencies 14 and 28MHz and can be operated with a comparatively flat SWR any place in the three bands. The band width of a high Q yagi may be limited to a portion of a band as the band width at resonance may be only 2.5%.

Compared with a tri-band yagi for 20-15-10m, which is generally a compromise antenna, the L-P should give the greater gain.

Of all the contacts made while testing these L-Ps during the past four years, not a single station worked (most using yagis for 20, 15, and 10) had a doublet for use as a "standard" or test antenna for comparison with his beam. Many have been most cooperative in rotating their yagis the full 360° to demonstrate the front-to-back, but none were able to demonstrate its forward gain. The front-to-back on some of the mono-band yagis was quite good, while others were very poor.

One MARS station worked had both a rotatable L-P and a yagi. He obliged by rotating the L-P 360° which gave a good demonstration of its pattern. When both antennas were beamed in this direction, the L-P showed greater gain; however, he did not have specifications on the yagi.

An advantage of having several fixed beams for various directions is that they can be selected instantly by a coax switch or relay. This allows for more accurate data in comparing antennas. Even under fading conditions a fair comparison can be made by switching rapidly and averaging the readings.

Compared With a Rhombic

Anyone having room for a rhombic certainly has room for several L-Ps for various directions and is then not limited to one direction as is the rhombic.

The TCI engineers (Technology for Communications International of Mountain View CA) advertise their "Extended

Aperture" L-P which is only 60.98m (200') in length and has a gain of 17 dBi. A rhombic to produce this gain requires a length of 518.29m × 228.66m (1700' × 750') width according to the TCI ads.

Further, the gain of a rhombic generally decreases at its low frequency end (less wavelengths per leg), whereas, the gain of the L-P is approximately the same over its bandwidth. If anything, at least from the tests here, the L-P seems to give slightly better gain at the low frequency cutoff end. The forward lobe of the L-P is generally wider than the rhombic, requiring less accurate aiming than the latter.

Compared with Phased Arrays

To date only two comparisons were made with phased arrays on 20; a 5-element Bruce and a 6-element collinear array mentioned earlier, both strictly single band antennas. Neither gave the performance of the L-Ps.

The SWR of Log-Periodics

As a general rule the SWR of a L-P does not exceed 2:1 over the band width for which it is designed, i.e., 14-28MHz. From the tests, the SWR over an entire band, 7.0-7.3; 14.0-

Table 1-1.

SWR Readings						
kHz	LP =1 7-element 20 & 15	LP =2 12-element 20—15—10	=11 17-element 20—15—10	kHz	=9 5-element 40 only	LP =15 5-element monopole For 80m only
14.0	1.1:1	1.4:1	1.4:1	3.5	NA	1.2:1
14.1	1.1:1	1.5:1	1.4:1	3.6		1.2:1
14.2	1.02:1	1.6:1	1.3:1	3.7		1.1:1
14.3	1.02:1	1.7:1	1.2:1	3.8		1.2:1
14.35	1.01:1	1.7:1	1.1:1	3.9		1.4:1
				4.0		1.25:1
21.0	1.01:1	1.1:1	1.3:1			
21.1	1.01	1.2:1	1.15:1	7.0	1.05:1	
21.2	1.05:1	1.3:1	1.05:1	7.1	1.05:1	
21.3	1.15:1	1.4:1	1.01:1	7.2	1.01:1	
21.4	1.25:1	1.4:1	1.02:1	7.3	1.1:1	
21.45	1.3:1	1.5:1	1.1:1			
28.0	NA	2.0:1	1.5:1			
28.2		1.5:1	2.0:1			
28.4		1.6:1	2.25:1	Also see SWR readings for mono- band L-Ps, Aug 1973 issue of 73 Magazine, p. 23 and 24.		
28.6		1.6:1	2.0:1			
28.8		1.8:1	1.3:1			
29.0		2.0:1	1.01:1			
29.2		1.6:1	1.5:1			
29.4		1.6:1	2.0:1			
29.6		1.4	2.0:1			
29.7		1.3	2.7:1			

Table 1-2.

Element Lengths and Element Spacing Distances

LP = & Length Bandwidth	LP = 1–38' 7-element 14–22 MHz	=2–70' 12-element 14–30 MHz	=4–40' 12-element 14–30 MHz	=7–50' 5-element 40 only	=11–102 17-element 14–30 MHz	Exp 25' 5-element 20 only
Element =1 (Overall Length)	36'	36'	36'	70'	36'	35'
2	32	32	32	64	34	33
3	28	29	28	56	31	28
4	24	26	25	49	29	24.5
5	21	22.5	22	40	26.5	20.5
6	18	20.0	20		24.0	
7	16	18.0	17.5		22.0	
8		16.0	15.5		21.0	
9		14.0	13.5		18.5	
10		12.0	12.0		17.0	
11		11.0	10.5		16.0	
12		10.0	9.5		14.5	
13					13.0	
14					12.0	
15					11.0	
16					10.0	
17					9.5	
Total Wire For Elements	175'	246.5'	231.5'	279'	345'	141'
Spacing =1 Distance	0'	10'	6'	14'	14'	7'
2	7.25	9	5.4	13	10	6.5
3	6.25	8.25	4.5	12	9	6.0
4	6.0	7.2	4.25	9	8.5	5.0
5	5.5	6.9	3.6		7.5	
6	4.25	5.7	3.5		7.0	
7		5.35	3.2		6.5	
8		4.8	2.8		6.0	
9		4.3	2.5		5.5	
10		4.0	2.0		5.0	
11		3.4	1.8		4.7	
12					4.2	
13					3.8	
14					3.5	
15					3.3	
16					3.0	
Boom Length	37.25'	68.9'	39.55'	48'	101.5'	24.5'
X2 Feeder Wire Required	74.5	137.8	79.1	96	203.0	49.0
+ Element Wire	175.0	246.5	231.5	279	345.0	141.0
Total Wire	249.5	384.3	310.6	375	548.0	190.0
Apex Angle	29° (α = 14.5)	22° (α = 11°)	36° (α = 18°)	32° (α = 18°)	16° (α = 8°)	32° (α = 8°)
Approx. Gain	8–10 dB	10 dB	8 dB	10 dB	12–13 dB	10 dB
For Bands	20 + 15	20–15–10	20–15–10	40 only	20–15–10	20 only

14.25 or 21.0-21.45 does not exceed 1.5:1. Table 1-1 gives some of the readings taken from several of the L-Ps tested.

Log-Periodic Site Selection

The first step is to determine if space is available for the L-P when beamed in the desired direction. The second step is to decide the desired band width or the bands it must cover and the gain desired. These will, of course, determine the size (length) of the L-P and if it will "fit" the space available.

The long rear element (reflector) must be at least 5% longer than the lowest cutoff frequency. The short forward element should be 50% shorter than the high frequency cutoff. The pages of math required for their complete design will not be presented here.

To simplify the design and eliminate the formulas entirely, Table 1-2 presents in tabular form some of the doublet type L-Ps (DLP) assembled and tested for the ham bands as mentioned earlier.

This tabulation gives frequency band width, element lengths and element spacings, overall (boom) length, apex, angle, etc. of each.

Similar information on the vertical mono-pole L-Ps for 40m and 80m is supplied by Fig. 1-25.

If space is available for a L-P at your QTH, at least one of these can be tried.

Figure 1-19 is a sketch illustrating four masts used to support a typical DLP for 20-15-10m. These masts can be inexpensive 12.20m (40′) collapsible guyed TV masts, power poles, towers, trees (as used here) or other supports if available.

Figure 1-20, illustrates two high and four stub masts for an inverted V-Log-P is called the "Λ-Log-P" configuration.

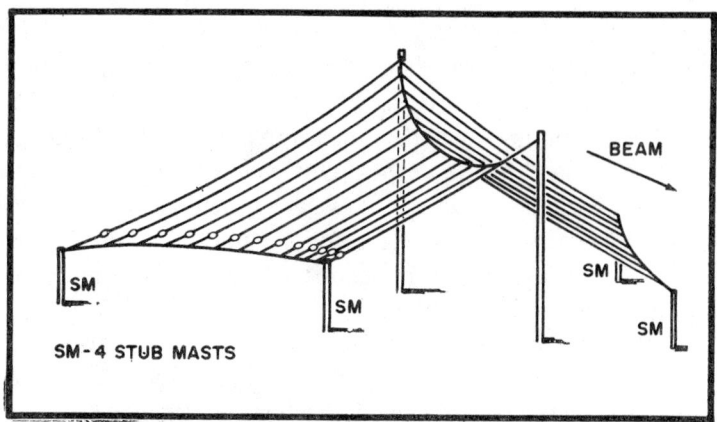

Fig. 1-20. Inverted Vee log periodic.

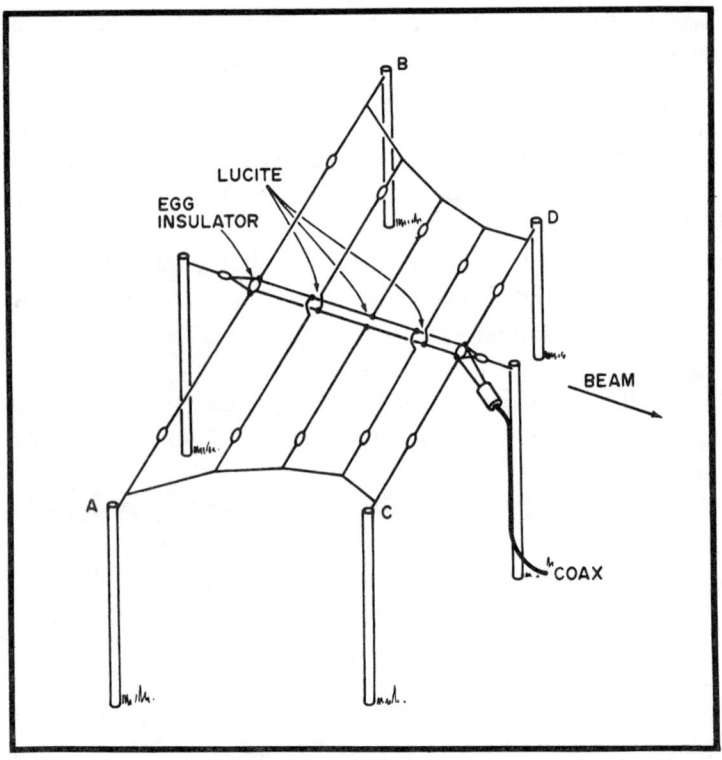

Fig. 1-21. Five element monoband log periodic—fine for any band 10 thru 80m.

Figure 1-21 illustrates a simple 5-element mono-band L-P which requires the least space. This is especially adapted for 40m.

Figure 1-22 illustrates an "acreage saver," using a DLP on edge in the vertical plane. This only requires one high and one lower mast and little width.

This one is only suited for the higher bands due to the rear mast height. The vertical DLP will usually have a lower angle of radiation than an equivalent horizontal DLP. It will generally not be too good for short-haul on 20m or 15m but might be better on longer, multi-hop circuits. The one tested worked extremely well on 10m.

Being vertically polarized, it is more subject to man-made QRM. This type is only suggested as a space saver or possibly mounted on the roof of a building where length may be available but with insufficient width for a four mast horizontal DLP.

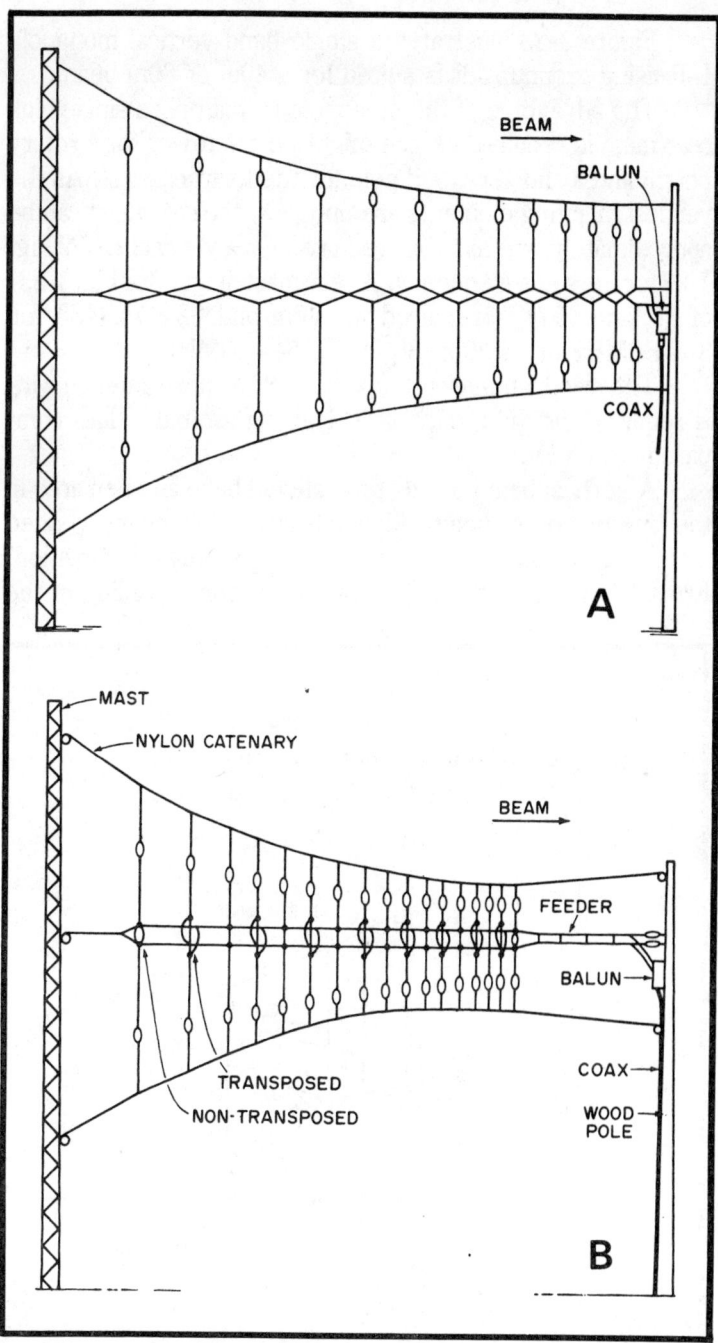

Fig. 1-22. Vertical dipole log periodic—acreage saver model.

Figure 1-23 illustrates a single band vertical monopole L-P using ground radials suited for a 40m or 80m beam.

The advantage of the monopole is that only a single high rear mast is required (which might be the tower for a rotary beam) and a shorter wood pole for the forward mast. As the vertical radiating elements are only ¼λ, the rear mast can be approximately one half that required for a vertical DLP, Fig. 1-22 for the same frequency. A rear mast height (for Fig. 1-23) of 15.24m (50′) is required for 40m and 22.87m (75′) for 3.8-4.0MHz or 24.39m (80′) for 3.5—4.0MHz.

The disadvantage is that at least 30% more antenna wire is required for the monopole L-P using ground radials compared with a DLP.

A vertical beam of this type should have an open area in the direction of the beam. Aiming toward a hill, heavy wooded area, etc., should be avoided due to its low angle of radiation. From the tests made, a two or three story dwelling in the

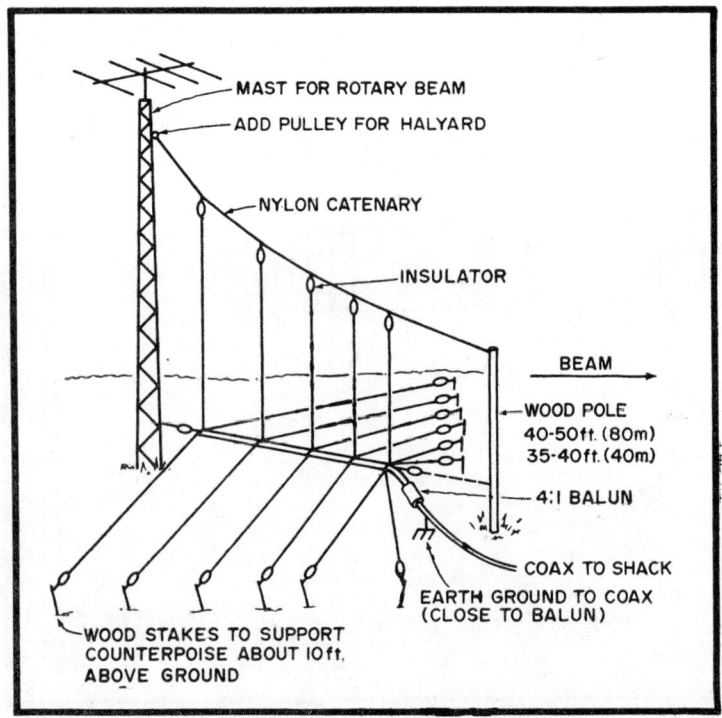

Fig. 1-23. Single band vertical monopole—for **40 or** 80m. About 10 dB gain.

beam's path seems to give about 5dB attenuation. No doubt the plumbing, electrical wiring or air conditioning ducts either resonate or give sufficient screening to cause this attenuation. It is, therefore, suggested that vertical beams be used only on open terrain, having good ground conductivity. Avoid trees or other obstacles in the path of the beam.

The ideal location for a vertical beam of this type would be at a coastal area as near the shore line as possible with the beam aimed seaward toward a DX continent. Those lucky enough to have such a location would no doubt have excellent results with a monopole L-P having a 10dB gain on 40m or 80m. One aimed across a lake might also be good.

A vertical monopole for both 40m and 80m of the "skip band" type is not out of reason but would require at least 45.73m (150′) in length by 42.68m (140′) or 6,042.44m^2 (21,000 sq. feet) of open space which is quite an area except for one lucky enough to live on a ranch or farm.

The following is a step-by-step procedure for assembling simple, inexpensive 2:1 bandwidth DLPs for 20-15-10m, single band L-Ps for 40m or 20m and 40m or 80m vertical monopoles.

Log-Periodic Assembly Procedure

After determining if there is sufficient area for the L-P when aimed in the desired direction, it is suggested that a scale drawing be made showing the proposed mast locations for the L-P as it will be when suspended from the masts. By drawing this to scale, it is quite easy to determine any needed or unknown dimensions.

Next procure the necessary material for the L-P selected. Figure 1-24 illustrates the construction or assembly of a typical DLP and Fig. 1-25, the monopole L-P configuration.

Note that for the long rear element (#1) and the short forward element of a horizontal DLP, small ceramic egg type compression insulators are used as these two end elements carry most of the load or strain of the center 2-wire open feed line and its center insulators or spacers. The latter are home made from .64m (¼″) thick Lucite or Plexiglass. This can

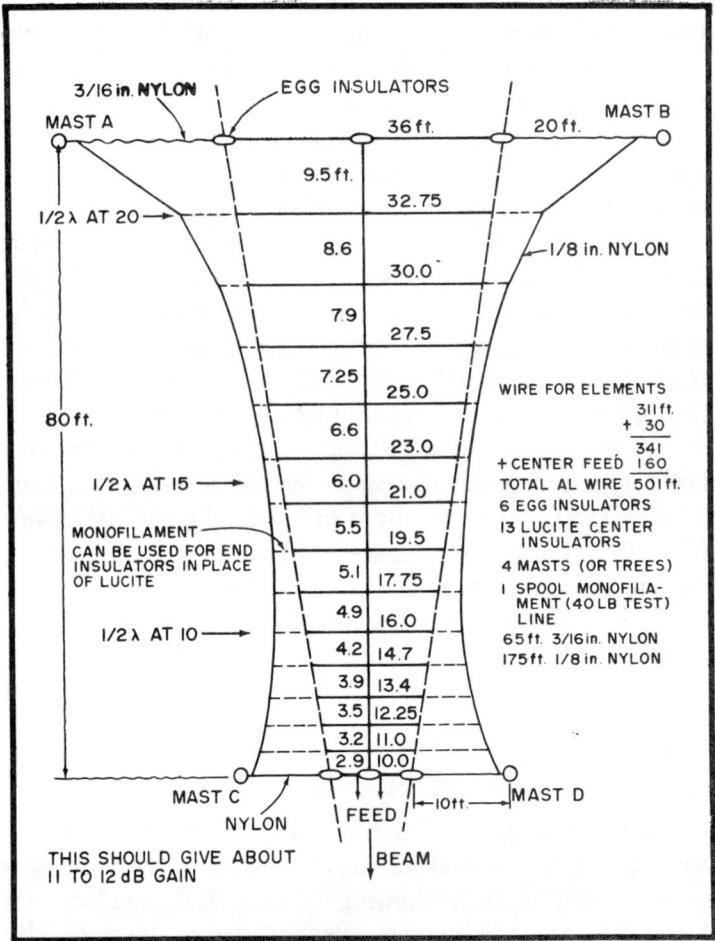

Fig. 1-24. 15 element 20/15/10m periodic.

usually be purchased at hardware, building supply or radio stores.

The Lucite is cut into strips 1.59cm wide × 15.24cm long (⅝" × 6"). These are then drilled to make three type insulators for the L-Ps, which are:

1) End insulators for all elements (except the front and rear as mentioned above). Two holes are drilled in this type.

2) Center insulators for the DLP center feeder which serves as the center insulator for all elements (except front and rear), also supporting and spacing 10.16cm (4") the 2-wire center feeder. 4-holes drilled.

3) Center insulator for the monopole L-P. Same as the DLP type except these have an extra center hole for securing to the ¼λ vertical elements. For this type the two outside holes are for securing the ¼λ ground radials or counter-poise.

The hole spacings for above are illustrated in Fig. 1-26. These are all the same size to simplify production.

Lucite is used for these as it is difficult to locate a ceramic insulator of this type. The Lucite is light in weight, easy to cut

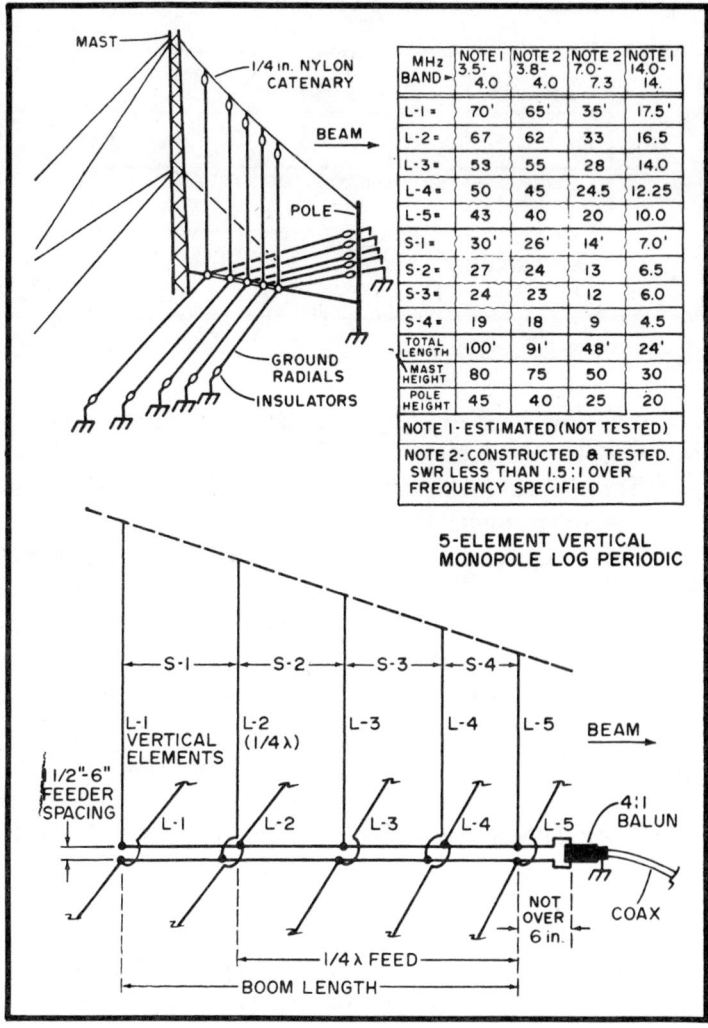

Fig. 1-25. 5 element vertical monopole log periodic.

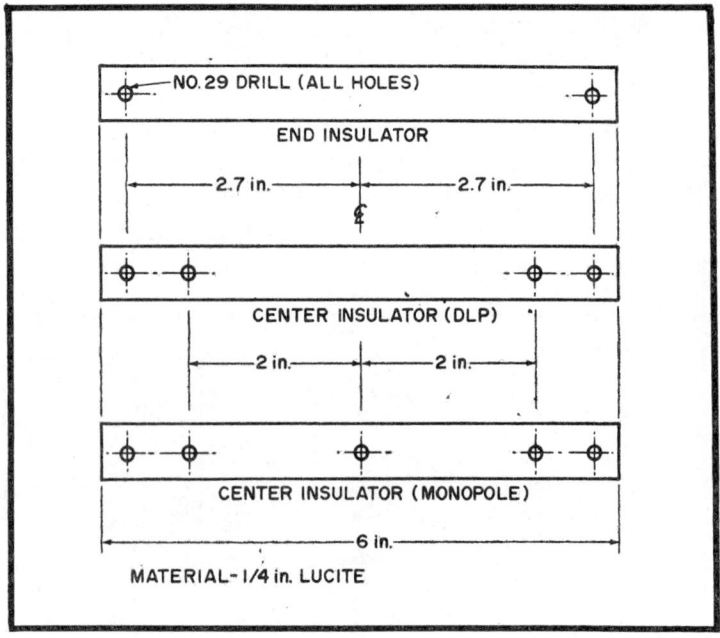

Fig. 1-26. Hole spacings for the insulators.

and drill, low loss and less expensive than commercial insulators. They average 10 to 20¢ each.

The importance of transposing between elements can not be stressed enough. This is accomplished either by crisscrossing the feeder as illustrated in Fig. 1-16A or by transposing the feed to the elements as illustrated in Fig. 1-16B. Both work equally well in providing phase reversal to alternate elements. The latter method is better suited for wire beams from a construction standpoint as shown in Figs. 1-21 and 1-25. This method has been used for all but one L-P. It is the method generally used for the large commercial L-Ps.

An L-P is in effect a multi-element end-fire array and *must have a phase reversal between adjacent elements* as with any end-fire array (example, the "ZL Special" or the "W8JK.") If there is no phase reversal between elements, you do not have an L-P.

Briefly, an L-P is similar to a yagi except all elements are driven. The "active" section of an L-P consists of a rear driven reflector, a driven or "active" ½λ radiator, and a number of driven forward directors. It must, therefore, function as an end-fire array. If the adjacent elements are not approximately 180° out of phase, there will be no forward lobe or gain.

Several have written that their L-Ps were non-directional and gave no gain. After checking, it was found they failed to transpose.

Antenna Wire

Because the forward and rear elements and the 2-wire center feed line are the only portions requiring a strain type wire, these should be #7/22, #7/24 or #14 copper or copper clad.

All of the other elements can be #16 soft drawn bare copper, enameled or tinned (hook up) wire. This can be purchased economically in 304.88mm (1000') spools. Even #18 has been used here which seems entirely satisfactory; at least to 500W. This saves weight and cost.

Since an L-P has a lower Q than a yagi, there is not the high rf current in the elements. The yagi generally requires tubing whereas wire is entirely satisfactory for an L/P. Wire is used for the large commercial or military fixed L-P antennas (reference A, B and C). Further, since there are several "active" elements per band, the rf current is no doubt distributed over several elements, therefore, wire is entirely satisfactory.

Soft drawn wire is suggested for all elements except #1 and the short forward element since there is practically no pull on the remaining elements. Being soft drawn, the wire will not tend to coil up or kink as does hard drawn or some of the copper clad. There is enough tension on the forward and rear elements to prevent this problem.

After all material has been collected, and the Lucite insulators fabricated, proceed as follows:

1) First assemble the two wire center feeder.

Select two sturdy posts, trees or other supports with about 1.53cm (5') greater separation than the required length of the center feeder for the L-P selected. Secure one end of the pair to or around the post at a height of approximately 1.83 (6') above ground level. Now thread the center Lucite insulators on the 2-wire feeder at the free end. This end may now be secured to the second post or tree. Stretch the two wires so they will be parallel and separated about 20.32cm (8") at the support ends. They will tighten to 10.16cm (4") separation after the center insulators/separators are spaced. They should be at about shoulder height to make for easy assembly. If necessary, two turnbuckles can be used temporarily at one end to tighten the two parallel wires and to adjust them for equal tension.

Now slide the center insulators (spacers) and distribute along the feeder in their approximate locations as given in Table 2. Starting at one end mark or indicate the location where the 2-wire open feeder will be attached to the center of the long rear element #1. A piece of 2cm (¾") masking tape can be used on each of the two wires to indicate this starting point, which should be about 30.48cm (12") from one of the end supports. The #1 element will be located at this starting point.

Now measure from this point with a steel tape the first spacing distance, S1 which will separate Elements #1 and #2. The first Lucite center insulator will be located at this point (location of the second element, #E2). This insulator is held in place between the 2-wire feeder by means of a few turns of 2cm (¾") masking tape served on either side of the Lucite insulator on both wires. Allow a slight distance or "play" on each side of the insulator so the tape will not be snug against the insulator. The wires should be able to turn free in the insulator holes. This helps keep the 2-wire line from twisting after the antenna is completed. The masking tape hardens after a few days in the weather and prevents the center insulators from sliding on the wires, which would alter the correct spacing of the elements.

Next measure the spacing distance, S2 and secure the next center insulator. Continue measuring and securing the insulators until all are in position. Then measure the last spacing distance and mark with tape as was done for the starting, #1 element. This last marking will be the location of the shortest end element (egg insulator) and will also be the feed point to the L-P.

The distance from the back side marking to the last forward marking will be the overall length (boom length) of the

L-P and will total the spacing distances, S1 + S2 + S3...etc. It is suggested that this total length of the center feeder be measured to make certain no errors have been made in any of the spacing distances. This total length is given in Table 1-2.

2) The next step will be cutting the various elements (or doublets) to length; L1, L2 etc. It is suggested that the rear element #1 and the short forward element be cut last as these will not be connected to the feeder until all of the other elements are cut and secured to the center insulators; thus leaving the feeder attached to the supports for convenience until all except the forward and aft elements are in place, connected and soldered to the feeder.

In addition to the actual element lengths, allow seveal centimeters for connecting to the end insulators and about 25.4cm (10″) extra for the center connections from the element center ends to the 2-wire feeder, as *every other element is transposed* as illustrated in Fig. 1-16B and 1-24. By using a continuation of the element centers, it eliminates an extra splice.

An odd number of elements is recommended since this allows the 2-wire feeder to be connected directly (non-transposed) across the center (egg) insulators of the end elements.

Also note that the rear of the center feeder is "fanned" or separated at the rear element. This helps in keeping the two feeder wires separated on the longest rear (S1) span, especially important for lower frequency L-Ps. This precaution helps prevent the two feeder wires from becoming twisted or from touching during a high wind. Additional Lucite spacers between S1 + S2 and possibly S2 + S3 may be necessary for 40m, or even 20m L-Ps. This can usually be determined after the L-P is finally assembled at the 1.83 m (6′) level.

3) After the elements are cut to the various lengths, they can be attached to the center Lucite insulators, starting with element 2. The connections from the elements to the feeders can be made after all elements (except the rear and forward elements) are secured to the center insulators. *Note that every other element is transposed*, i.e., Element 1, non-transposed; #2, transposed; #3, non-transposed...etc.; or *all even*

number elements transposed and uneven numbers non-transposed.

Figure 1-26 illustrates the Lucite center insulator, the transposed and non-transposed method of connecting the element center ends to the feedline and the method of connecting the feeder to the short forward element and the long rear elements which use the egg strain insulators.

4) After the elements (except forward and rear) are attached to the center insulators and in turn connected to the feeder, all joints can be soldered while the center feeder is still elevated 1.83m (6').

The ends of the center feeder can now be removed from the 1.83m (6') supports and lowered to the ground. The feeder can now be attached to the rear and forward elements and soldered. Spread the complete L-P on the ground at its approximate location (when aimed in the desired direction) between the four masts (DLP type) from which it will be suspended.

Nylon Catenary Support Lines

The DLPs used here are supported by two catenary side lines. These are stretched between masts A-C and B-D and the L-P suspended between these. Nylon line, .32cm (⅛") is used. .48 cm (3/16") nylon is used for supporting the long rear element, #1 and the short forward element as shown in Figs. 1-19, 1-21 and 1-24. Nylon does not shrink when wet or stretch when dry as does most rope. Further nylon will not rot and should last several years. After four years in constant use here none of the nylon line has broken.

The next step is to suspend the L-P between the two catenary side lines.

At this point the L-P has been assembled and is spread out on the ground between the four masts or other supports, aimed in the beam direction. It should now be raised 1.83—3.05m (6-10') above ground level and suspended at this height between the masts to be used in its final full height position. By using these masts, all angles and distances will be the same as when the L-P is hoisted to its maximum height.

The long rear element, #1, and the short forward element are attached to the .48cm (3/16″) nylon line which supports the rear element between supports A & B. The short element is stretched between C & D.

The .32cm (⅛″) side catenary lines or bridles are now stretched between A & C and B & D. Actually these are supported A-B and C-D, however, these splices will be near the masts; the .48cm (3/16″) lines carry all the load and will be tied to the mast halyards.

Next, add the Lucite end-insulators to all elements except #1 and the short forward element. These use the egg strain insulators.

Now, starting with element #2, tie short lengths of #18 (165 lb test) nylon cord to the end insulators. These will in turn be tied to the side catenary lines, A-C and B-D. Element #2 will then be suspended between the side bridles.

When first tieing these element support cords to the catenaries, make a knot which can be easily untied. It may be necessary to adjust the tension on the various elements several times before they are correct and the catenary lines start taking their proper "suspension bridge" shape as shown by Fig. 1-19; 1-21; or 1-24.

Elements #1 and #2 should be parallel, by making certain that their end spacings are equal to the center spacing, S1. After element #2 has been attached and adjusted parallel with #1, proceed to suspending and adjust element #3 and the following elements, #4, #5, etc., until all are suspended between the side bridles. As these are attached, the catenaries will start taking on the shape of a commercial L-P.

Adjusting the tension of the elements between the side lines is the only "cut-and-try" procedure required for the L-P assembly. When constructing your first L-P it may require several tries but it will soon assume the correct shape illustrated by Figs. 1-19; 1-21; or 1-24.

Note: All elements other than the rear #1 and the short forward element will have some sag. This does not seem to affect the operation. If the elements are pulled too tight between the side support lines (to try and level the elements), too much strain will be placed on the side lines, possibly requiring larger line and even sturdier masts.

There will also be some sag of the center feed line sagging toward the center. This shows no ill effect in the L-P's operation. Some sag or "give" in all elements (except the long #1 and the short forward element) is desirable. If all lines are too tight, they might break during heavy icing conditions.

None of the L-Ps here have come down over the past four years. During this time there have been three heavy ice storms. The L-Ps sagged almost to the ground from the ice build-up. As soon as it melted they returned to their normal height. They have also withstood several high winds without damage.

After all element support cords (#18 nylon) have been adjusted (and readjusted) several times so the sag of these are approximately the same, all elements parallel, and the side lines appear identical and have a similar catenary "curve" as in Fig. 1-19 cords can be secured permanently to the side lines.

I suggest that a few turns of 2cm (¾") masking tape be served on the .32 cm (⅛") side lines on either side of the #18 nylon support cords. This will prevent the latter from sliding out of place along the side lines after the antenna has been raised.

Before raising the L-P to normal height on the masts, an SWR should be run while the antenna is still 6 to 10 ft. above ground. Proceed as follows.

Feeding the Log-Periodic

The simplest method of feeding the L-P is to connect the high impedance balance winding of a 4:1 broad band balun at the feed point (short element end). The coax is then connected to the balun. Two other feed methods will be presented later but the 4:1 balun method is the easiest for running the initial SWR before raising the L-P to full height.

A low powered transmitter or transceiver should be placed on a box or table directly under or a short distance in front of the short element feed end. Connect a short length of coax from the 4:1 balun to the SWR meter and another short length to the transmitter or transceiver.

An SWR run should be made over each of the bands for which the L-P has been designed to cover. Readings should be taken at least every 100kHz over each band. Record these for comparison with a second SWR run to be made after the L-P has been hoisted to full height and the final length coax used between the antenna and the shack is positioned.

While the L-P is still at a workable height it is interesting to check the element ends for rf voltage on each of the bands. Either a small ¼ watt neon or a "sniffer" can be used. This test will give one a better idea as to the operation of the L-P.

If the SWR readings are 2:1 or better, the L-P should be O.K. after it is raised to full height. Generally the SWR readings will improve after being raised higher above ground. They should then be similar to the SWR examples given by Table 1-1.

Other Feed Methods

The feed method mentioned above using a 4:1 balun directly to coax is the simplest and is recommended; however, two other feed systems can be used:

1) Tuned open line from shack directly to the L-P feed point. This, of course, requires a tuner at the shack which must be returned when changing bands. The tuner with open line is O.K. for a mono-band L-P but is a nuisance when more than one band is used.

2) 300Ω TV flat line can be used from the L-P feed point to the shack, then the 4:1 balun and coax to the set. This is the method used here. Since trees are used as "masts," RG-8/GU or RG-11/U coax is too heavy, causing the L-Ps to sag. The 300Ω TV line seems entirely satisfactory for low power "bare foot" operation. Further the TV line has extremely low loss if properly terminated and is quite inexpensive for long runs. Some of my L-Ps use over 107m (350′) of TV line between the L-P feed point and the 4:1 balun.

After the final method of feed is selected, it can be connected permanently to the L-P feed point.

The beam is now ready to be hauled up to maximum height by the mast halyards. After the L-P is in place, another SWR should be run over each band and compared with those run at the lower level. They should not exceed 1.5:1 over any band (or any frequency within its band width, if necessary test equipment is available to make measurements outside the ham bands).

A doublet at the same height and broadside to the L-Ps beam should be used as a "standard" or test antenna for comparing gain in the forward direction.

Monopole Log Periodic Assembly

The assembly and erection of the monopole L-P configuration is similar to the DLP. Fig. 1-23 illustrates the general construction for either a 7.0-7.3 or 3.5-4.0MHz monoband monopole L-P. Fig. 1-25 gives element lengths and spacing distances for 40m and 80m.

A single catenary line is run from the high rear mast to the shorter forward mast, .64cm (¼") nylon line is suggested. The 5 vertical elements are suspended from the support line. Note the "suspension bridge" shape of the catenary illustrated by Figs. 1-17 & 1-23.

The short forward mast should be a wood pole or any other non-metallic support since it is directly in the line of fire of the vertical beam.

Note that the ground radials decrease in length from the rear end (below the longest rear vertical reflector, element #1.) to the #5 forward element, the radials being the same length or slightly longer than their ¼Ω vertical elements.

The radials should be about 3.05m (10') above ground to allow access under them. Although the radials can slant down from the center feeder, the ends should be high enough to prevent contact as some are quite "hot" with rf.

The 2-wire feed line is identical to the DLP type; however, the elements connected to and supported by the Lucite center insulators (Fig. 1-26) are arranged differently in that the two outside holes are for the two ¼λ side radials and the center hole is for the ¼λ vertical element. Actually the center insulator and the 2-wire feeder are suspended by the 5 vertical radiating elements and they in turn by the single catenary line. Fig. 1-25 illustrates these elements, showing the jumper connection between the two side radials. Transposition or the "criss-cross" feed is accomplished as illustrated in Fig. 1-25.

The suggested method of feed is by the 4:1 balun, then to coax. Be sure the coax shield is grounded to an earthground as near the balun as possible.

For these mono-band monopole L-Ps, the #2 or ¼λ "active" radiator is approximately ¼λ from the balun feed point. This ¼λ line provides a matching stub bètween the low impedance feed point of the #2 element and high impedance at the feed point which is probably in the order of 200-300Ω, making a good match to the input of the 4:1 balun.

Summary

Anyone having observed the gain of the L-Ps discussed here will agree as to their effectiveness. When using the 17 element 20-15-10m West beam, (L-P #11) on 20m, W6's often report "strongest W4 on the band at this time." Considering that many of the other W4's are using the legal limit with rotary beams, a report of this type is encouraging.

AN EASY WAY TO DECIBELS

One of the most often used yet little understood terms in amateur radio is the decibel, commonly known as the dB. Introductions to this not-too-complex topic are generally presented with the formula:

$$dB = 10 \log \frac{V_2}{V_1} \text{ for power and}$$

$$dB = 20 \log \frac{P_2}{P_1} \text{ for voltage}$$

Each has its use in amateur radio. However, the amateur not versed in mathematics and logarithms, when confronted with these formulas, usually backs off. Most hams, therefore, are content with the knowledge that an increase in the number of decibels is desirable and that a loss in the number of decibels is not.

This knowledge suffices for all practical purposes. However, considering the many times decibels are used in the operation of a station, it behooves the amateur to become more familiar with this elusive term. The amateur may consider increasing the power of this station or changing antennas. How many decibels will this change represent? Will it pay to increase transmitter power or install a more efficient antenna system? This discussion will help to answer these questions without the use of complex mathematics.

Decibels as such are not an absolute quantity. They refer to a change. This change expresses ratio whether it be current, voltage, or power. The ratio means nothing until some reference level is established. Once the reference is set this reference is considered zero (0) dB. For example, some communications engineers may use 6 milliwatts (.006 watt) as reference while others will use 1 milliwatt (.001 watt). The reference level or zero dB, when used in amateur radio, is whatever the amateur starts with.

For the purpose of illustration, assume Joe Ham has a 10 watt rig which is operating under ideal conditions (no losses).

It must be realized that there are always losses. However, for the purpose of facilitating the understanding of decibels all conditions will be considered ideal. The power input to the final (the 10 watts) is the reference level or zero dB. Any station changes to increase power will be referred to this starting power. If the station was running a 100 or 1000 watt transmitter, this would be the 0 dB or reference level.

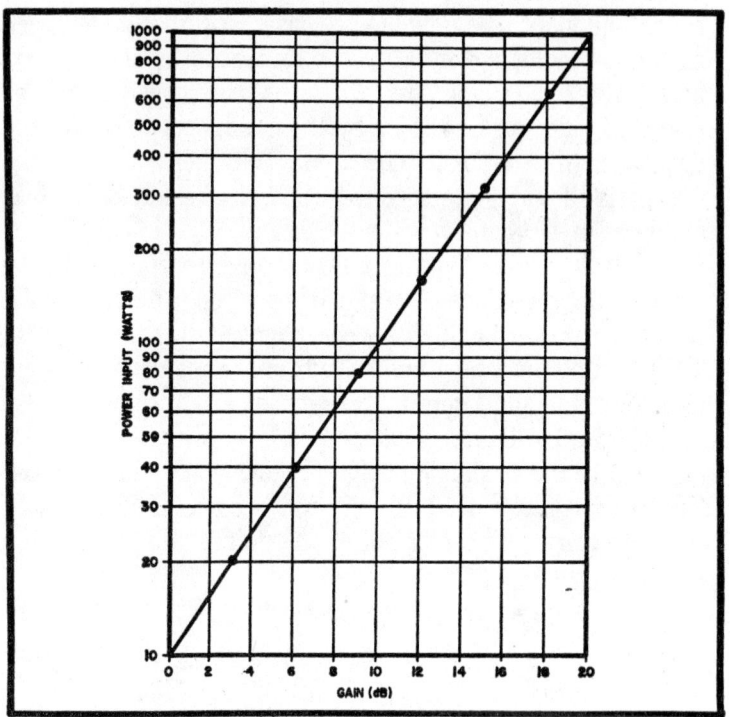

Fig. 1-27. A gain in transmitter power brings a gain in decibels.

Any contemplated change in the transmitter power will result in a change in decibels or a change in the power ratio. Joe ham decides to increase his power from 10 watts to 20 watts. What will the result be in decibels? Referring to Fig. 1-27 will disclose that the increase will be 3 dB. Observe that to obtain an additional 3 dB gain the 20 watts must be increased to 40 watts. Therefore, an important point to remember is: each time the power is doubled the gain is increased by 3 dB. Conversely a 3 dB loss reduces the power by ½. Knowing these facts any multiple of 3 dB can be calculated. Accordingly,

dB gain	0	3	6	9	12	15
watts	10	20	40	80	160	320

dB loss	0	3	6	9	12	15
watts	10	5	2.5	1.25	0.625	0.3125

let's see how this works out for an arbitrary value of 15 dB. In order to extend our ability to calculate dB's it will be advantageous to review some basic arithmetic. When a number is multiplied by 10 you affix a zero to the number. Multiplying by 100 you affix two zeros, etc. This is known as shifting the decimal. Dividing by any multiple of 10 moves the decimal in the opposite direction. The ability to shift the decimal is all that is required to be able to calculate decibels in multiples of 10.

Calculating decibels in muliples of 10 is simple if you consider the reference as 0 dB, if it is desired to determine the increase in power for 10 dBs we've changed the reference 0 to 10 by putting the digit 1 before the zero. Essentially this digit (1) tells you to add 1 zero to the reference. Therefore, the 10 watts with a 10 dB increase becomes 100 watts. For 20 dB the digit before the 0 is 2. This digit (2) tells you to add 2 zeros. The 10 watts with a 20 dB gain will become 1000 watts. In order to clarify this let's utilize this information to make a chart.

dB gain	0	10	20	30	40	50
watts	10	100	1000	10000	100000	1000000
Zeros added	0	1	2	3	4	5

To consider losses. It is best to consider that the digits before the zero indicates the number of places the decimal must be moved to the right. Observe the chart below.

dB loss	0	10	20	30	40	50
watts	10.	1.0	0.1	0.01	0.001	0.0001
Places decimal moves	0	1	2	3	4	5

Thus far we are able to calculate decibels in multiples of 3 dB and 10 dB. Combining these two facts with addition and subtraction we are now able to determine the gain or loss for any number of decibels. To ascertain that we do, we'll find the value for any number of decibels between 10 and 20 still using the original 10 watts as zero reference. Starting with 10 dB gain adds a zero to our reference, increasing the 10 watts to 100 watts. Each additional 3 dB gain doubles the power.

Decibels	0	10	13	16	19
watts	10	100	200	400	800

Now starting at 20 dB the 10 watts with two zeros becomes 1000 watts. Each 3 dB reduces the power ½.

Decibels	0	20	17	14	11
watts	10	1000	500	250	125

The only values we have not determined are values of decibels for 12, 15, and 18. These values are multiples of 3 and can be calculated by utilizing the method described previously.

The practical application of decibels in terms of power, can now dictate the advantages or disadvantages of station changes. Increasing transmitter power is a one way street. The old adage; you can't work what you can't hear is applicable. Therefore, when looking for dB gains; travel the two way street via the antennas. Improving antenna capabilities increases the number of decibels in both the transmitted power and the received signal.

When antenna gain is specified by the manufacturer, the gain is based on what the beam will do in comparison to a dipole under the same conditions. If a beam rated at 8 dB gain is replaced by one having 15 dB the gain is 7 dB. Stacking antennas increases the gain 3 dB. Raising the antenna an additional 30 feet will result in a 10 dB gain. These facts are presented so that you can start using this method of calculating decibels in terms of power the easy way.

Chapter 2
Non-Directional Antennas

VERTICAL ANTENNAS

Obviously, how well an antenna performs is determined by the percentage of the power fed into it when it radiates in the desired direction. *Direction* in this connection means both the compass direction and the angle above the horizon at which the radiation takes place. The latter is important, because practically all radio communications over distances much in excess of 50 miles on the frequencies between 1.5 and 30 mc are accomplished by signals radiated by transmitting antennas at angles above the horizon striking the ionosphere between 65 and 250 miles above the earth and being refracted back to the earth miles away.

For example, between the U.S.A. and Europe, signals arrive at angles between 10 and 35 degrees on 7 mc 99% of the time. On 14 mc, the arrival angle is between 6 and 17 degrees 99% of the time. Nine degrees is the median angle for DX signals on 10 meters. Higher angles of radiation are, of course, useful over shorter distances, particularly on the lower-frequency amateur bands.

With these facts in mind, refer to the vertical radiation patterns of the simple horizontal and vertical antennas shown

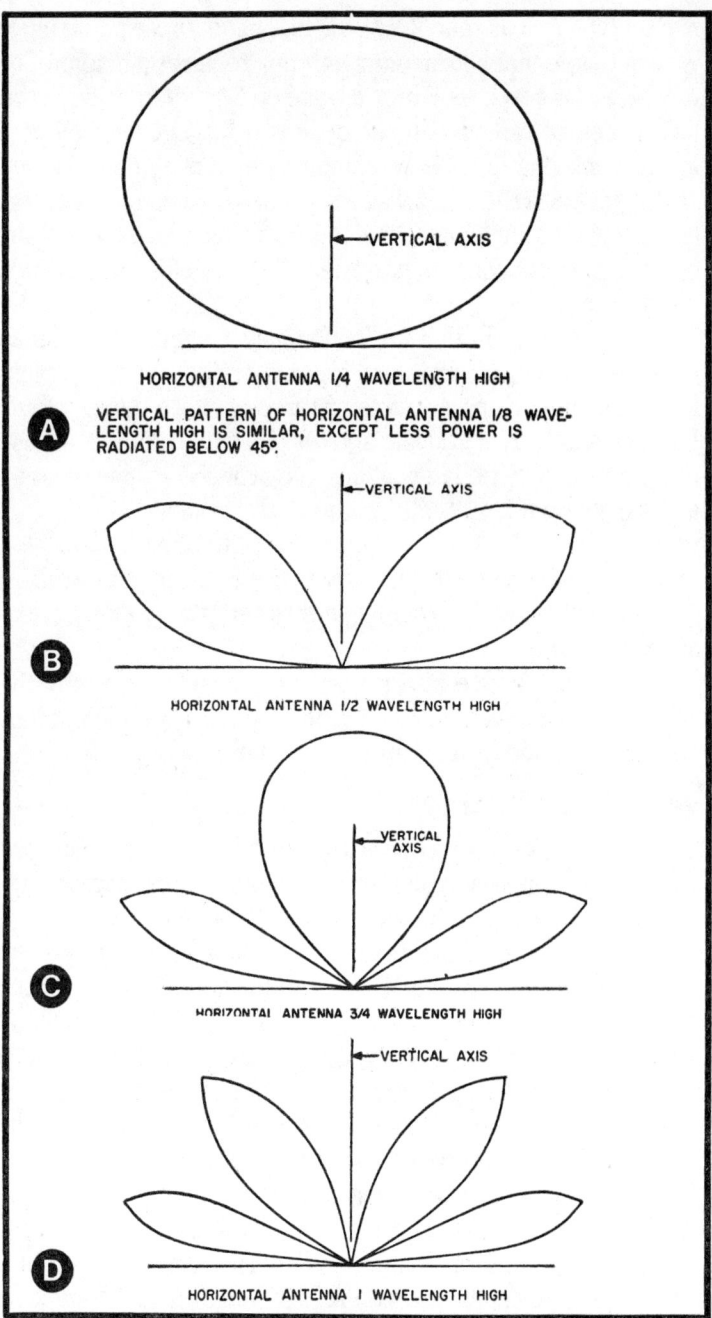

Fig. 2-1A, 1B, 1C and 1D. Radiation patterns in vertical plane of horizontal antennas ¼, ½, ¾ and 1 wavelength high respectively. Perfect ground assumed.

in Fig. 2-1A to 2-1D and 2-2A to 2-2D. From these patterns, it is easy to see that a horizontal antenna requires a height of 65 to 70 feet (½ wave on 7 mc, 1 wave on 14 mc) to achieve the low angles of radiation most desirable for DX work. But a vertical antenna ¼ to ⅝ wavelengths long is a powerful low-angle radiator. (Even a ⅛ wave vertical is a low-angle radiator, but the losses introduced by the necessary loading coil and other factors touched on later decrease the efficiency of such short antennas.)

Incidentally, Fig. 2-2A-2-2D apply to vertical antennas operated against an artificial "ground plane." In its simplest form, a ground plane consists of three or more (usually four) ¼-wave wires arranged like spokes of a wheel under the base of the antenna. Strictly speaking, the artificial ground must be at least ¼ wavelength above the earth to function as a true ground plane. At lower heights, the effective ground establishes itself somewhere between the earth and the artificial ground; nevertheless, these lower quasi-ground planes usually work quite well.

From the above information, it is easy to see why the vertical antenna has the reputation of being a good DX antenna. But how does it live up to its promise?

What Amateur Users Say

In the following paragraphs, the opinions expressed are based upon personal experience and observation and conversations with hams who had taken some pains to achieve reasonable results from their antennas, whether they were vertical or horizontal, as well as on the footnoted and other articles.

On the 3.5 to 4-mc band, one W8 reported, "The vertical worked fine on DX-I got S9 reports from South Africa, but my reports from U.S. stations were definitely weaker than with the old horizontal antenna. As I work the U.S. every day and South Africa about twice a year, I put the horizontal back up." Another ham questioned said, "After trying a vertical, I can say that the worst horizontal antenna I every used will outperform any vertical antenna made." A third ham said, "I must be satisfied with my vertical; I took down all my other antennas." On the average, a 3.5-mc vertical antenna is superior to a

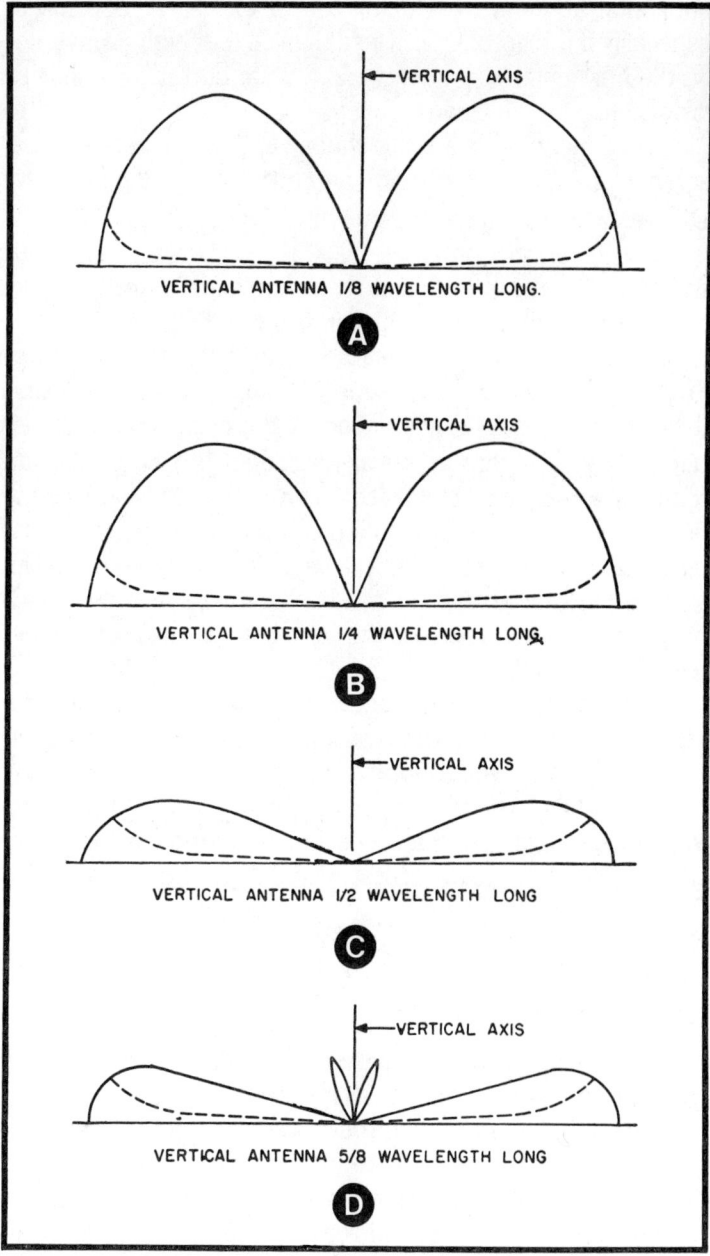

Fig. 2-2A, 2B, 2C and 2D. Radiation patterns in vertical plane and all compass directions of vertical antennas ⅛, ¼, ½ and ⅝ wavelengths long, respectively. Antennas mounted a few inches above earth or ground plane. Power at angles below dotted lines absorbed by ground losses.

horizontal antenna over distances in excess of 850 miles, especially if the horizontal is less than 60 feet high. Conversely, the horizontal may be several S units better over shorter distances.

On 7 mc there is little to choose between a good vertical antenna and a good horizontal antenna for working long-haul DX—provided that the horizontal is 67 feet high. But compared to lower horizontal antennas, the vertical is a better DX antenna. On the other hand, the horizontal usually works much better than a vertical over shorter distances.

Even on 14 mc, a low horizontal antenna usually outperforms a vertical over "short-skip" distances. But, once again, the vertical usually forges ahead of the horizontal over distances beyond 1,000 miles, at least until the horizontal antenna is a minimum of 50 feet high. And occasionally, the vertical will even outperform a horizontal beam on DX—just often enough to thrill the vertical owner and shake up the beam owner. On the 21 and 28 mc bands, there is little choice between the two types—if the horizontal is at least 45 feet high.

Incidentally, a vertical antenna is fine for working mobile stations within ground-wave range. Conversely, cross polarization effects between horizontal and vertical antennas can reduce signal strengths as much as 23 db over these distances, although the difference is seldom this great. Don't worry, however, about cross polarization effects on signals that travel through the ionosphere. The trip so mixes up the polarization that there are equal amounts of both in received signals, without regard to their original polarization.

Lowering Ground Resistance

Returning to earth-mounted verticals, you could obtain a good ground connection in the center of a salt march by dropping a length of wire in the water. In rich, permanently-moist soil, driving an eight to 12 foot pipe about an inch in diameter into the earth will produce a fairly low-resistance DC or low-frequency AC ground return. In dry, sandy, or rocky soil four additional pipes in a 10 foot square around the first one, all five connected together with heavy wire is recommended. Such an installation is good for lightning protection, but it isn't a particularly effective rf ground.

Actually, rf currents are introduced into the earth for many feet around an antenna. As a result, these currents must travel long distances through the earth to reach the ground-return point of the antenna system. Furthermore, because of rf "skin" effects, the effective ground resistance increases with frequency.

To obtain a low-resistance, rf ground, in addition to the ground rods, four or more heavy wires should be buried a few inches in the ground like spokes of a wheel around the base of the antenna and tied together at the center. For best results, each buried radial should be at least ¼ wavelength long at the lowest operating frequency of the antenna. Also, the more radials buried the better the results, although the rate of improvement goes down after about 12 are installed.

Before treatment, the rf resistance of an average earth ground will be 50 ohms higher. With the installation of a buried radial system, this resistance can be reduced to less than five ohms. As a ¼ wave vertical antenna has an effective radiation resistance of approximately 32 ohms, lowering the ground resistance from 50 ohms to five ohms will increase the radiating efficiency of the antenna from 39% to 86%.

An even more dramatic improvement is obtained by reducing ground losses when a shortened antenna is used. For example, a ⅛ wave, loaded vertical (33 feet long on 80 meters) has a radiation resistance of about 10 ohms. With such an antenna, reducing the ground resistance from 50 ohms to five ohms will raise the antenna efficiency from an anemic 17% to a respectable 67%.

Short radials will still bring down the ground resistance. Increasing their number will help compensate for their lack of length. In one installation, sixteen 25 to 40 foot radials decreased the ground resistance from 50 ohms to 7.6 ohms at 4 mc.

But even if all the local losses resulting from an imperfect ground were eliminated, this doesn't mean that the resulting antenna would necessarily be as good as one located where natural ground losses were less. Actually, the characteristics of the earth for miles around an antenna can affect radiation characteristics.

Signals radiated at very low angles from an antenna graze the earth's surface, and over the best possible earth, energy radiated at angles below 3½ degrees is absorbed within a few miles. Over a lossy earth, energy at angles up to 10 degrees can be absorbed in this manner. The dotted lines at the bottom of the curves in Figs. 2-2A-D show the effects of this attenuation over an average ground. Fortunately, the power loss in this manner is not too high, simply because practical antennas just don't radiate a very great percentage of their power at very low angles. Nevertheless, the effect is present.

No dotted lines are shown in Fig. 2-1A to 2-1D because, at the heights shown, horizontal antennas just don't radiate any appreciable power at angles below five degrees.

Practical Vertical Antennas

Probably the simplest effective vertical antenna is the ¼ wave coaxial-fed one illustrated in Fig. 2-3. Fed with 50 ohm coaxial cable, the feedline swr will be approximately 1½ to 1 at the antenna's resonant frequency. If the antenna is being operated as a "ground plane," dropping the ends of the radials to produce a 30 degree angle below the horizontal (eight feet for a 16½ foot radial) should bring the line swr down to near 1 to 1.

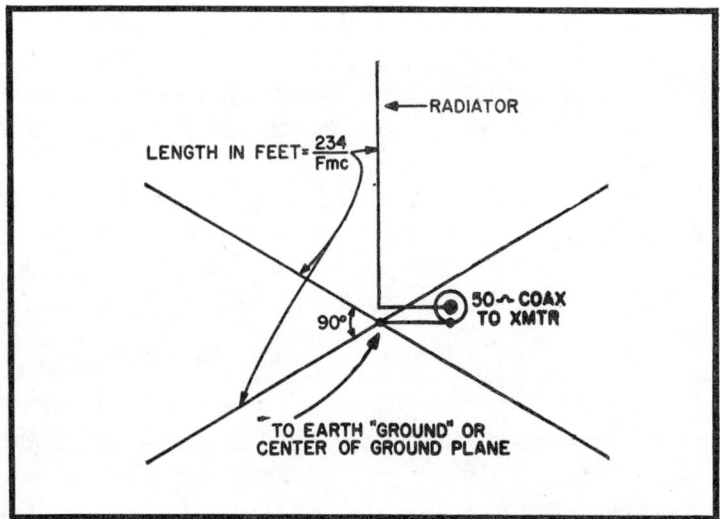

Fig. 2-3. Quarter-wave vertical antenna.

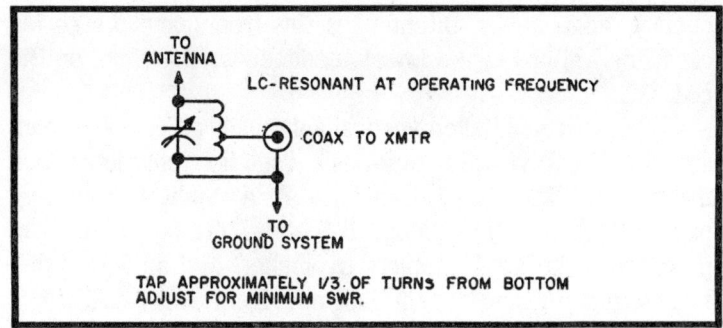

Fig. 2-4. Connection of a tuned circuit between the base of the antenna and ground.

Connecting a tuned circuit between the base of the antenna and ground as detailed in Fig. 2-4 will permit using the antenna on twice the frequency, also with low feedline swr.

Referring to Fig. 2-5 an antenna 5/8 wavelengths long at the highest frequency will perform efficiently over a 4 to 1 frequency range and fairly efficiently over an 8 to 1 range.

Careful positioning of the taps on the loading coil will produce minimum swr on all except the second highest fre-

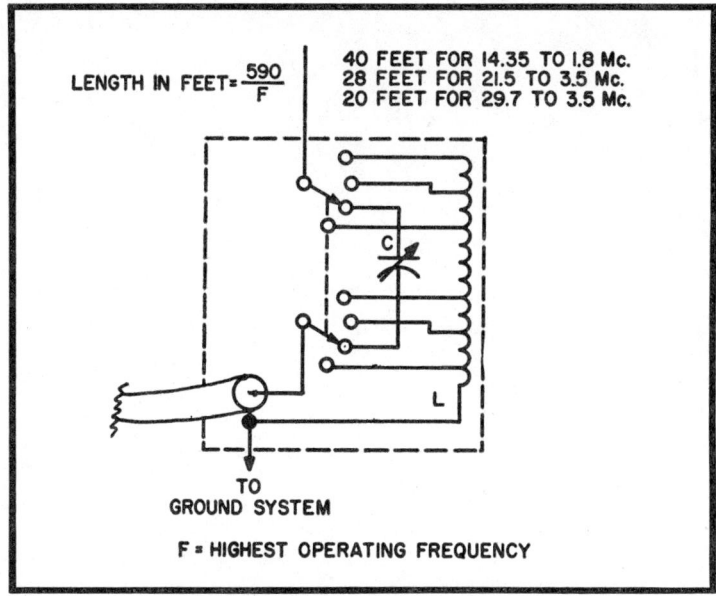

Fig. 2-5. Simple 3 to 5 band vertical antenna.

quency range of the antenna. At this frequency, the series capacitor will produce a lower feedline swr than taps on the coil.

A properly installed vertical antenna is an excellent ham antenna. However, its low-angle radiation characteristics make it somewhat of a DX antenna. As a result, its operating performance may not please 3.5 and 7 mc operators who prefer to make solid contacts over short and moderate distances to squeezing out the last mile from every call.

A MULTIBAND GROUND PLANE

This antenna is simple to construct with a little plumbing ability or help. The top section is an old CB whip antenna. These are easy to come by. The whip is bolted to a one-half inch pipe cap by drilling a ⅜ inch hole in the center of the cap (Fig. 2-6). This assembly is installed on an 86 inch section of one-half inch pipe. A coupler from one-half inch to one inch pipe is used to add the remaining 12 inch section of one inch pipe. The overall length of the vertical section is about 16 feet, but length is not critical.

Prepare the dielectric union by welding four 3/16 inch eye bolts to it for attachment of the ground radials.

The mast is 1¼ inch pipe. The higher it is the better. The plane of the radials must be at least ¼ wave length above ground to be a true ground plane. The radials serve as guy wires and should be at a 45° angle to the mast.

By feeding the antenna with 450Ω open wire and using a match box and an swr indicator, this antenna will perform well on 15, 10 and 20 meters. SWR is usually less than 1.3 to 1 on these bands and the antenna can be used on 40 meters with an SWR of only 3 to 1.

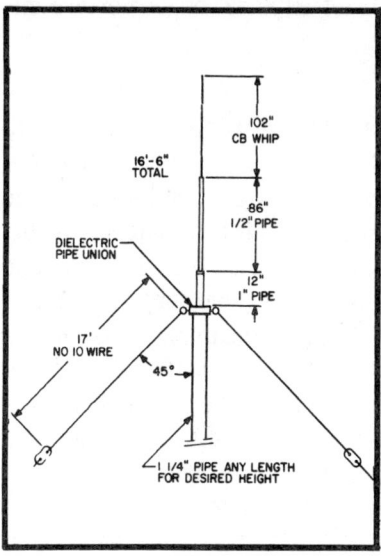

Fig. 2-6. Length of elements not critical. Feed with 450 ohm open wire for use on 10, 15, 20 meters.

USES FOR THE ⅝TH WAVE VERTICAL

In fixed-station usage and particularly for VHF mobile use, the ⅝ λ vertical has various advantages over a ¼ λ vertical or whip. The advantages are discussed and a simplified matching system is presented.

Many amateurs use quarter-wave verticals in both fixed-station and mobile use because of the low radiation angle and constructional advantages that they possess combined with a direct match to a coaxial transmission line. Not every situation will allow extension of the antenna length to approximately twice its quarter-wave length, but, if it can be done, additional gain can be achieved and matching the ⅝ λ vertical to a coaxial transmission line need not be complicated. In a vhf mobile situation, even a construction advantage will be achieved since, as is discussed later, the ⅝ λ vertical mounted in the same position as the regular automobile radio antenna will equal or outperform an awkward to mount roof-top quarter-wave ship.

Why a ⅝ λ long antenna should show a "gain" over a ¼λ long antenna while both antennas still exhibit omnidirectional radiation patterns is a subject that many amateurs still find somewhat confusing. So, a brief explanation of vertical antenna gain might be in order.

Gain of Vertical Antennas

The word "gain" itself is often misused with respect to antenna systems. It is defined as the relative amplitude of the maximum radiation from an antenna as compared to the maximum radiation from a reference antenna. The key points are that the maximum radiation determines the gain figure and the figure can vary widely depending upon what antenna is used for a reference antenna. Whether the maximum radiation takes place in a direction or at an angle that is useful in a given communications situation is another question. It is quite possible that an antenna with a higher gain figure will perform poorer than an antenna with a lower gain figure.

A half-wave horizontal dipole is frequently quoted as the reference antenna for directive antennas, such as beams, quads, etc. However, it is not the most basic reference antenna used and is a particularly awkward one to use when speaking of the gain of a vertical antenna. The most basic antenna used is a so-called isotropic radiator. Such an antenna does not exist in reality, but in theory it is an antenna in free space which radiates equally in all directions (its radiation pattern is a perfect sphere which surrounds the antenna).

By looking at the radiation patterns shown in Fig. 2-7 for various vertical antennas, it should be clear why these antennas have gain as compared to an isotropic antenna. If the same input power is used for the isotropic antenna and the vertical antennas, and all the antennas radiate all the power fed into them, the vertical antennas must radiate more energy in some directions than the isotropic radiator in order to account for all the power fed into the antenna.

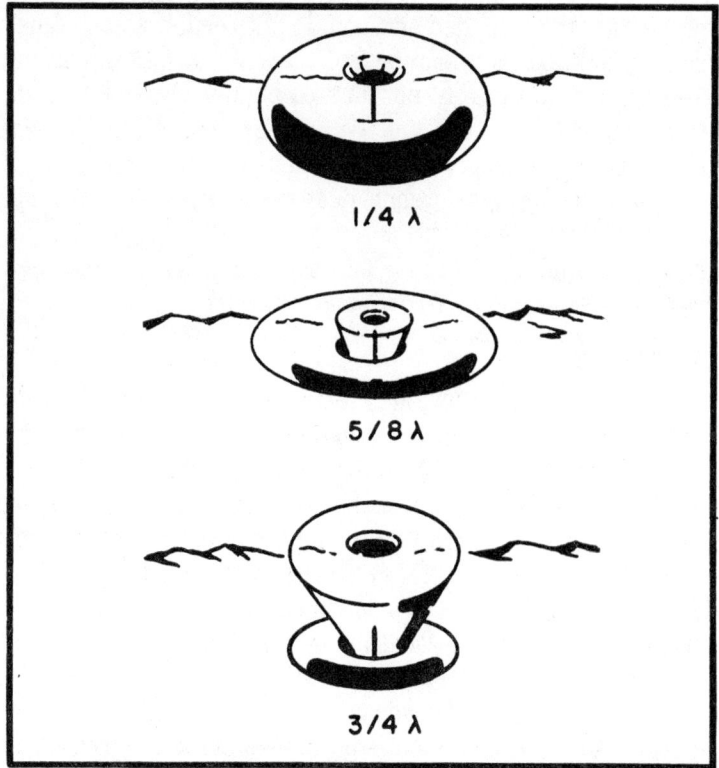

Fig. 2-7. Pictorial representations of the radiation pattern of various length vertical antennas over a perfect ground.

Compared to an isotropic radiator, the direction of maximum radiation from a ¼ λ antenna shows a gain of about 1.6dB and that from a ⅝ λ vertical is about 2.5dB. Fig. 2-8 may provide a clearer illustration of the maximum radiation from the ¼ and ⅝ λ verticals. The figure shows relative field strength versus the radiation angle measured from the ground plane surface, the ¼ λ vertical presents a single, constantly decreasing radiation amplitude as the radiation angle increases. The ⅝ λ vertical presents also a constantly decreasing radiation amplitude; however, with a relatively greater amplitude at the extreme low radiation angles and with a secondary peak centered around 50°. If the graph included the ¾ λ vertical, it would show approximately, the relative amplitudes of the low angle radiation and 50′ radiation from the ⅝ λ vertical interchanged. High angle radiation takes place from the ¾ λ vertical because the currents in various sections of the directly extended antenna do not reinforce each other such as to keep the radiation angle low. By means of phasing networks, it is entirely possible to build a vertical antenna many wavelengths high but with extremely low angle radiation. One prime example of such an antenna is a commercial uhf-TV transmitting antenna built by RCA. The horizontal radiation pattern from this antenna is omnidirectional, but yet it has a 60 dB gain because all the radiation is concentrated within an extremely narrow band around 1.

Ground Effect

The solid lines in Fig. 2-8 represent the radiation that takes place with a perfect ground surface. This condition can be approached fairly well on the vhf bands where the antenna is placed over a large metal surface, such as an automobile body, however, it is rare on the high-frequency bands unless an extensive ground radial system is used. The dotted lines in Fig. 2-8 represent the loss which takes place in extreme low angle radiation over a fair to poor ground surface.

An interesting point to note is that, although both the ¼ λ and ⅝ λ verticals suffer considerable low radiation angle losses, for the ⅝ λ vertical it is not so severe at the low angles below 5. For DX operation where one-hop F layer propagation is used, the ⅝ λ vertical usually shows an apparent gain over a

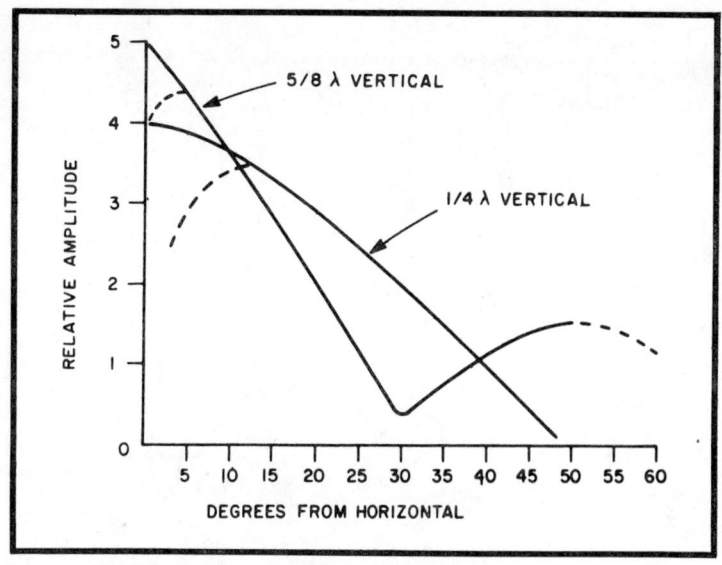

Fig. 2-8. Graphic representation of the vertical radiation pattern from ¼ and ⅝ λ verticals. The dotted lines show the approximate effect of ground losses.

¼ λ vertical that is considerably greater than the approximate 1dB difference theoretically indicated.

Although the increase in antenna height from ¼ λ to ⅝λ, where feasible, represents one of the least expensive ways to gain effective dB's either for long-haul DX or extended ground wave coverage on the vhf bands, many amateurs are hesitant to try this approach because a ⅝ λ vertical does not directly match a coaxial transmission line. However, it is quite simple to match the antenna to a coaxial line for single band operation without the need for an antenna coupler or any expensive components.

Figure 2-9 represents the terminal or base resistance of a vertical antenna while Fig. 2-10 shows the reactive part of the base impedance. These graphs may appear somewhat complicated but lead to a very simple matching means for a ⅝ λ vertical (or for most other length verticals as well). The graphs are drawn for various A/D ratios. (Antenna length divided by antenna element diameter.)

An example should make the use of the graphs clear. Suppose that it is desired to use a ⅝ λ vertical (of about 52" long) on 2 meters which is constructed of tubing having a 4

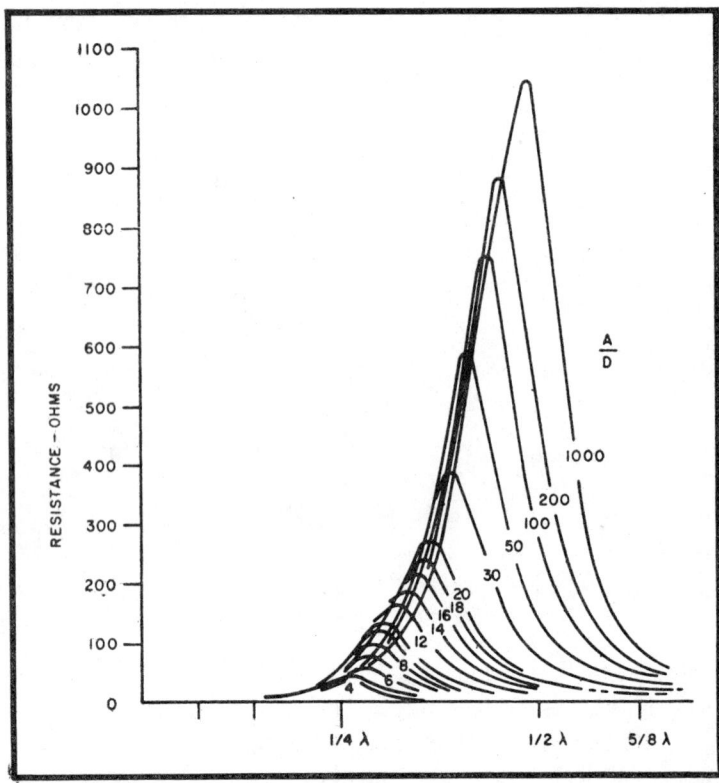

Fig. 2-9. Vertical antenna terminal resistance as a function of length. A/D is the ratio of antenna length to antenna diameter (d).

mm. diameter (which would be the typical, average diameter for a telescoping automobile whip). This results in an A/D of about 200-250. From Fig. 2-9 it can be seen that the resistive portion of the vertical's impedance is about 60 ohms and could present an almost perfect match to a 52-ohm coaxial cable. From Fig. 2-10 the reactance is noted as being 180 ohms capacitive. Once this reactance is cancelled, therefore, an excellent match to the transmission line will result.

There are several ways by which the capacitive resistance can be cancelled. One of the simplest ways is shown in Fig. 2-11. A shorted coaxial stub is connected to the base of the antenna and its length chosen to present an inductive reactance that just cancels the antennas capacitive reactance.

Determining the length of the stub can be done either by cut-and-try or by formula. For those who don't mind wasting a bit of coaxial cable, the cut-and-try procedure involves simply making the stub initially ¼ λ long at the operating frequency and with an swr meter in the transmitter coaxial line at the antenna base, cutting back the stub until unity swr is achieved. Note that every time the stub is cut back, the inner conductor and outer shield must be shorted together. The starting length

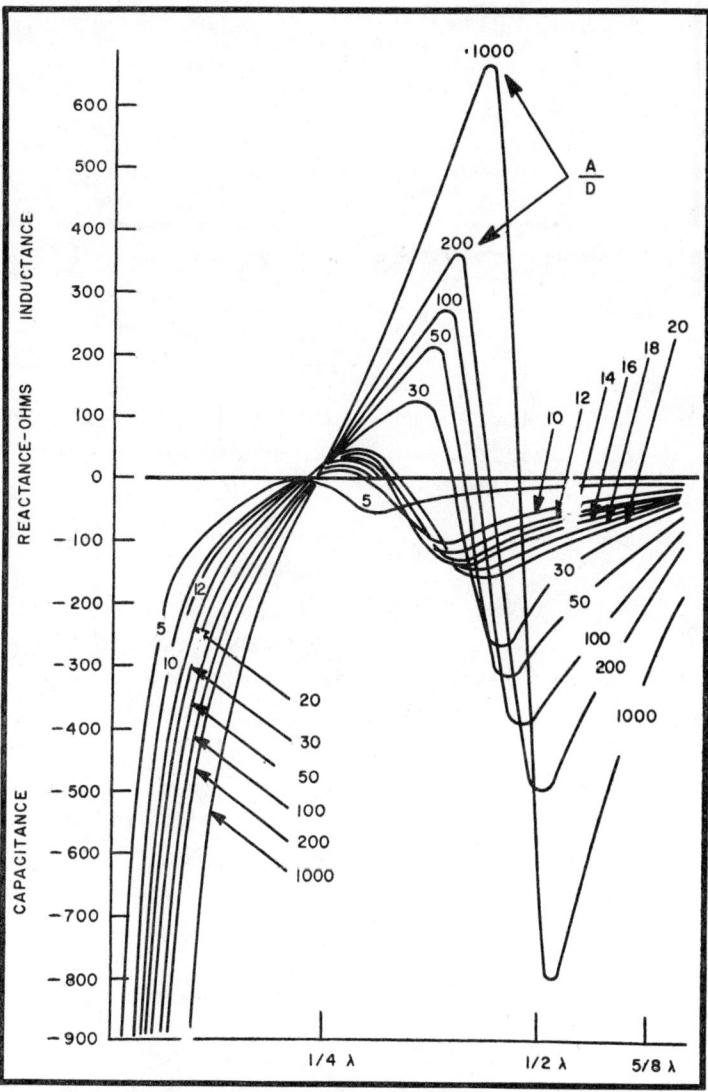

Fig. 2-10. Terminal reactance of a vertical antenna as a function of its length. A/D is the length/diameter ratio, both expressed in the same units.

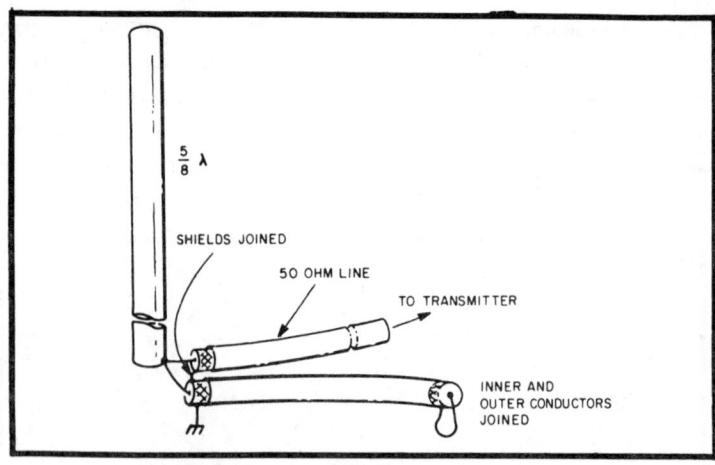

Fig. 2-11. Use of stub to match a coaxial line to a ⅝λ vertical.

of the stub would be, in inches

$$\frac{2808}{f(mc.) \times V.}$$

V is the velocity of the coaxial cable used, usually about .66.

The stub length can be determined by formula to a very good degree of accuracy by the formula:

$$L \text{ stub inches} = \frac{32.8 \text{ VL}}{f \text{ (mc.)}}.$$

V is again the velocity factor and L is an angle whose tangent is the value of inductive reactance desired divided by the impedance of the cable used for the stub. For example, the 2 meter vertical L would be (using a 52 ohm cable for the stub

$$\tan \frac{180}{52} = 3.6 \equiv 75°$$

Placing this value in the formula would yield

$$L = \frac{32.8 \,(.66)\,(75)}{144} = 11¼'' \text{ stub length.}$$

If one desired to use an actual coil instead of a stub, the required inductance, in microhenries, would be

$$L = \frac{X_L}{6.28 \text{ (f mc)}} = \frac{180}{6.28 \,(144)} = .2 \text{ rh}$$

A coil of this value can be formed but unless the equipment to measure its inductance is available, the stub method of matching will be found to be far simpler and quicker.

Mobile Application

The ⅝ λ vertical is useful in any situation where the greatest amount of omnidirectional low angle radiation is desired without resorting to a complicated antenna structure. However, for mobile applications on the vhf bands, the ⅝ λ vertical has particular appeal.

It can be mounted on the automobile in the same location as the regular automobile whip. In fact, it can be constructed from any ordinary telescoping automobile antenna since its length would be about 52″ on 2 meters. The sections of the automobile whip should be soldered together at the telescoping joints to insure good electrical contact. Figure 2-12 shows

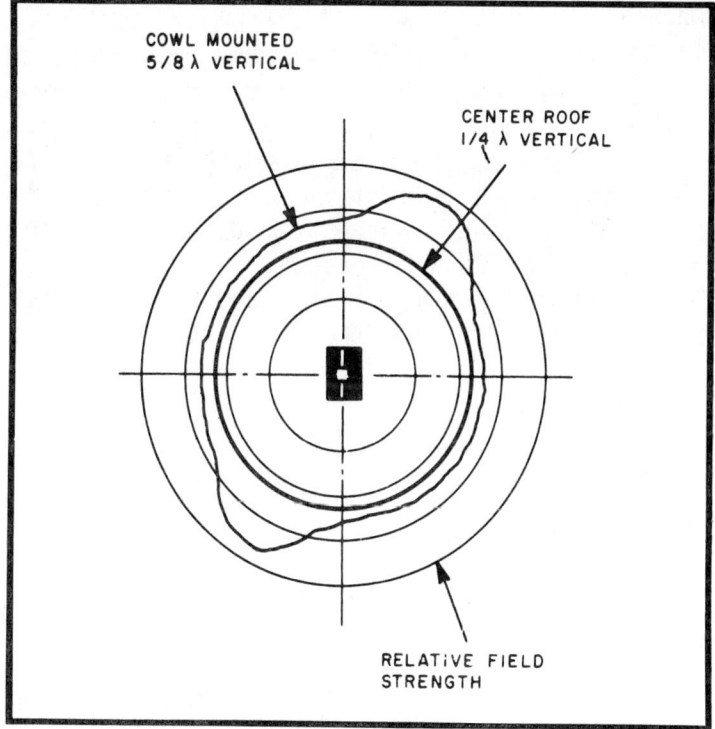

Fig. 2-12. Tests by many manufacturers have shown that a simple to mount ⅝λ cowl antenna will equal or outperform a roof mounted ¼λ antenna.

a comparison of the radiated field from a ¼ λ whip mounted directly in the center of the car roof and that of a ⅝ λ vertical mounted in the same position as the regular automobile radio antenna. The ⅝ λ vertical is never worse than the very-awkward-to-mount ¼ λ vertical and shows usable gain over the ¼ λ whip over a fairly broad portion of its pattern.

Other advantages of the ⅝ λ vertical are that it offers a much closer match to a 52-ohm coaxial cable than a ¼ λ vertical and its bandwidth is broader. The latter results due to the stub matching system since the reactance of the antenna and the stub change in opposite directions (but at different rates) with changes in frequency. At least until the rate change becomes excessive, this means that the reactances cancel each other and the swr will remain very low over almost any band when the antenna is designed for the band mid-frequency.

Summary

The ⅝ λ vertical offers one of the simplest ways to extend the performance of a vertical antenna system where low-angle radiation is desired. On the high-frequency bands for DX purposes, the real gain from such an antenna can far exceed the expected gain since it seems not to suffer the severe loss of extreme low angle radiation that occurs with a more ground-dependent ¼ λ vertical.

On the vhf bands for mobile operation, it offers wide bandwidth performance combined with a simple and inconspicuous installation.

SIMPLE GROUND PLANE ANTENNA

One of the classic vertical antennas for VHF work is the ground plane, and one of the classic problems is how to assemble it mechanically. Lacking machine tools to make special parts, this can seem formidable, but the exploded drawing shows how to assemble a sturdy ground plane quickly from parts available in any well stocked hardware store (Fig. 2-13). The size of pipe and fittings to be used is not critical, and

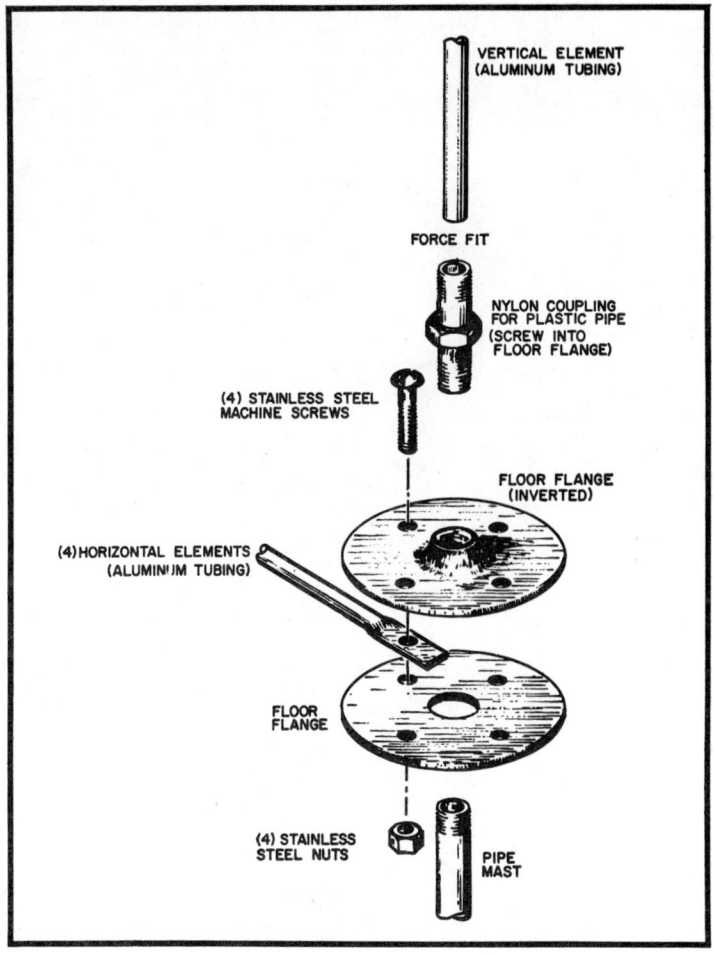

Fig. 2-13. Assembling a sturdy ground plane.

can be determined by what is available. For a 50 MHz antenna a reasonable size of ¾ inch pipe and the aluminum tubing of such diameter to make a firm drive fit into the nylon coupling. For a high wind area the pipe might be a 1 inch or one could use ½ inch to make a light portable job. In addition to quick assembly this construction permits quick disassembly merely by unscrewing the pipe mast and loosening the four bolts that hold the horizontal elements in place. For optimum match to a 52 ohm line the horizontal elements may be bent down about 30 degrees. This is best checked with a standard SWR bridge.

AN EFFICIENT 75 METER MOBILE ANTENNA

The success or failure of mobile operation depends entirely upon the efficiency of the antenna installation. In mobile operation, one must compete with extra-ordinary noise levels, QSB, and high-power fixed stations. Considering all of these factors, the highest efficiency mobile antenna possible is essential for good, solid communications. The physical configuration of this antenna was chosen with regard to the highest possible performance.

The Whip

The top section whip is adjustable in length. It is a two-section telescoping unit with a setscrew on the side for locking the length after adjustment has been made. The base of the whip is threaded and will mate with the top of the loading coil. At the top of the whip is a large corona ball about ½ in. in diameter. The extended length of the whip is 54 in. and the collapsed length is 43½ in. This gives 12½ in. of adjustment. See Fig. 2-14.

Capacity Hat

The capacity hat (Fig. 2-15) is of the homebrew variety. It consists of three equally spaced (120 degrees a part)

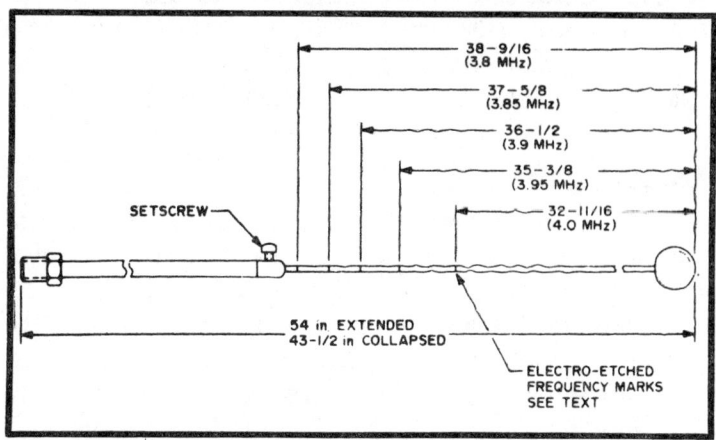

Fig. 2-14. Top section whip.

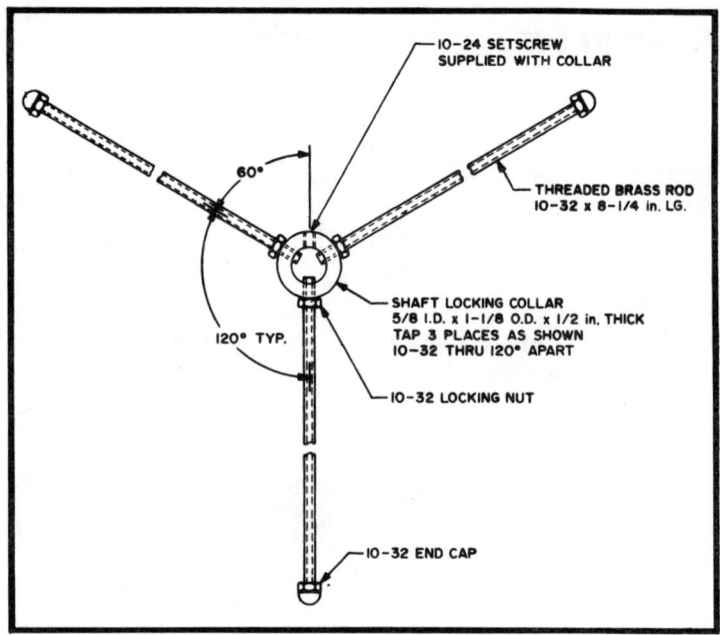

Fig. 2-15. Capacity hat detail.

threaded brass rods 8¼ in. long. The use of this capacity hat above the loading coil effectively lengthens (electrically) the top section whip. The hat increases the effective capacity of the top section whip to ground. This also allows for the shorting out or removal of some turns on the main loading coil which is desirable. The overall effect of the capacity hat is to bring the rf up into the upper portions of the antenna and away from the car body which has detrimental effects on performance such as absorption and detuning of the antenna.

The Loading Coil

The loading coil is the heart of the loaded low-frequency whip antenna. Its purpose is to cancel the negative reactance offered by the short antenna to the tuning device. The shorter the whip antenna (compared to a quarter-wave antenna), the greater the negative reactance offered to the tuning device. The loading coil adds an equal and opposite amount of positive reactance to the antenna circuit. This will neutralize the negative reactance of the short whip, leaving only the radiation resistance of the whip and the inherent losses of the loading coil as the terminating load for the tuning device.

Unfortunately, most loading coils are far from perfect. They exhibit resistance as well as reactance. The ratio of reactance to the rf resistance is given by the symbol Q. A coil having a reactance of 1200Ω and a resistance of 10Ω is said to have a Q of 120. If this coil is used with an antenna having a radiation resistance of 8Ω, the coupling efficiency is then $8/(8+10) \times 100$, or 45%. Over half the output power of the transmitter is lost in the loading coil. If the radiation resistance of the antenna is only 4Ω (as it may well be with base loading), the coupling efficiency is then $4/(4 + 10) \times 100$, or 28%. This means that 72% of the transmitter output power is lost in the loading coil. This power can only be dissipated in the form of

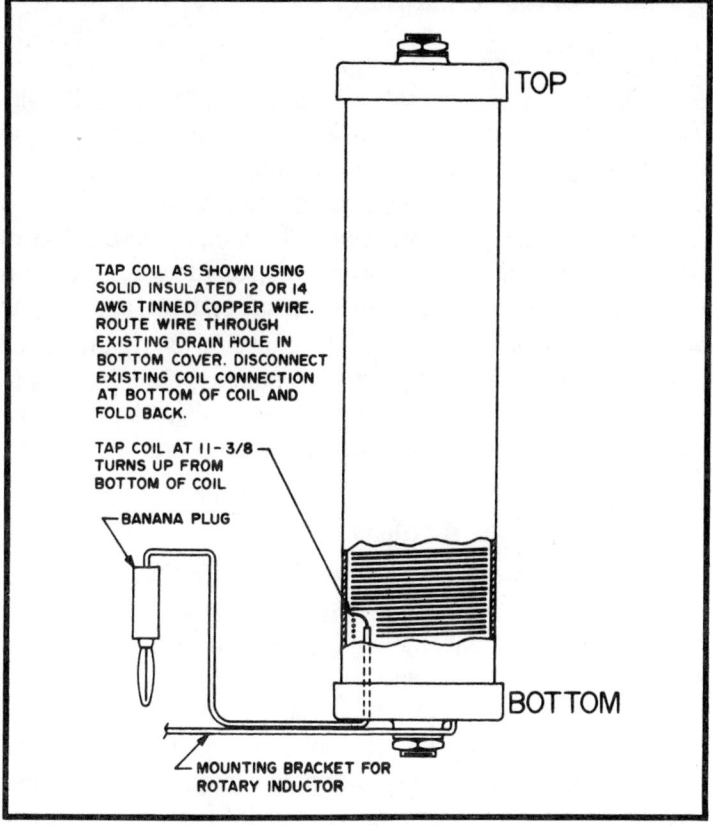

Fig. 2-16. Loading coil detail.

heat generated in the loading coil itself. For this reason, center loading was chosen. Improving the efficiency of the loaded whip antenna means increasing the radiation resistance by using either a center- or top-loaded whip antenna. The center-loaded version is the most practical method to use from the standpoint of available commercial parts. The longer, thinner coils, rather than the shorter, fatter ones, consistently give higher field-strength readings. This would indicate that the coil is actually radiating energy in conjunction with the whip antenna. With this type of coil, overall antenna efficiency should be on the order of 60 to 70%. The coil chosen was a Master-Mobile Ultra Hi "Q" 75/80 meter loading Coil (Fig. 2-16). It has well-spaced coil turns (about ⅛ in.) to eliminate arcing at high power. The coil is 13½ in. long × 3 in. in diameter and contains 95 turns.

The Rotary Inductor

The rotary inductor is a Master-Mobile Micro-Z unit. The basic rotary coil measures 1 × 3 in. long. The housing measures 3 × 4 in. It is designed for manual rotation and has a turns counter (numerals marked on one end cap). The tap to the coil is accomplished by a grooved wheel, mounted so as to slide on a bar. The grooved wheel rides on the coil. There is a coupling capacitor mounted between one terminal and one end of the rotary coil (Fig. 2-17). The capacitor is removed and discarded and the input and output contacts are made continuous through the coil via the roller contactor. The rotary inductor is mounted to one side of the main center line of the antenna.

Operation of Loaded Antennas

One of the penalties that must be paid for high Q and high efficiency in the antenna system is that the tuning of the antenna becomes extremely sharp. It is necessary to adjust the loading accurately, and to adjust it for each significant change in operating frequency. To put this in practical terms, if the frequency is shifted 5 kHz without reloading, no appreciable loss will result; 10 kHz and the output and the plate current of the transmitter will start to drop; 15 or 20 kHz and the performance will seriously suffer. Obviously, some means

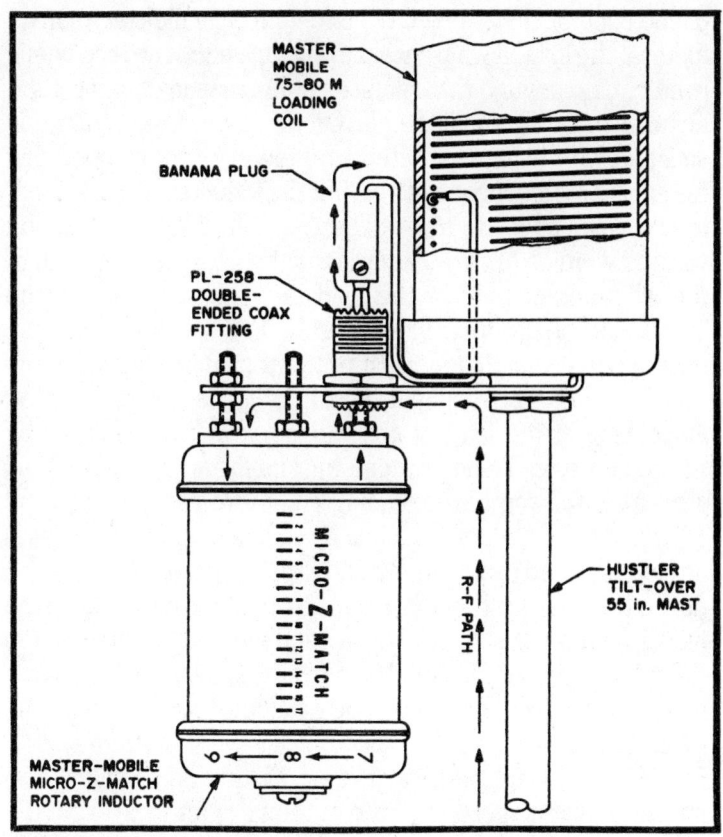

Fig. 2-17. Rotary coil detail.

must be provided for adjusting the antenna loading or some other portion of the circuit to tune out the antenna reactance at each frequency setting.

On 40 meters, the problem is somewhat less pronounced. First of all, the loading coil is smaller, and the whip is longer in proportion to the wavelength than on 75 meters. Changes in frequency of 30 kHz may be made with no appreciable loss of antenna circuit efficiency. However, for changes of 50-100 kHz, some means of varying the antenna loading must be incorporated into the system.

Maintaining Antenna Resonance

On 10, 15 and 20 meters, the whip antenna (usually one electrical quarter-wavelength long) will usually tune the entire phone band or the entire CW band with little difficulty. On 40 meters, the loaded whip antenna will tune about half the phone band. Going beyond that requires some means of varying the inductance of the loading coil. On 75 meters, unless one is satisfied with being restricted to one very narrow operating frequency, some means of varying the inductance of the loading coil is an absolute must. The resonant frequency may be varied by either of two basic means: by altering the length of the whip antenna above the loading coil, or by varying the inductance of the loading coil itself. Both methods may be employed. To vary the length of the top whip section by any other means than doing it manually is very difficult. The inductance of the loading coil can be varied manually or by remote control. Tuning is done by running an insulated metal strip from the top of the loading coil down past the side and adjusting the length of this strip by rolling it up and down with an old-fashioned can opener key, or a sliding tap on the coil will do the trick. Or try running a brass slug up and down inside the inductor with the slug electrically and physically connected to the bottom side of the coil. Relays may also be used to switch in taps on the coil, or use a rotary inductor in series with the main loading coil.

To obtain high performance on 75 meters, one must first get the radiating portion of the antenna as far away from the metal body of the automobile as is practical to prevent body absorption and the detuning effect it has. A long mast is mounted on the rear deck adjacent to the corner of the rear window.

This tilt-over mast measures 55 in. long. With the extra heavy-duty stainless steel base spring, it brings the length of the base mast to exactly 5 ft. The total antenna length approaches 10½ ft. This puts the top of about 13 ft. in the air. This is a lot of antenna but to get out, a lot of antenna must be out in the air and be properly matched and tuned. While driving, in order to keep the antenna in a vertical position, which is also necessary to maintain antenna resonance, a large nonmetallic washer is slipped over the top section whip and rests on top of the loading coil. A small hole is drilled in one side of this washer. Through this hole tie some waxed nylon lacing twine as shown in Fig. 2-18 and run it up forward to a gutter clip on

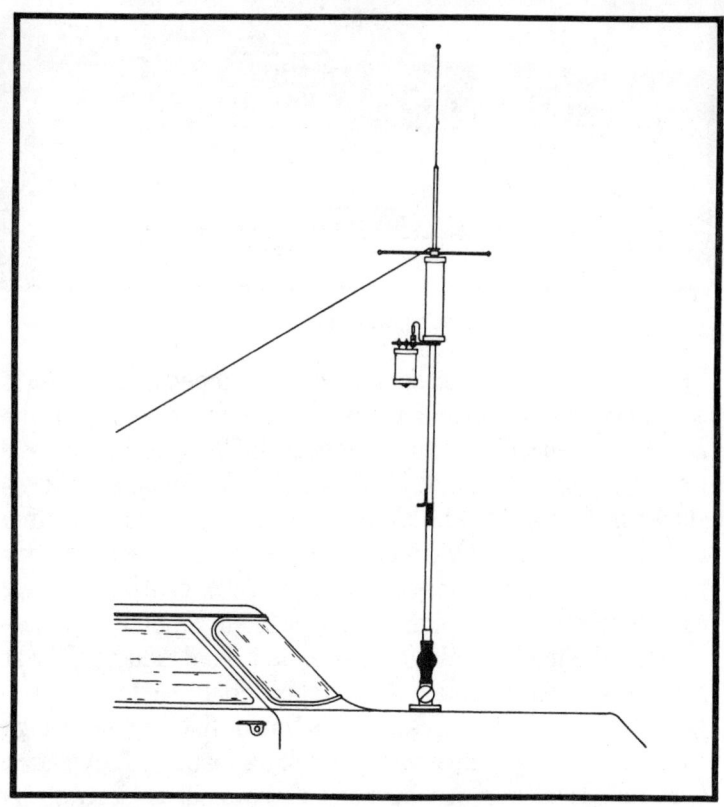

Fig. 2-18. Assembly and installation of 75m antenna.

the driver's side of the car. This is a very low-loss line and does not affect the performance of the antenna even when wet, because it is waxed.

Tuning the Antenna

Tuning this antenna is a relatively easy job using nothing more than the cathode current meter on the transceiver and a field-strength meter.

Most mobile antennas exhibit a feed-point impedance less than the characteristic impedance of the transmission line. The feedline is matched to the antenna by using a simple matching network. It is composed of a rotary inductor which will be called L_M and a capacitor, CM, which is used to shunt the feed point of the antenna to ground. The required values of C_M and L_M may be determined from the following formula. See

$$C_M = \frac{2\pi f \text{KHz} \ R_O \sqrt{\dfrac{R_A}{R_O - R_A}}}{10^3} \text{ pF and}$$

$$L_M = \frac{\sqrt{R_A(R_O - R_A)} \times 10^3}{2\pi f \text{KHz}} \ \mu H$$

Fig. 2-19. Equations for obtaining C_M and L_M.

Fig. 2-19. R_A is the antenna feed-point impedance and R_o is the characteristic impedance of the transmission line. If the antenna impedance is 20Ω and the line is 50Ω coaxial cable and the operating frequency is 4000 kHz, the inductance is as shown in Fig. 2-20. The chart at Fig. 2-21 shows capacitive reactance of C_M, and the inductive reactance of L_M necessary to match various antenna impedances to 50Ω coaxial cable. L_M can be duplicated by adding an equivalent amount of inductance to the loading coil. If L_M is to be a fixed inductor, C_M at least should be variable until the appropriate value of capacitance is found. When this value is found, fixed, high-voltage capacitors of the same value can be substituted. A rotary (variable) capacitor can be used if the shaft is secured by locking it in place so it will not change capacitance due to vibration.

First start tuning up the antenna at the low end of the phone band about 3.8 MHz. This is done by extending the top section whip up and out all the way and setting the rotary inductor so the maximum number of turns were in use. Inject

$$C_M = \frac{(6.28)(4000)(50)\sqrt{\dfrac{20}{50-20}}}{10^3}$$

$$L_M = \frac{\sqrt{20(50-20)} \times 10^3}{(6.28)(4000)}$$

$$= \frac{(1256)(10^3)\sqrt{0.6666}}{10^3}$$

$$= \frac{\sqrt{600}}{25.12} = \frac{24.5}{25.12} = 0.97 \ \mu H$$

$$= (1256)(0.816) = 1024 \text{ pF}$$

Fig. 2-20. Inductance equations.

LOAD RESISTANCE—OHMS
REACTANCE—OHMS
X_L
X_C
10
12
14
16
18
20
22
24
26
28
30
32
34
36
20 24 28 32 36 40 44 48 52 56 60 64 68 72 76 80

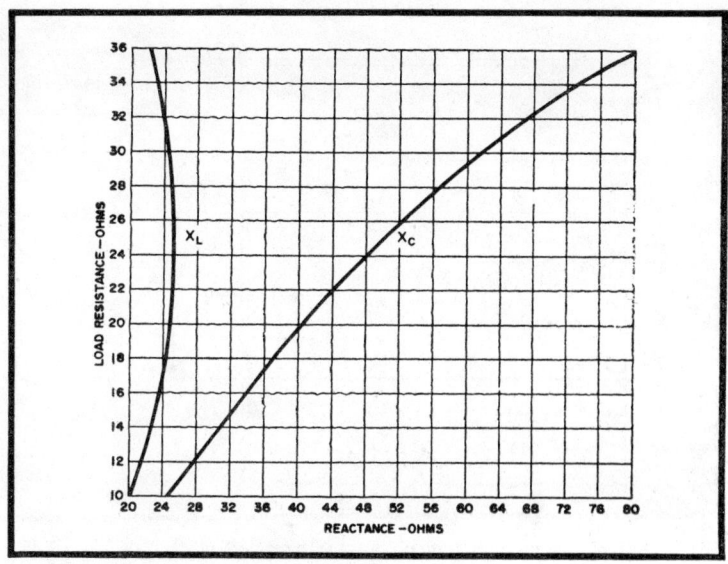

Fig. 2-21. Resistance-reactance chart.

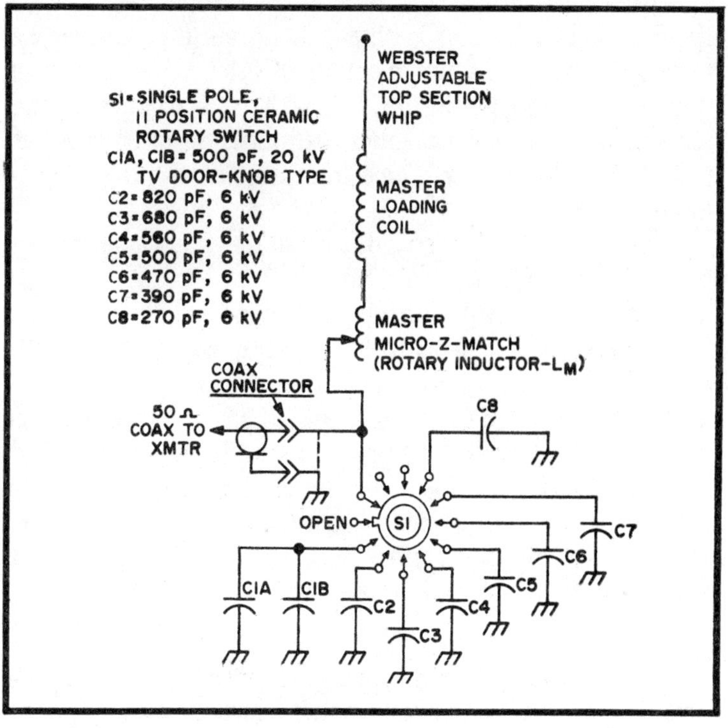

S1=SINGLE POLE,
11 POSITION CERAMIC
ROTARY SWITCH
C1A, C1B=500 pF, 20 kV
TV DOOR-KNOB TYPE
C2=820 pF, 6 kV
C3=680 pF, 6 kV
C4=560 pF, 6 kV
C5=500 pF, 6 kV
C6=470 pF, 6 kV
C7=390 pF, 6 kV
C8=270 pF, 6 kV
WEBSTER
ADJUSTABLE
TOP SECTION
WHIP
MASTER
LOADING
COIL
MASTER
MICRO-Z-MATCH
(ROTARY INDUCTOR-L_M)
COAX
CONNECTOR
50 Ω
COAX TO
XMTR
C8
OPEN
S1
C1A
C1B
C2
C3
C4
C5
C6
C7

Fig. 2-22. Schematic of antenna circuit.

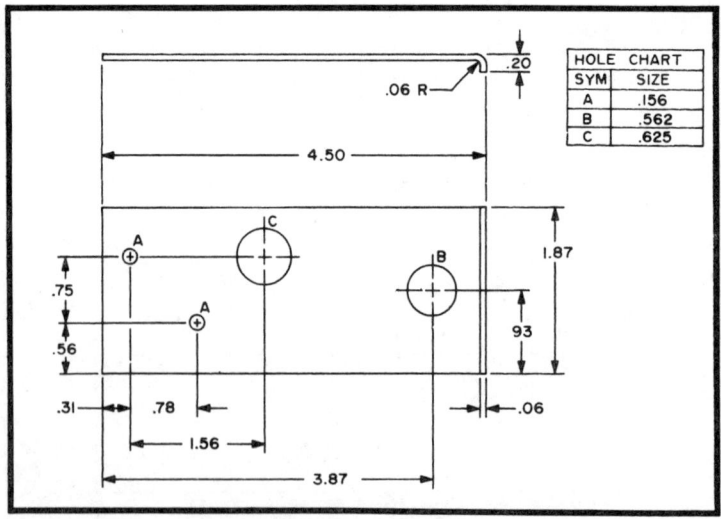

Fig. 2-23. Dimensions of mounting plate for rotary inductor.

just enough carrier to get some output, then begin running the tap up on the main loading coil until some output indication on the field-strength meter is obtained. Locate the point on the loading coil where maximum output occurs and then back up a turn or two and solder the tap in place. Now, start lowering the top section whip until maximum output is again obtained as indicated on the meter. Once this point on the top whip is located, mark it.

Now, turn the rotary inductor all the way in and retune the transceiver for maximum output. This process is repeated and other marks made for other frequencies.

In reference to the matching capacitor C_M, shunt the base of the antenna by using a ceramic rotary switch mounted on an angle bracket which in turn is mounted on one of the antenna body mounting bolts. This switch is a single-pole, multiple position switch. On this angle bracket mount the ceramic rotary switch, the matching capacitors and a coaxial fitting (Fig. 2-22). Any of 10 different capacitors, all of different values can be switched into the circuit.

Results

The results obtained with this antenna have been excellent. Ground-wave communication is excellent with little or no fade.

HI-Q 80-40 METER VERTICAL ANTENNA

80 meter operation has enjoyed considerable interest by all amateurs. The reliable but generally moderate transmission range has been welcomed by the "old time" amateur for trouble-free "local" round table "rag chews," whereas, the novice, because of FCC restriction, is relegated to this particular niche in the spectrum.

To secure maximum transferral of power, the time proven half-wave dipole has been the standard method of radiation; however, few operators are fortunate enough to construct this type of system due to the demanding erection area. As a consequence, the vertical base loaded antenna has been the logical substitute. Notwithstanding several inherent defects, the small space requirements and low angle of radiation justify its employment.

Construction

The radiator segment is constructed from two 12′ lengths of telescoped aluminum tubing. The lower base section is ¾″ diameter, and the upper section is ⅝″ diameter (both .058″ wall) and telescoped about one foot. The uppermost part of the base section was slotted lengthwise and a ⅞″ stainless steel clamp positioned to secure both elements.

The coil assembly is the most electrically critical component feature. A HI-Q ratio was chosen to insure sharp attenuation of spurious harmonics since the promixity of a number of concentrated city-located television antennas would normally have led to serious TVI complaints. The coil was constructed from #14 magnet wire stock. A tin can of 3½ inch dia. will serve as the coil form. Approximately 12-14 turns were close wound and removed from the form. Adjust 8 turns to occupy 1″ insuring that equal spacing has been provided throughout. In order to reinforce the coil, short lengths of polystyrene strips (¼″ × 1½″) were placed across the winding and glued with liberal amounts of liquid polystyrene or Duco cement for low loss. One strip can be made somewhat longer and a hole drilled to allow for subsequent securement to a 3½ × 5 × 7″

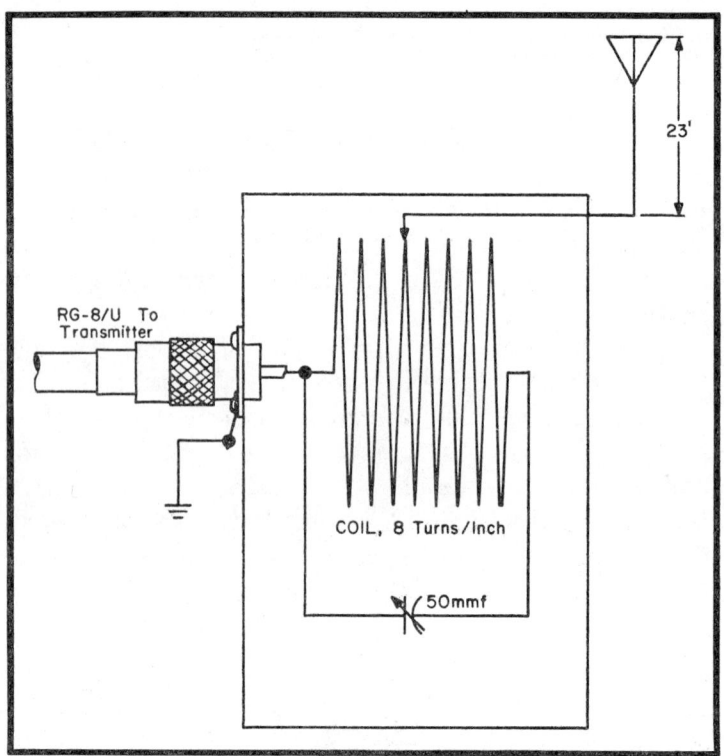

Fig. 2-24. Coil assembly.

plastic food container. The container will afford a measure of protection from direct moisture accumulation. One end of the coil is soldered directly to the center terminal of a mounted SO-239 connector. A moderately spaced 50mmfd variable capacitor is placed across the coil to allow for initial tuning and/or retuning due to minor frequency excursions.

A short length of 58U cable is soldered 10 turns from the input terminal and the other end is connected to the vertical antenna base. A grid dipper tuned to desired frequency will assist in adjusting the capacitor-coil factor once the entire antenna assembly is operative. A resistive bridge placed in the line will indicate the characteristic 30-35 ohm resistive load. The antenna is coupled to the transmitter via 50 ohm cable with a minimum SWR reflection. A reasonably non-floating ground is necessary at the transmitter and SO-239 connector for proper loading and propagation.

For those interested in tuning the vertical for 40 meter operation, either a remote or manual low-loss ceramic shorting switch might be integrated to shorten the coil length to 4½ TP from the input terminal. As a consequence, the assembly will be resonate at 7 MHz.

Although antenna placement is not especially critical, some care should be exercised to prevent resonating the vertical supports of neighboring TV antennas or similar power-absorbing objects.

The HI-Q characteristics of a coil result in a sharp dipper null at the operating frequency and will insure a maximum transferral of RF potential.

THE UMBRELLA ANTENNA

Short vertical antennas for the low-frequency bands are certainly nothing new. Inductively loaded mobile whips are a common form. They take up very little space and perhaps for the mobile situation are the only practicable antenna form. However, considered for usage in a fixed station situation where even a moderate amount of space is available, it appears foolish to accept the limitations of such a form. The limitations of the loaded mobile whip—poor efficiency, very narrow bandwidth and an awkward value of terminal impedance—arise because of the form of loading used. A small loading coil is required because of space limitations and in order to provide even usable efficiency the coil must be of high "Q" with resultant narrow bandwidth.

If the space limitations did not exist, it would be desirable to make the loading coil as large as possible to increase efficiency. Also, it would be desirable to include some form of capacity or "top-hat" loading since the reactive effects of the inductive and capacitive loading will act to maintain antenna resonance over a greater bandwidth. It is rarely possible to do this in a mobile situation but it is possible in a fixed station situation. A form of antenna is described which combines a very efficient method of combined inductive and capacitive loading while still requiring very little space compared to any conventional antenna of full-size dimensions.

Umbrella Loading

The basic form of the umbrella antenna is shown in Fig. 2-25. The vertical mast is relatively short (10 feet on 40 meters, 20 feet on 80 and 160 meters) and the wire web on top introduces inductive loading by linearly extending the length of the mast and capacitive loading due to the capacity effect between individual web wire and between the wires and the mast. A side view of the antenna is shown in Fig. 2-25B. The dimensions shown were not randomly chosen as one might at first imagine. They are a compromise between a number of factors. If the web wires are brought closer to the mast, the

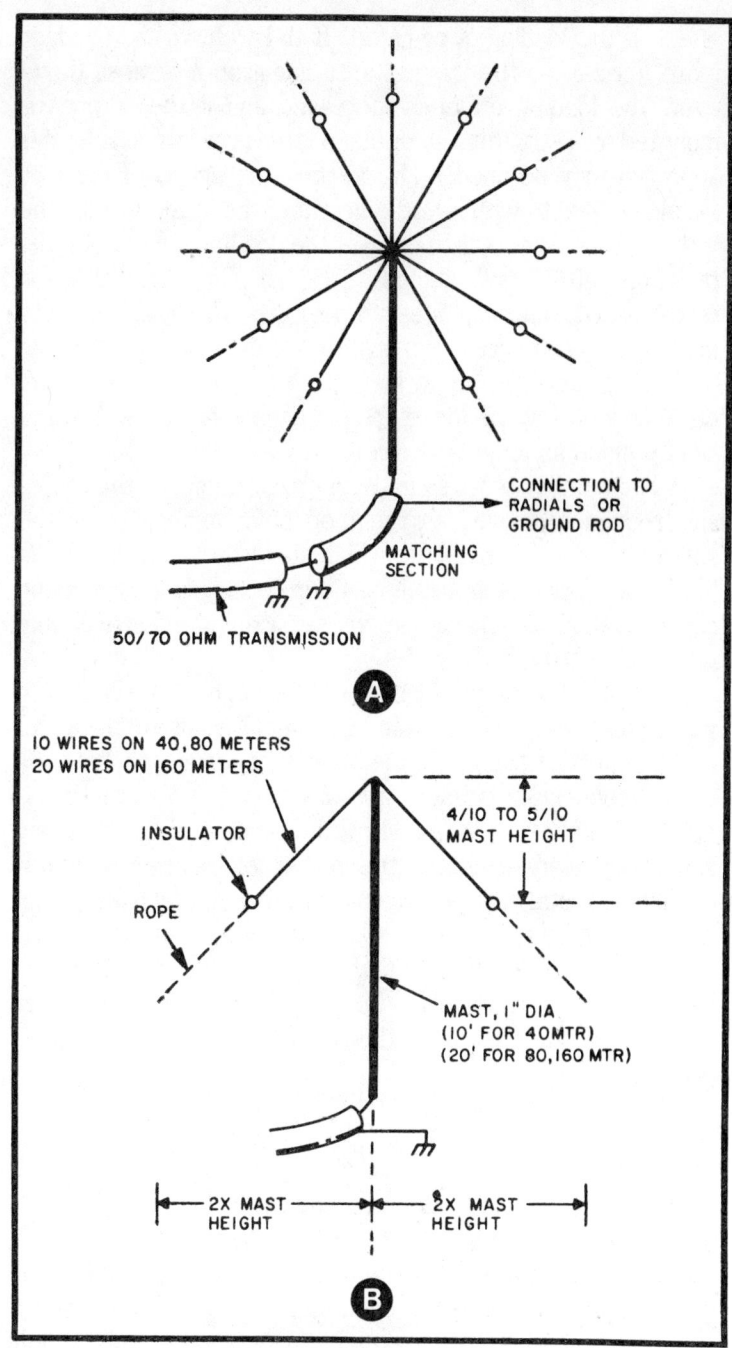

Fig. 2-25. Basic dimensions of the umbrella antenna.

effect of the loading is reduced. If the web wires are made more horizontal, the ground area required increases (however, the loading effect is increased and if the antenna is mounted on a structure such that a large area is available, this approach may be used.) The vertical projection of the web should be held to within ½ the height of the mast. Making the web wires longer will increase the loading effect but the radiation pattern will change such that high angle radiation results and the antenna does not perform as the equivalent of a full-length quarter wave vertical. If one does not object to the high angle radiation or, in fact, prefers it for short-medium distance work on the lower-frequency bands, the web wires can be made as long as desired.

The number of wires in the web will influence the resonant frequency as well as the feed point impedance of the antenna. Six is the minimum which should be used and probably 10 or 12 is a reasonable number considering loading and constructional complexity. Increased loading effect will take place up to at least 20 wires.

The feed point impedance is not as high as one is used to with a purely inductively loaded antenna. Because of the effect of the capacity loading, the feed point impedance when a good ground connection is used will vary between 5 and 8 ohms. Any one of the conventional methods used to match a low impedance load—such as the driver element on a multielement parasitic beam—to a coaxial cable may be used and so they will not be shown here. One slightly different method to match the antenna by means of a quarter-wave transmission line transformer is shown in Fig. 2-26. Three lengths of coaxial cable are paralleled to form the required simulated low impedance transformer section.

Construction

The main vertical section of the antenna was made from a standard 10 foot aluminum TV mast. The mast was mounted to a ground stake by two standoff insulators (Birnbach No. 448 pillars with tube clamps at each end). A home-brew standoff can be easily fabricated from a block of wood or polystyrene and cutting a hole at both ends for insertion of two adjustable

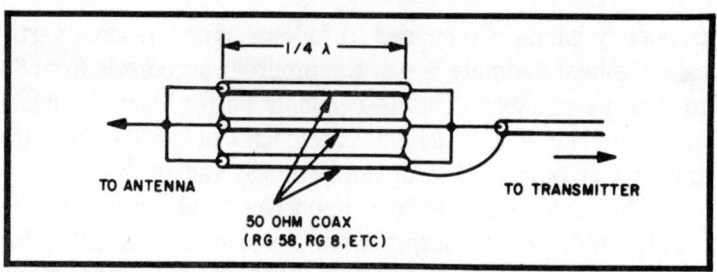

Fig. 2-26. Transmission line matching transformer for use between base of umbrella antenna and 50/70 ohm coax line.

hose clamps. Since the base is at a low voltage point, the quality of the insulator used is not critical.

The ten wires comprising the top web were fastened to the mast by means of an ordinary ground lug. Hook-up wire was used to construct the web although it is suggested that stronger wire be used—common TV guy wire would be an excellent choice, for example, for rugged installation.

As with any antenna being worked against ground—be it full-size or a loaded type—the quality of the ground plane has an important effect upon antenna performance. In moist soil a ground rod driven several feet into the ground may suffice but otherwise a cluster of 10 to 12 radials buried several inches and extending at least to the point where the extension of the web wires touches ground is very desirable.

The test antenna was constructed following the outline dimensions given in Fig. 2-25B to resonate on 40 meters. After installation and testing it was found that resonance was very slightly below 40 meters. The situation was corrected by placing the ground terminal for the web wires slightly closer to the mast. This was done experimentally while checking the transmission line SWR. There are several variables involved in determining the exact resonance of the antenna and either one can vary the angle of the web wires to "fine-tune" the antenna or these wires can be firmly placed on either a small inductor or capacitor used at the base of the antenna for final adjustment.

Summary

Antenna efficiency is at best a difficult thing to measure or estimate even under the most ideal conditions on low-frequency bands. Compared to full-size quarter-wave verticals, the best estimate is that the umbrella antenna is from 60 to 70% as efficient. This is certainly better than the usual loaded mobile whip which has efficiencies of 2-5% (100 watts transmitter output, 2 to 5 watts actually radiated).

The bandwidth depends upon the band for which it is constructed. On 40 meters when resonance is adjusted for 7150 kc, the antenna will satisfactorily cover both the CW and phone band edges with an SWR of form 2:1 to 2.5:1. On 80 meters either the entire phone band can be covered or the CW portion; the bandwidth is about 250 kc overall. No tests were made on 160 meters but the coverage should be in the order of 75 to 90 kc which is certainly adequate for most uses on that band.

If the antenna is constructed for 40 meters, a double resonance will be found to occur on 21 mc—the same as for any monopole or dipole operated on odd multiple harmonic frequencies. The antenna should be quite efficient on 3rd harmonic operation but the problem is one of the resultant radiation pattern. The pattern appears to produce mostly high-angle radiation which, of course, is useless for DX purposes. Nonetheless the antenna may still be useful as an auxiliary antenna on this band.

The umbrella antenna appears at first glance to be a rather simple and elemental type of loaded antenna. Actually, it is not when one considers its advantages in terms of preserving a low radiation angle and achieving quite good efficiency—all within reasonable dimensions. This form of antenna may not allow an apartment dweller to put out a booming signal on 80 meters but it should certainly permit someone with a moderate amount of space to considerably improve his signal on any low-frequency band as compared to using a whip or randomly placed and tuned length of wire.

MULTIBAND VERTICAL ANTENNA

Many hams would like to install a vertical antenna but are stumped for a suitable design that will work for all bands and be structurally strong. Here is one such design that will be very successful on both counts.

The schematic of the antenna is shown in Fig. 2-27. The trap is resonant at 7.1 mc. Because the impedance of a parallel tuned circuit is infinite at resonance, the trap serves to disconnect the portion of the antenna above the trap from the portion below the trap at 7.1 mc (40 meters). The mechanical length of the lower section is not a classic ¼ wave length long at 40 meters because the effective diameter of the antenna approaches that of a 5 inch diameter cylinder. The capacitance of such a body with respect to earth greatly reduces the mechanical length required to obtain the desired electrical length.

On 80 meters the trap is inductive and this inductance plus the total length of the antenna serve as a ¼ wave radiator. On 20 meters and above, the trap is capacitive in character and therefore shortens the electrical length of the antenna. The feed impedance varies from 22 ohms to 196 ohms and RG-8/U was selected for the feed line.

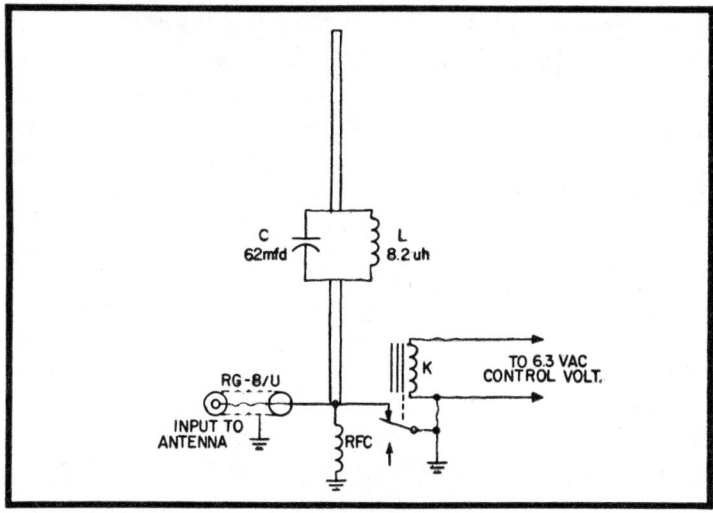

Fig. 2-27. Schematic of multiband vertical antenna.

The VSWR on 40, 20, 15 and 10 meters is less than 2 to 1 across each of the bands. On 80 meters we run into the perennial problem of bandwidth with small diameter radiators. It is impossible to cover more than about 75 kc and remain below 2 to 1 on VSWR. The VSWR on 80 meters will be about 3.8 to 1 over the entire band and less than 2 to 1 over the 70 to 80 kc bandwidth just mentioned.

Note the rf choke from the base of the antenna to ground. This choke bleeds off electrostatic charges and therefore reduces the possibility of damage to radio equipment connected to the radiator. It is wound on a piece of 1 inch solid polystryene rod and mounted at the base of the antenna. It is connected to the antenna and ground with copper braid made from the braided shield of RG-58/U coax.

The grounding relay is wired in fail-safe fashion to ground the antenna when it is not in use. This relay is mounted inside a 3 × 4 × 5 inch Mini-Box and mounted at the base of the antenna. Copper braid is used to connect the relay from ground to antenna. The box can be easily waterproofed by sealing all seams with liquid rubber. This liquid rubber is obtainable at any hardware store.

The antenna ground system consists of seven radials of aluminum stranded wire buried about 6 inches beneath the earth's surface. A sod edger shovel can be used to slice a slot in the ground. The aluminum wire should then be laid in the slot and the earth tamped shut. Each of the aluminum ground radials should be clamped to an aluminum pipe driven 3 feet into the ground at a distance of 6 inches from the base of the antenna. The use of an aluminum strap prevents problems from dissimilar metals when clamping the radials.

A practical design of the antenna element is shown in Fig. 2-28. The base of the lower antenna section is mounted on six 1¾ inch high ceramic honeycomb stand-off insulators. Since the load on these insulators is both compression and shear, they are of good quality. The insulators are screwfastened to a fir wood block which, in turn, is secured to a one cubic foot block of concrete in the earth. For the design shown, a large 1 quart soft drink bottle will make an excellent insulator. To install this insulator, the bottle should be coated generously with petroleum jelly. After digging a 1 foot cubic hole in the

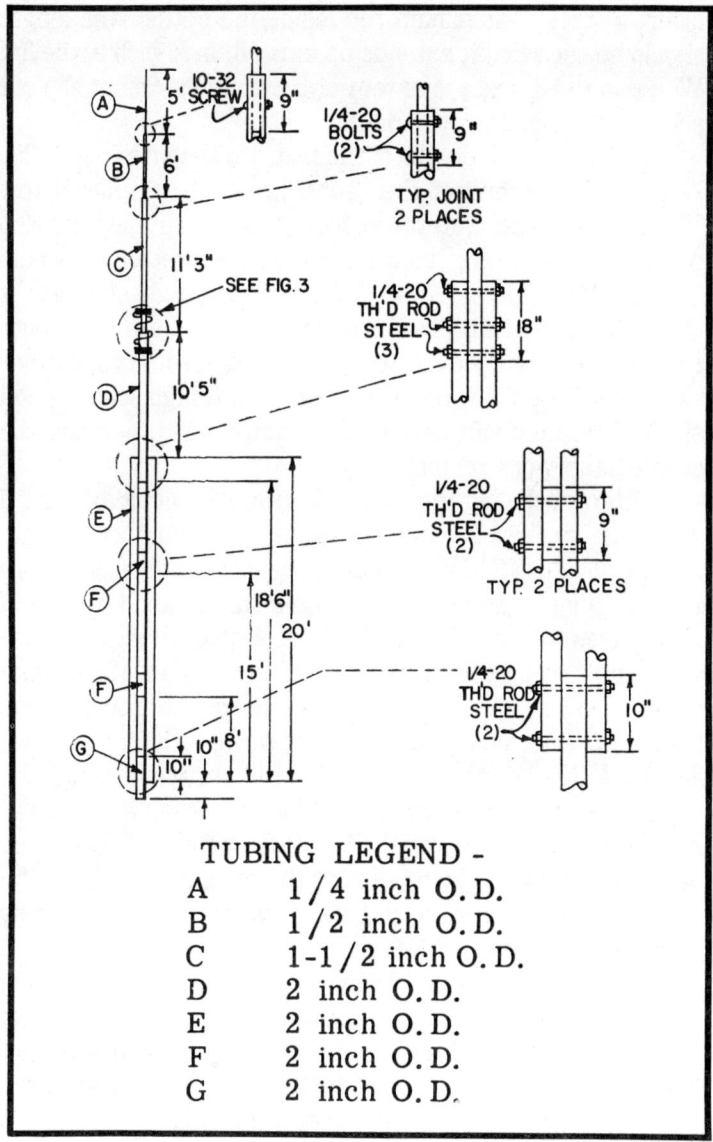

Fig. 2-28. Construction of antenna.

ground, place the bottle on a brick in the hole. Pour cement into the hole and around the bottle. When the cement is just about firm, *carefully* rotate the bottle to free it slightly from the concrete. When the cement is hard, removal of the bottle will be possible should it ever need replacing. As a safety feature,

131

insert a steel or aluminum rod inside the bottle. The length should be such that it extends up into within ¼ inch of the lip. With this rod in place, the foot of the antenna cannot slip out should the insulator ever break.

The structural details of the trap are shown in Fig. 2-29. Notice the hole through the center line of the trap insulator. This hole prevents water from lodging in the antenna and also prevents split tubing in climates where the water would freeze. The insulator is phenolic resin reinforced cloth and is obtainable from many plastics supply houses. Seasoned white oak could also be used if it were first coated with transformer varnish or boiled in wax for 3 or 4 hours. A high school wood shop or a friend can usually be counted on to perform the simple lathe work required.

After machining the plastic insulator it should be placed inside the smaller diameter tubing. Using a number 41 drill, drill about 10 or 12 holes in the tubing and into the insulator to at least ½ inch depth. Remove the insulator and tap each of these holes with a 4-40 tap. Drill out each of the holes in the aluminum tubing with a number 32 drill to clear a 4-40 cadmium or nickel plated machine screw. Reinsert the insulator and fasten it to the tubing with one or two ⅜ inch long screws, lightly. With a ¼ inch drill, drill a hole all the way through the diameter of the aluminum tube and insulator in two places as shown in Fig. 2-29. Remove the insulator and, using a ½ inch drill, enlarge the ¼ inch holes in the aluminum tubing only. Assemble the insulator inside the smaller tubing permanently with 4-40 screws ⅜ inches long.

Referring to Fig. 2-29 again, locate two ¼ inch holes in the large aluminum tubing as shown. Drill *only* the lower hole marked "A" and drill this one *only* through one wall of the tube. Now place the smaller tube inside the larger tube and when the lower holes of each are lined up across the diameter, drill a ¼ inch hole through the other wall of the large tube using the ¼ inch hole in the insulator as a guide. Remove the drill and place a ¼ inch machine bolt into this hole. Measure the location of the other through the hole in the insulator from the lip of the large tube. After carefully locating this position on the outside of the large tube, drill a hole through the large tube, through the insulator hole already drilled and through the far wall of the

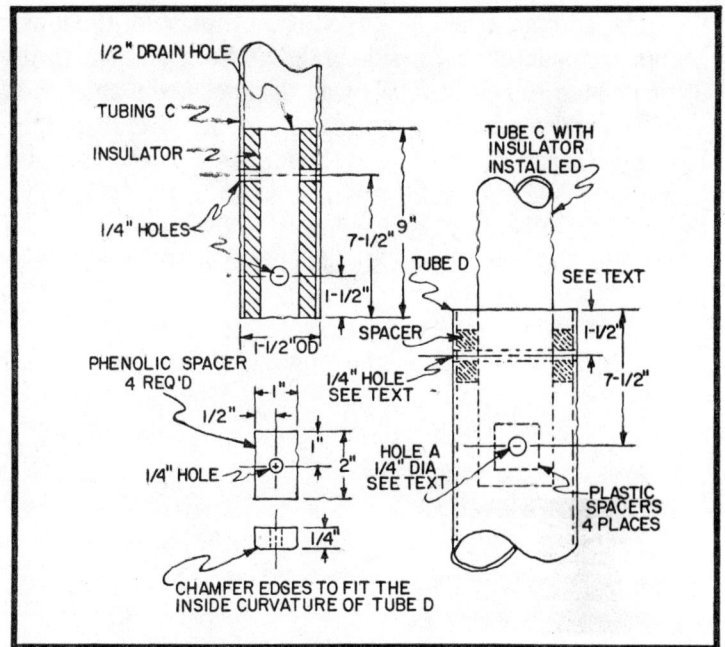

Fig. 2-29. Structural details of the trap.

large tube. Remove the ¼ inch machine screw and pull the small tube out. Clean up the burrs on the inside of the large tube.

Make 4 spacers as detailed in Fig. 2-29. Fasten these to the outside of the small tube with *one layer* of plastic electrical tape. These spacers must be in line with the through holes of the insulator.

Carefully re-insert the small tube inside the larger tube and place ¼ inch cadmium or nickel plated machine bolts through the assembly. The joint is mechanically strong and electrically forms a capacitance between large and small tubing.

The coil is wound with #10 wire and supported with aluminum straps around the large and small tubing. The wire is prevented from shorting against its neighboring turns by means of a split insulator made from 2 strips of plastic. The two strips are clamped together and a hole for each turn drilled through the parting line. This labor can be avoided by using ready made coil stock.

The antenna could be guyed from the top of the lower section or mounted on the side of the house. By using 2024S-T4 aluminum tubing of 0.043 wall thickness, the antenna is very light and yet doesn't require any support other than at the base and a bracket to the house about 8 feet up from the base. Using an SW12 meter lengthen or shorten the uppermost length of the tube to obtain resonance at the desired point on 80 meters. The other bands should exhibit an SWR of 2 to 1 or less.

DIPOLE INSTALLATION

If you're a beginning ham operator or SWL, why don't you try a twinlead dipole? The half-wave dipole offers good performance from a design as simple as a random long-wire antenna, but without the longwire's grounding problems. Its relatively narrow-band performance is usually no inconvenience, and it can be tuned up without a transmitter or SWR bridge.

The Twinlead Dipole

Two pieces of 300-ohm twinlead will make a very fine folded dipole, usable for receiving and low power transmitting applications. One piece is to cut slightly under a half-wave length at the operating frequency and hang parallel to the ground. The other piece is attached to its center to serve as feedline, as shown in Fig. 2-30. The feedline may be any convenient length.

The 300-ohm antenna and feedline impedances are properties of the folded dipole antenna.

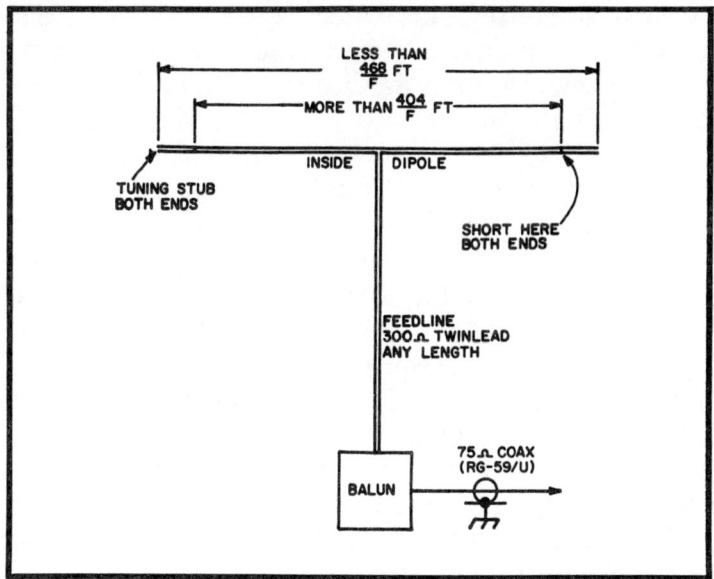

Fig. 2-30. This is the complete antenna system you're going to build.

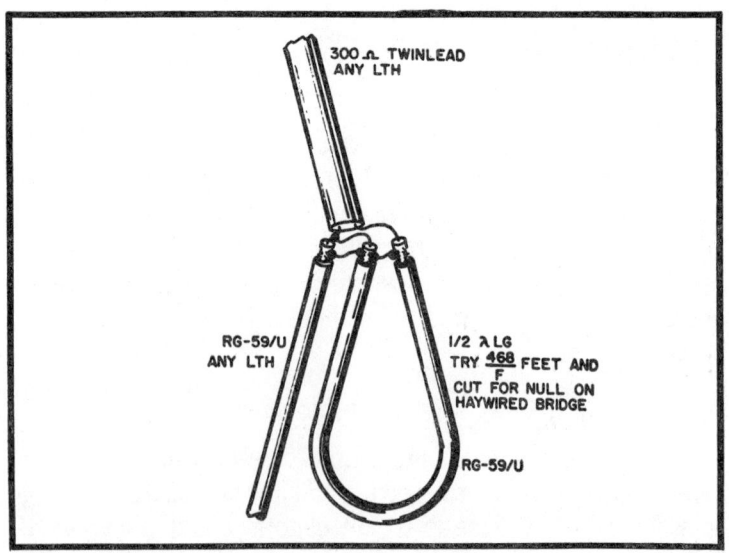

Fig. 2-31. This coax balun is as simple as possible. Avoid shorting between center and outer conductors!

Many radio receivers will take 300 ohm balanced or 75 ohm unbalanced inputs, but most amateur transmitters are designed to feed a 50 or 75 ohm load. These two load systems cannot be connected directly, one to the other, but a transformer or an electrical circuit which seems to act like a transformer may be used as an almost perfectly efficient adapter. From the unwieldly term "balanced-to-unbalanced" came the modern word "balun."

Three types of baluns are used in most amateur work. All are equally appropriate for transmitting and receiving, if they are heavy enough to handle the transmitter's rf. They are the toroidal transformer, the interwound coil, and the coaxial cable varieties. Only the coaxial cable balun, shown in Fig. 2-31, requires tuning.

The easiest and least expensive way to get a balun is to make it from coax cable. The exact length of cable required depends upon how fast the rf travels through it, so, if the balun is made up from a piece of coax perhaps 5% too long, simply use the haywire bridge described later to zero in on the proper dimensions. This eliminates difficulties with a piece of cable whose specs may not be quite right.

The dipole's "radiation pattern" indicates which way the rf goes once it leaves the dipole. For receiving purposes, efficiency is best in the directions that get the most rf when transmitting. The dipole's radiation pattern varies sharply with height above ground, and the highest possible installation is not necessarily the best. Figure 2-32 shows the striking variations in radiation pattern as the dipole's height is increased from 0.2 wavelengths above ground to 0.5 wavelengths.

The dipole's electrical properties also depend upon distance from ground. Figure 2-33 shows how the input resistance varies with height. For best results try for some height near one giving a 300-ohm input resistance.

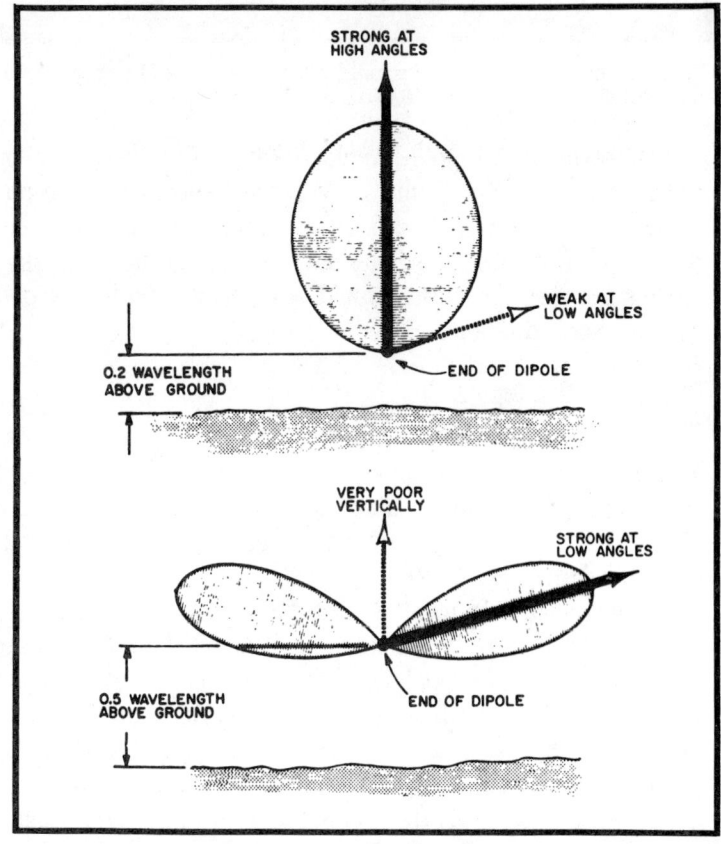

Fig. 2-32. The dipole has two basic radiation patterns. Here they are, as seen from the dipole's end.

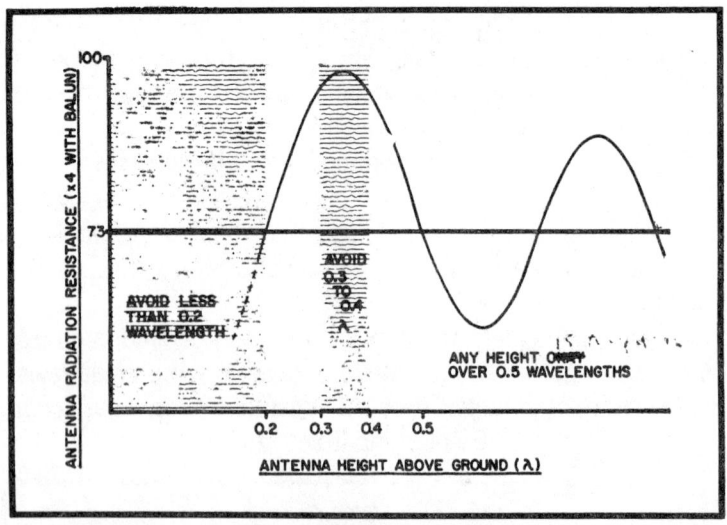

Fig. 2-33. The dipole's input characteristics depend upon height. Here are the two important numbers again: 0.2 and 0.5 wavelengths.

Although the dipole is said to show zero response or radiation off its ends, quite good signal reports may be obtained in these directions. This is because the dipole really does radiate off its ends, but only steeply up into the sky. Signals in these directions would be improved with a second better-oriented dipole.

Putting Up the Dipole

Trees, drain pipes, lightning rod systems, and towers, will all affect the folded dipole. The usual result is a measured resonant frequency lower than calculated. The dipole seems a little too long. And various nearby structures may also electrically unbalance the dipole.

Suppose L is in feet and F is the operating frequency in MHz. Cut a strip of twin-lead at least 404/F feet long, short both ends and attach the feedline to its center. Add tuning stubs to each end, both stubs equally long, for a total length of 462/F to 468/F feet. The twinlead makes good stubs. The dipole, minus stubs, is too short, with stubs fully extended it is too long. See Fig. 2-34. By folding the stubs back, it is possible to tune or retune the dipole to frequency, and balance out uneven effects of nearby structures.

Assemble the antenna in any convenient way similar to the handbook and magazine descriptions. Nylon line will do a fine job and the insulators can be omitted for low power. Seal twinlead openings and splices since moisture leakage may change the twinlead characteristics or corrode the wire. Tie the tuning stubs back against the dipole as shown in Fig. 2-34A and pull it up into the air.

Testing and Tuning the Dipole

With the twinlead dipole in place and the transmission line strung to the vicinity of the receiver or rig, everything that could affect its tuning or input characteristics has come into the picture.

Compare the dipole with a resistor at various frequencies, and specially note the frequency which gives the best similarity. The haywire bridge shown in Fig. 2-35 does this.

It's a simple Wheatstone bridge with a diode detector. Assume that the antenna terminal has a 75 or 82 ohm composition resistor across it, the same value as R3. Then the LH and the RH component strings are 2:1 voltage dividers, points A and B see the same rf voltage, in phase, and there is no DC fed to the meter.

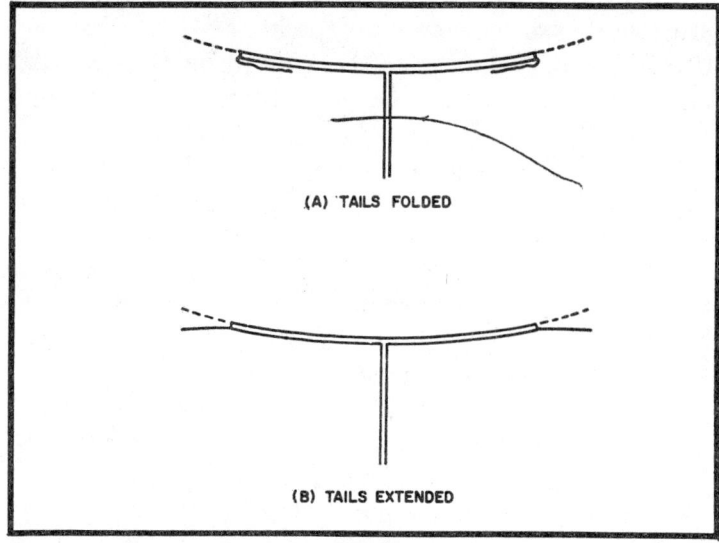

Fig. 2-34. How the tuning stubs are placed for highest (A) and lowest (B) resonant frequency.

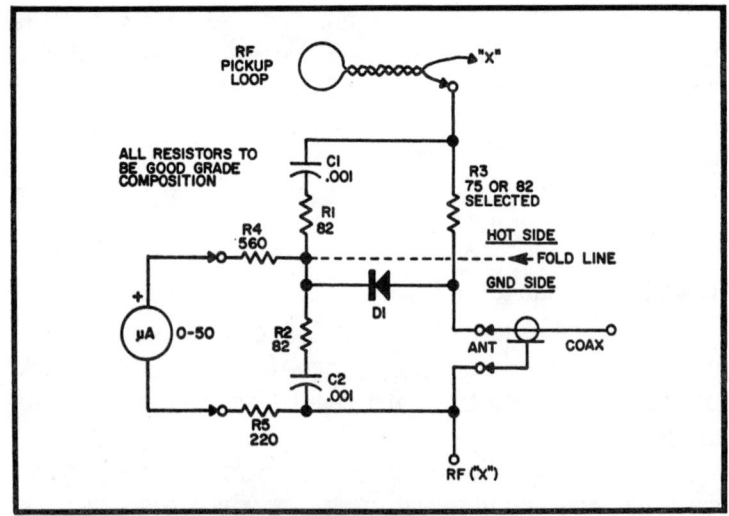

Fig. 2-35. Schematic of the haywire test bridge.

If the antenna resistance is replaced by any other value, or if inductive or capacitive reactance is added, the balance is disturbed and a meter reading is obtained. This circuit tells when the antenna terminals look most like a resistor, which is enough to get by.

The bridge circuit will give the best results if it is built of properly chosen components. Capacitors C1 and C2 are not critical, but should measure the same value on a capacitor tester. R1 and R2 are also ballpark accuracy if they measure the same on an ohmmeter. R4 and R5 are uncritical isolating resistors. Choose a good new resistor for R3. All resistors should be composition variety. D1 is any known, good germanium rf diode.

Assemble the parts with short leads in the relative positions of Fig. 2-35. Then, fold the entire assembly, like a book, around the diode, bring the two outside rf points together, and solder them to the rf input line.

Feed enough rf into the bridge to bring the meter to full-scale with an open circuit across the antenna terminals. Without changing anything, place a resistor equal to R3 across the antenna terminals, and the meter reading should go very close to zero. It should stay there, independent of large changes of rf frequency.

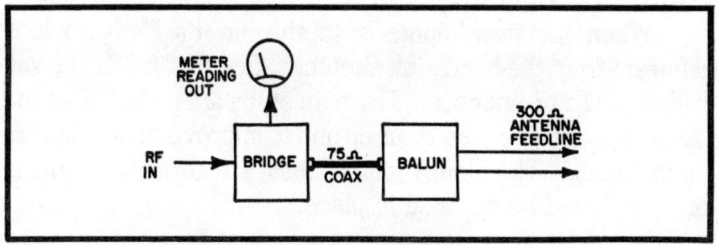

Fig. 2-36. Block diagram of the antenna test setup. A signal generator could replace the GDO. See how this fits into Fig. 2-30.

Connect a 300 or 330 ohm composition resistor across the balun's twinlead side, set up the bridge, and check for drop from full meter deflection to near zero. When using a coax balun, the null will be frequency dependent and belongs on the chosen operating frequency. Double check by finding the signal on a receiver.

Now test the antenna (Fig. 2-36). As before, adjust the rf source and bridge for full scale deflection, attach the balun and antenna and find the best null frequency. Repeat the test with the stubs fully extended, to find the lowest operating frequency. The target frequency should be between these extremes.

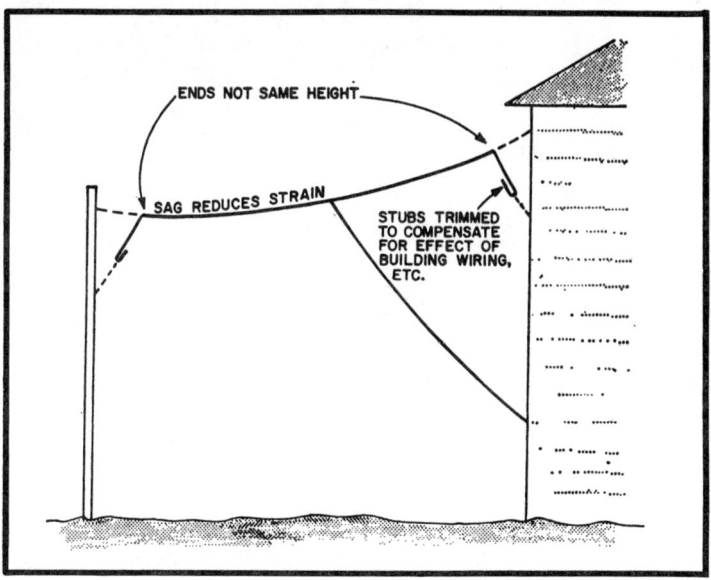

Fig. 2-37. With the new dipole in the air, the sag reduces load on the supports.

When the new dipole is in the air it's likely to look different from the handbook sketches. See Fig. 2-37. The sag reduces load on supports. The trim stubs are pulled off to the sides to guarantee their location and to improve their influence upon tuning. The floppy cords that are probably hanging around should be secured in place.

THE DOUBLE INVERTED VEE

A beam costs about $200 and a quad over $100. Then there is the problem of what to hold it in the air with. A tower is the most popular device, ranging in price from perhaps $50 for a used tower up to many hundreds of dollars, depending on the type desired. The more expensive types might be crank-up, non-guyed (with tilt over action). Of course a heavy duty rotor such as a Ham M costs around $130 plus cable, freight, etc., and the cost never seems to end.

This antenna, which may also be built for other bands if desired, is known as a *double inverted vee*, gives good directivity and power gain in the direction chosen, but also allows signals to be heard and worked from the sides and back. Construction is relatively simple and cost can be held to a minimum depending on how it is constructed. The antenna will give a much lower angle of radiation and thus a better signal to DX areas not normally workable with simple antennas such as dipoles.

When finished the antenna looks like the outline of a tent, Fig. 2-38. The lower this antenna is placed to the ground the shorter the elements become due to ground effects. This can be determined by experiment with an SWR meter and cut and try, the easiest method being to allow a foot or two of the element to hang down beyond the end insulator, where it may

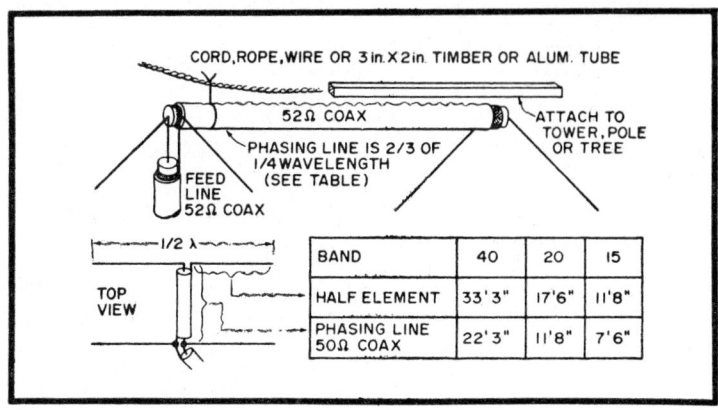

Fig. 2-38. The antenna resembles the outline of a tent.

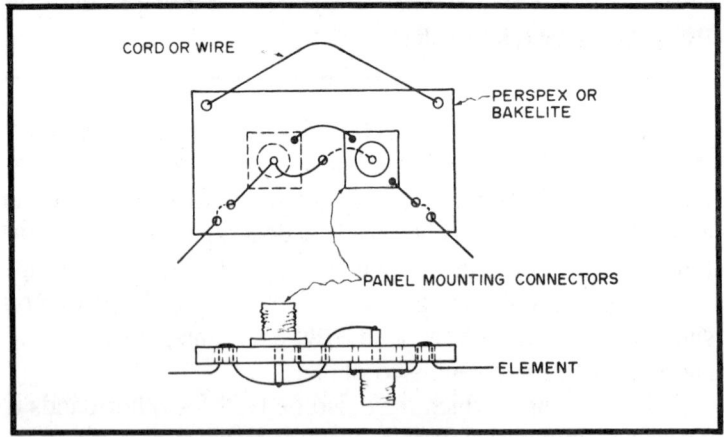

Fig. 2-39. Connections are made at the feedpoint.

easily be trimmed. This saves unfastening insulators each time. Fig. 2-39 shows how the connections may be made at the feedpoint. Alternatively the elements and coax may be soldered together at the appropriate points using egg insulators or similar supports. Theoretically a 1.1 balun should be used at the feed-point, but it does not appear to make such difference.

The antenna will work well with 75 ohm co-ax but the SWR will be slightly higher, although not excessive (less than 1.5 to 1). With 52 ohm co-ax SWR should be near unity, depending somewhat on height and surrounding objects.

If no co-ax is available a twisted pair of wires will serve the same purpose as 72 ohm co-ax and should substitute quite nicely. Another possibility is 75 ohm twin lead, which will make the whole structure lighter.

More elements can be added for higher directivity and gain. Element ends are insulated and tied off on bushes, trees or stakes in the ground. The beautiful thing about this antenna is that it is highly transportable, fitting into a box when traveling to a Field Day site and easily erected in a matter of minutes in emergency conditions. Note when more than 2 elements are used the element length, co-ax length, and spacing are exactly the same. Just add them on.

BALANCED DIPOLE (WITHOUT A BALUN)

Dipole antennas and the method of feeding them are well known among amateur radio operators.

Here is another method of feeding dipole antennas that results in a balanced antenna with a low swr, and it is fed directly with RG-8/U coax. A balun is not necessary to produce a 1-1 match.

Present day transceivers and transmitters are designed for 52 Ohm output impedance. With this fact in mind, this antenna was designed to a circuit with a feed point of 52 Ohms.

Construction is very simple and can be checked out easily and quickly. A great many circuit changes and tests have been made to arrive at the design shown in Fig. 2-40.

The coil in the circuit is not used as a loading coil but is used in the antenna circuit to determine the 52 ohm feed point.

Coils for low and medium power are made from ⅛" copper tubing or No. 9 copper wire. Coils for high power are made from ¼" copper tubing. All coils are spaced the diameter of the tubing and all the coils are self-supporting.

The coils are supported by a center insulator or to an insulator that is fabricated from a durable insulating material.

The length of wire needed for each half of the dipole is found by the formula

$$\frac{238}{F_{MHz}} = \text{Length (ft.)}$$

No. 12 solid or stranded copper wire may be used.

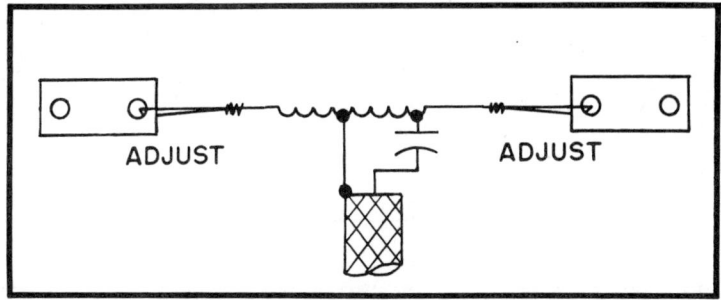

Fig. 2-40. Construction design of a dipole antenna.

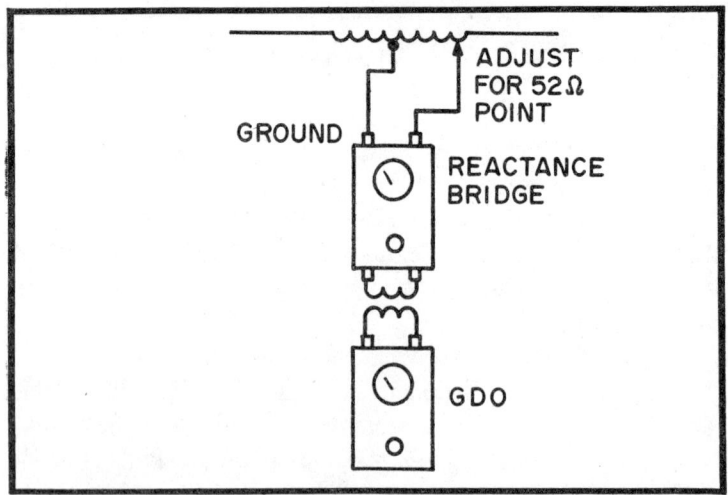

Fig. 2-41. The reactance bridge and gdo are connected.

A convenient method to determine the antenna frequency is to suspend the antenna about six feet above the ground, supporting the antenna coil on a wooden platform on top of a six foot step ladder. Using the platform for a work table, set a gdo to the desired frequency, bringing it into inductive relationship with the antenna coil. Adjust the length

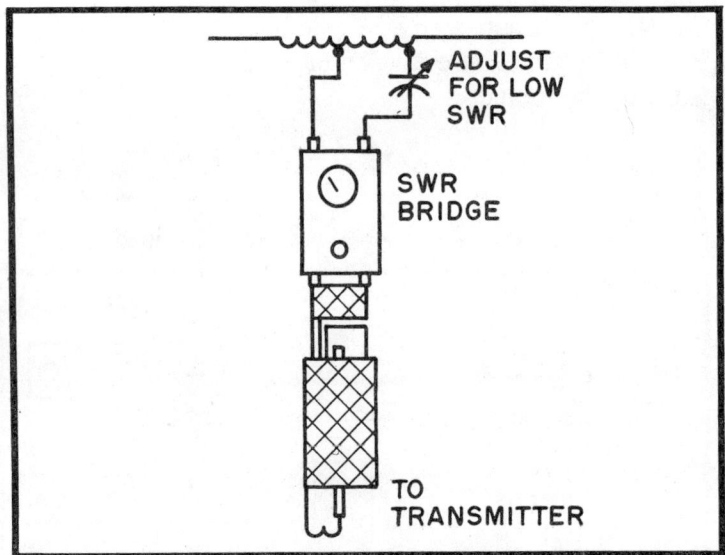

Fig. 2-42. A variable capacitor is placed in series with one side of the circuit.

of each element until resonance is obtained, keeping each end the same length.

Locate the center of the coil and solder on a tab there. Connect the reactance bridge and gdo as shown in Fig. 2-41. With the reactance set at 52 Ohms and the gdo set at the desired frequency, slide the terminal from the reactance bridge along the coil until 52 Ohms is indicated on the bridge meter. At this point solder on a tab.

Before any checks are made of the swr, a variable capacitor is placed in series with one side of the circuit as shown in Fig. 2-42. This capacitor is used in the circuit to tune out the reactance, and should have a capacity of 7 pF per meter. The capacitor may be left in the circuit or replaced with a fixed unit of the same value.

Attach feed line and swr meter as shown in Fig. 2-43 and adjust the capacitor to the lowest reading swr. Use low power for this operation.

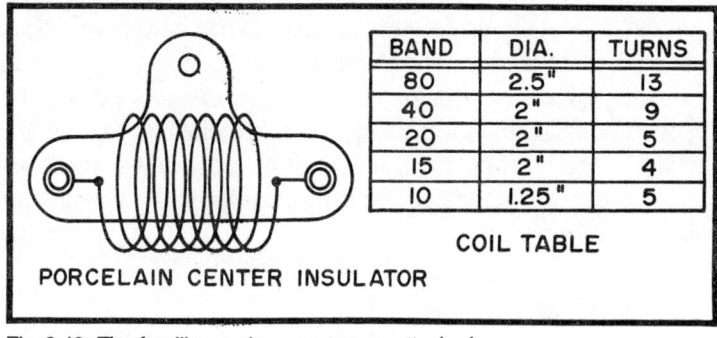

Fig. 2-43. The feedline and swr meter are attached.

Precautions

- Be certain test instruments are calibrated correctly.
- Do not allow the leads from the coil to droop down along the coil, let them come straight out for an inch or two.
- In construction of the coils be sure the spacing of wire or tubing does not change.
- If swr is high do not start to cut the feedline, but check the work and adjustments.
- Seal the ends of the coax so moisture cannot enter.

TRAPLESS MULTIBAND ANTENNAS

The harmonic relationship of the high-frequency amateur bands has over the years generated a multitude of multi-band dipole designs. That is, a dipole form which can be fed with a single feedline (preferably coax) and operated on several or preferably all bands 80—10 meters without any tuning adjustments. "Trap" type dipoles have been most popular over the past several years and any number of variations of trap antenna designs are available. In general, trap type dipoles when constructed of high quality components give a good account of themselves. Their disadvantages are usually sharp resonance within each band so that it is difficult or impossible to obtain optimum performance in both the CW and phone portions and the physical loading of the antenna structure due to the trap components. A number of amateurs have been trying to get away from the use of traps by finding the right combination of dipole element lengths so one can construct an efficient multiband antenna solely from wire elements. The following is a description of some of these designs which have proved popular and interesting. One can duplicate these antennas directly, if desired, but they also provide some very useful ideas for individual experimentation with antenna designs.

G5RV Multibander

The antenna design of Fig. 2-44 is often called the G5RV antenna. By a proper combination of choice for the antenna's flat-top elements lengths as well as for the length of the 300Ω feedline, resonance can be achieved on all of the amateur bands from 80 to 10 meters. The impedance present at the end of the 300Ω line is about 50—60Ω and a regular coaxial cable of any desired length can be used after this point. Although not absolutely necessary, it would be well to use a 1:1 balun for connection between the coax and the 300Ω line (note again that a 1:1 balun should be used and not a 1:4 balun). The antenna is slightly "short" on 80 meters and the 300Ω line section serves as a form of matching stub on this band and a

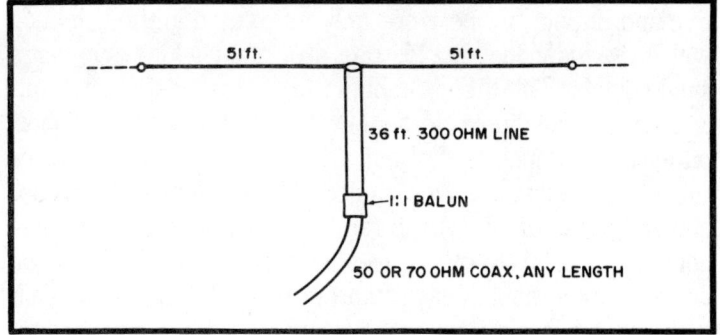

Fig. 2-44. The multiband 80-10 meter antenna.

combination of stub and/or impedance transformer on the other bands. An SWR maximum of 2:1 can be achieved across most bands with a minimum of about 1.3:1 at the best frequency within the band. One should take a bit of time to properly trim the 300 line section for the best SWR, particularly on the 20 and 15 meter bands. A few feet, plus or minus, depending upon the installation environment, can make a considerable difference in the SWR on 20 and 15 meters. Generally, as far as the cost of construction goes as related to performance, the G5RVC design is about the best design available.

DJ4BQ Double-Dipole

An only slightly more complicated design is the DJ4BQ multi-band double-dipole as shown in Fig. 2-45. This antenna operates on every band from 80 through 10 meters. One dipole (the longer one) operates on 80, 20 and 15 meters while the other dipole (the shorter one) operates on 40 and 10 meters. So, on every band only one dipole at a time is opera-

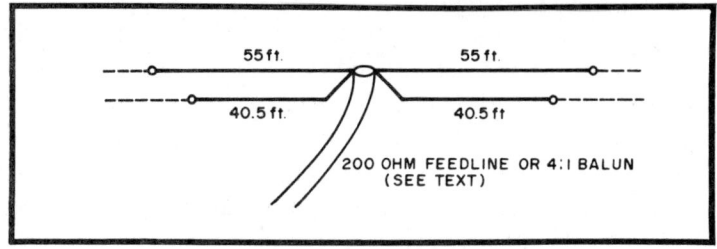

Fig. 2-45. The DJ4BQ double dipole for 80-10 meters.

tive and, in fact, if one were only interested in the bands on which the individual dipoles operate, one could put up just a single dipole. The theory behind the operation of the antenna is to choose the dipole element lengths just to be slightly short on each band such that about a 200Ω feed point impedance results on each band. This is accomplished by the dimensions shown for the dipole. If the feedpoint impedance is matched correctly, an SWR of no greater than 1.5 to 1 should be achieved over most of each amateur band. The dipole which is operative on each band can, of course, be fine trimmed for almost a perfect 1:1 SWR in any specific portion of a band. The matching of the 200Ω feed point impedance can be done easily with a 1:4 toroid balun working from a 50Ω coaxial line. The 1:4 balun can be connected directly to the feed point of the antenna and coax used to the transmitter or 200Ω open wire line constructed for a light-weight and more economical approach and used as a feedline with a 1:4 balun at the transmitter. The developer of the antenna recommends spacing the dipole wires at least 6 in. apart to prevent interaction of the elements. The flat-top could be constructed from 300Ω twin-lead with each wire in the twinlead forming one of the dipole elements. Heavy duty twinlead with copperweld wires should be used (Belden 8230). Some experimenting with the elements lengths will be required because of the close spacing of the elements.

A Modern Windom

Some old timers may still remember the classical Windom antenna shown in Fig. 2-46 and named after the amateur who developed it in the 1930's. It is simply a half wave antenna feed 14% off of the center point with a single wire feeder of any length. It operates only on the even harmonically related bands (80, 40, 20 and 10 meters for the basic 80 meter antenna; 40, 20 and 10 for a 40 meter flat-top, etc.). It enjoyed great popularity in its day as a simple but very efficient multi-band antenna and a theoretical analysis of its construction proved the soundness of its design. By studying the current and voltage relationships which exist along the flat-top on each band, it can be seen that the feed point chosen 14% off center

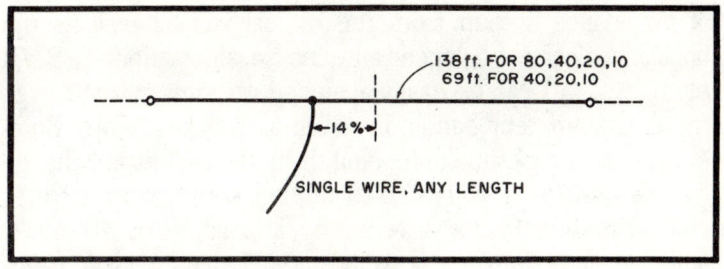

Fig. 2-46. The Windom antenna of the 1930's.

does indeed provide the correct matching point on even bands for a single wire feeder. The characteristic impedance is 500Ω at the feedpoint. The era of TVI pretty well killed the Windom because of the rf that the feedline brought into the shack which in turn made efficient transmitter shielding and filtering almost impossible. For a short while in the mid-1950's a variation of the Windom became popular where the flat-top portion of the antenna was broken and fed by a 300Ω line at approximately the same point as the original single feed line was connected. The 300Ω line could be any desired length but at some point a 4:1 balun had to be used to bring the feedline impedance down to 75Ω for a coaxial cable feed. The requirement for this multiband balun, which was not a simple thing to construct before the event of toroids, caused the antenna to fall into oblivion.

Recently, a modification of this antenna has come to light. Instead of using 300Ω twinlead, it goes directly from a coaxial cable through a 1:6 balun transformer connected in the antenna flat-top as shown in Fig. 2-47. A 1:6 balun can be wired using one of the toriod balun kits in much the same manner as the common 1:1 or 1:4 baluns. The balun serves the purpose

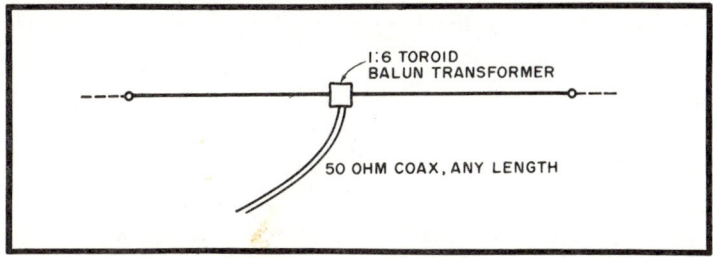

Fig. 2-47. The modernized Windom. Same dimensions apply as shown in Fig. 2-46. Balun is made using toroid kit.

of impedance stepup from the coaxial line as well as the unbalanced coax to balanced antenna transformation. An SWR of under 2 to 1 can be easily achieved with this antenna over the entire amateur bands on which it operates. Note, however, that this method of feeding the antenna does not change its basic nature. It will operate only on even harmonic bands and 15 meters is not covered by a basic 80 or 40 meter antenna. One simple way to include 15 meters would be to erect parallel to the Windom a regular half wave 15 meter dipole connected in parallel to the coaxial feedline on the low impedance side of the balun transformer.

Performance and Construction

The trapless types of antennas described here, once properly adjusted and checked for SWR, will generally perform as well on the low frequency bands as a trap antenna and usually better on the 20, 15 and 10 meter bands. The latter effect is due to the fact that on the higher frequency bands, the full antenna is still used while in the trap antennas, the traps are arranged so that the antenna remains a half-wave dipole on each band. Trapless antennas on 20, 15 and 10 start to exhibit some gain and hence directivity.

Naturally, the performance of any antenna depends upon its height above ground. However, this also effects the impedance developed at the terminals of an antenna. Since the electrical height above ground changes as the antenna is used on different bands, the feed point impedance is also changing on various bands. A bit of patience is required to carefully check the SWR on each band before rushing to put the antenna into operation. Some time spent in trimming the flat-top element lengths, or the feedline in case of the G5RV antenna, will pay dividends over the long run with far better SWR performance on each band.

Many amateurs will undoubtedly think of using these multi-band antennas as V types from a tower support. Basically, the antennas should work fine in this manner as long as the SWR is properly controlled. The Windom antenna has also been reportedly used with the shorter leg vertical and the longer leg arranged at a right angle to it with satisfactory results.

TEN METER FOLDED DIPOLE ANTENNA

The length of a folded dipole can be found using the formula 468/f which will give the length of the dipole in feet. The length of 16 ft 6 in is a good compromise for the whole of ten meters. The advantages of using a folded dipole are numerous: It is a broad band device. This is good for ten meters where the operating range is from 28.0 MHz to 29 MHz. The next advantage and most desirable characteristic is that it does not accept power at twice the fundamental frequency. This means that it should attenuate the channel two harmonic providing there is no capacitive coupling at the tuning end. Cutting the feeders 22-32 feet long or the length 52-64 feet long will also help.

The folded dipole does not accept power at even multiples of the fundamental, because the folded section acts as a continuation of a transmission line. The folded dipole is better than a single dipole because the current in the two conductors flows in the same direction and acts as two conductors in parallel and the current, therefore, in each conductor is divided. Thus the feed line sees a higher impedance because it is delivering the same power at half the current and the impedance is about four times greater at the feed point than a regular dipole. This enables the use of 300 ohm twin lead to feed the antenna.

The folded dipole antenna mounted in a vertical position with the bottom 12 to 24 inches off the ground offers a low angle of radiation and probably a lower angle than most beams, but without the gain. The lower angle is really an advantage when reaching out for the extreme DX. The all around pattern is also an advantage for SSB round table discussions.

For the investment, the vertical folded dipole is a good solution for those who do not want to put up a beam for the short period of ten years is open.

Construction

Make a trip to the lumber yard and buy a 20 foot 2 × 2 and give it an undercoat and three coats of Z-Spar boat paint.

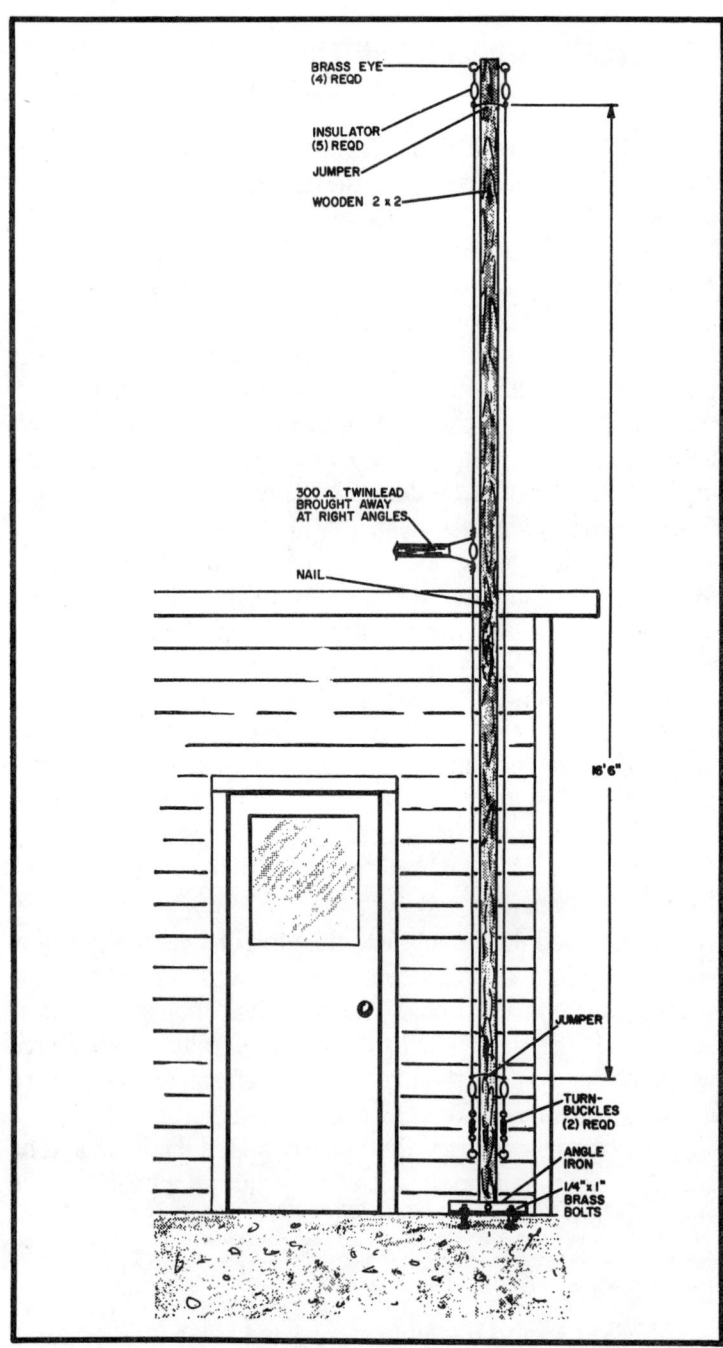

Fig. 2-48. The folded dipole antenna mounted in a vertical position.

Put two brass screw eyes at the top of the pole and two at the bottom a sufficient distance away to accommodate the length of wire and insulators plus turnbuckles. String the wire on each side of the 2 × 2 between the insulators and tighten the turn buckles. Solder jumpers on each end. The center insulators where the feedline is attached should be a short one. Wires can be soldered from it to several 6-32 machine screws on a piece of lucite which has been screwed to the 2 × 2 to help take the strain when the 300 ohm line is pulling. The 300 ohm twin lead should come away from the antenna for about ten feet. It can be held by end slotting some sticks which are then nailed to the house or garage to support the line.

The antenna mast can now be nailed to the eaves of the house or garage. The bottom of the 2 × 2 can be bolted to an angle iron or aluminum four inches long which in turn can be bolted to the concrete with ¼ × 20 bolts set in the cement. An easy way to do this is to star drill ⅜ holes about 1 inch deep and put the head of the bolt in the hole. Heat some sulphur in a can with a torch and pour it in the hole, and it will be secure.

Since most transmitters use PI-Network outputs, it will be necessary to use an antenna tuner. Link coupling to the tuner and the condenser adjustments will provide a means of obtaining zero SWR throughout all of the ten meter band as far as the transmitter is concerned.

For the few dollars investment in wood, paint and wire compared with the beam, tower and rotor, this is a remarkable antenna.

ALL-BAND ROTATABLE DIPOLE

Want an antenna with 8 to 10 dB gain, small dimensions, all band operation, low angle radiation, low SWR, and a low price? No such animal.

This antenna is simply a dipole for 40-10 meter work; on 80 meters, it can be loaded with a little experimentation. Get this antenna up 50' off the ground on a rotator and it will perform well.

The coils, L_1 and L_2, are B&W #3022 miniductors. Insulators are Johnson type 65. The two 16' elements are ½" diameter electrical thin-wall tubing, either aluminum or steel conduit. The wooden support is a 2 × 2, about 6 feet long, and the mast is a TV mast, preferably leading to a rotor at the bottom.

Mount all of this as shown in Fig. 2-49 and that's it. The antenna should be fed with 52 or 75 ohm coax and the coils tapped for the lowest SWR on the preferred band. If tapped for 40 meters (usually about 10 turns onto the coil from the element) the antenna works fine on 40 and 15, and sometimes even 80 meters. Tap for 20 (low end of the band, about 1 turn of coil; high part of band, bypass the coil completely) and it works well from 20 through 10 meters. For 10 meter operation, the lowest SWR will occur when there is no inductance at the antenna (bypass the coil).

Check the SWR while the antenna is still on the ground if desired, as very little, including transmission line length, will effect it to any great extent after it is raised. The SWR on all bands will be better than 2:1 at resonance, and SWR's of 1.1:1 at resonance are not unusual. The SWR will not go above 2.5:1 anywhere in any band if the antenna is built and installed correctly. An antenna tuner will provide nearly instant bandswitching without touching any taps and a pi-net in the transmitter will also help.

Incidentally, this antenna works quite well on 80 and 75 meters by experimenting with the tapping on the inductor, which for these bands, will be near the bottom of the coil (from the element) for maximum inductance.

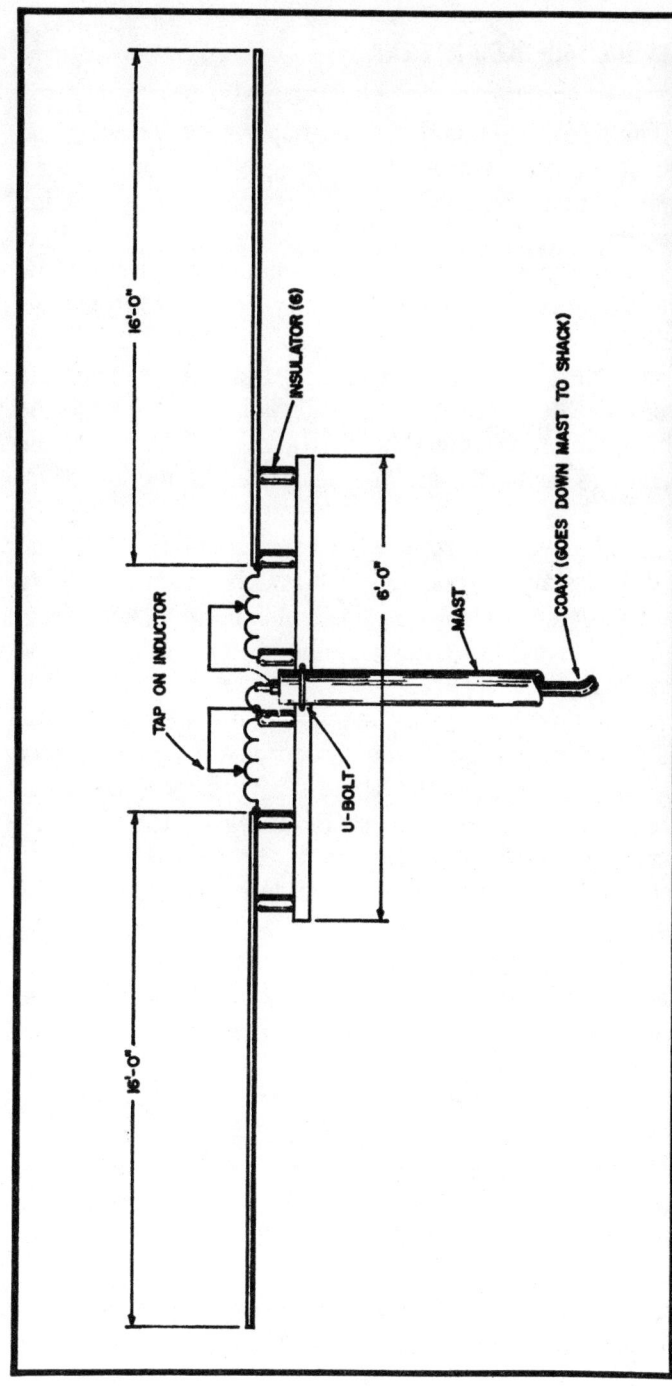

Fig. 2-49. Construction of the all band rotatable dipole for 80 through 10 meters. When properly adjusted this antenna provides excellent results without a large cash outlay.

A BALANCED DIPOLE ANTENNA

The dipole antenna described displays the following features: rf balance to ground; lightning protection; low loss ground system; multi-band capability; and simple erection procedure.

The principal radiation from such a dipole is in a plane perpendicular to its direction; the polar plot of field strength is essentially a circle tangent to the surface of the ground. Thus, the radiation is concentrated at the high angles, near the vertical. The desirable low-angle radiation will increase as the height of the antenna is increased. In addition, the losses in the antenna and in the ground are decreased as the height increases.

An effective compromise between operating results and cost of the installation normally calls for a height of about 40 to 60 feet. Greater height is desirable but expensive, whereas at lower heights the efficiency and the pattern shape become unacceptable.

It turns out that the five-section telescoping TV masts, available in the popular TV supply stores are satisfactory for supporting the center of the dipole at a height of 45 feet. This mast can be assembled on the ground and walked up or hoisted to a vertical position. If clamped against a house or other structure at about the 15 foot level the mast need only be guyed at the top.

The two halves of the 80 meter dipole serve as top guys in, say, the north-south directions, and the two halves of the 40 meter dipole (with a common coaxial feed line) double as east-west guys.

For ease of erection the mast is pivoted at its base. This base support is a 5 foot length of 1-½" diameter galvanized steel tubing projecting 1 foot above the ground. A 5/16" bolt through the post and mast provides the pivoting shaft. The side of the mast is hacksawed diagonally at the base to clear the supporting post.

The coaxial feed line for the dipoles is inside the mast; this decreases wind problems and also serves to shield the outer sheath of the coax from unwanted rf coupling.

A balun coil arrangement is frequently used to assure balanced currents at the center of a dipole antenna. In this instance a novel method is used which not only provides the required balance to ground but which also offers a low-resistance DC current path to ground from both sides for lightning protection.

The mast itself forms the outer conductor of a coaxial section having the sheath of the feedline at its inner conductor. The length of this coaxial section inside the mast is approximately one quarter wave length. Therefore, if the shield is shorted to the mast at its base, the impedance viewed from the top will be high so that it can be connected across the center feed point of the dipole without appreciable loading. This unorthodox arrangement is completed by connecting the center conductor of the coaxial feedline directly to the top of the mast. One half of the dipole antenna is connected to this common point, the other half to the shield. The arrangement provides DC current paths to ground for lightning protection of both sides of the dipole and it achieves balance of antenna currents through its balun action.

The rf transmitting currents in either half of each of the dipoles are conveniently indicated by means of four flashlight bulbs, not shown in the figures. Each bulb is shunted across a 3" length of the antenna wire just outside of the top insulator. A 3" length of #18 copper weld wire is used from each side of the bulb so that the latter is soldered in place at a corner of a 3" equilateral triangle. This area of pick up loop gives adequate brillance for tests without burning out the bulbs. The equal brillance of the two 80 meter bulbs, for example, provides an excellent indication of the balun action of the unorthodox mast connection used.

In order to insure a stable dipole pattern with symmetrical ground reflection and to reduce ground losses, a grid of nine parallel ground wires, each approximately 200 feet long, spaced ten feet apart, was buried under the dipole parallel to its plane. A single cross-buss was placed across the middle of the grid and soldered at the cross-over points. The base support of the antenna mast was connected at the center of this ground system.

The clamping hardware supplied with the mast was discarded. The mast was extended horizontally on blocks on the ground and the overlapping sections were secured against vibration by use of #18 self-tapping screws. The mast was painted to discourage rust.

The antenna wires and the coaxial feed line are supported at the top of the mast by a cylindrical plastic insulator. This insulator, which combines high strength with low wind resistance, is made by sawing off the grooved male portion of a "T" fitting for plastic water pipe. There is a size which is a snug push fit over the 1-¼" o.d. mast. The grooves serve to anchor the wires and they also provide an improved leakage path for rf currents.

The details of the insulator assembly are shown in Figs. 2-50 and 2-51. All wires are #12 stranded copperweld. The "D"-shaped wire loops pass through the insulator, one above the other, as shown. There are no screw connections to loosen or corrode and, of course, all connections are twisted and soldered. A radiator hose clamp holds the insulator in place. The RG-8/U coaxial feed line is supported from the "D"-loops, centrally within the mast. (This feed line is cut to be a multiple of a half wavelength—approximately 82' 2" for a

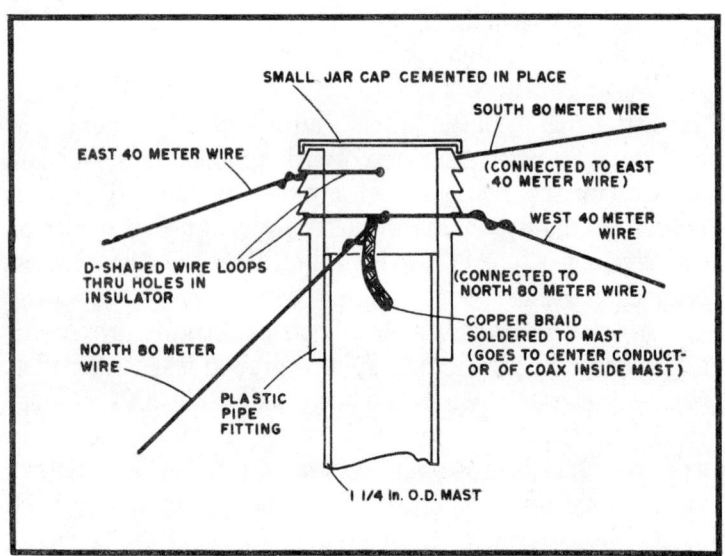

Fig. 2-50. Top insulator—side view.

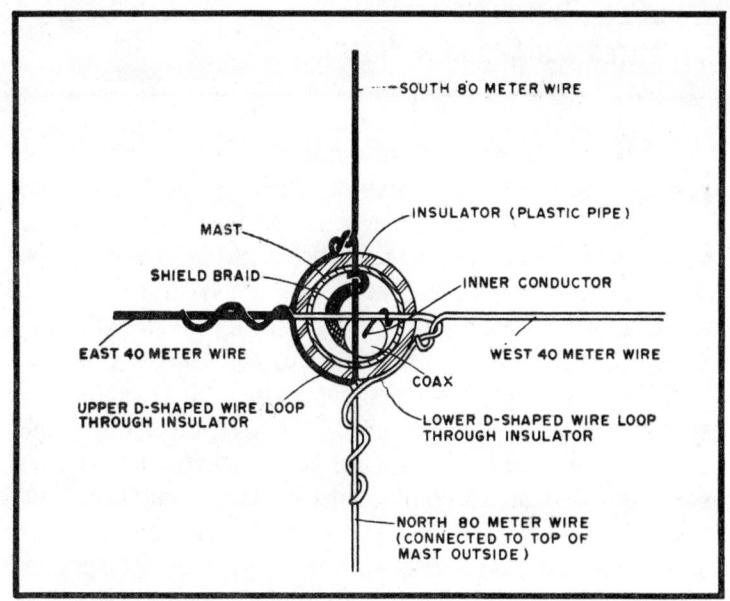

Fig. 2-51. Top insulator—top view.

3.955 MHz.) A noise bridge is used for this and all other resonance measurements.

The mast was hoisted to the vertical position using an inexpensive 4-pulley rope hoist attached at the 16' height. After erection the mast was clamped to the house roof at the 15' level using a steel strap and lag screws.

The four halyards at the end of the antenna wires are of ⅛" nylon rope. These are supported by 2-½" diameter galvanized posts seven feet high. These posts are each located at a distance of 80 feet from the base of the mast and are connected to the ground system. The ropes are secured by swivel snaps attached to the tops of the posts, for ease of lowering for antenna adjustments.

The dipole was adjusted to resonance, using a noise bridge, by adding equal lengths at either end. Resonance was achieved at 3.955 MHz with a length of 120'9" and an input resistance of 50Ω.

The SWR of the antenna is quite low (approximately 1.01:1) and its response is sufficiently broad to cover the 80 meter phone band without appreciable decrease in signal strength.

OLD ANTENNAS AND NEW BALUNS

Anyone who has been in amateur radio for ten to twenty years will remember the days of elaborate "wire" antennas. Newcomers can also glance in some of the old antenna manuals and find them replete with "wire" antenna designs. Wire antennas as the name indicates, are simply more elaborate antenna forms than a simple dipole which provide some gain and directivity and which were usually constructed from wire hung between the necessary supports. The advantages to such antennas was primarily cost, since relatively high gains could be achieved for the cost of additional antenna wire. All sorts of collinear arrays, broadside arrays, curtains, etc., were developed and used successfully. The problems associated with such antennas were many and one of the primary ones was the often awkward feed point impedances and the requirement for a balanced feed. Open wire lines had to be used to feed the antennas at impedances ranging from 150-600Ω and then the balanced open wire line converted via an antenna tuner to an unbalanced coaxial feed. For these and other reasons, elaborate wire antennas have fallen into disuse. Nonetheless, for the amateur who has the necessary space and is primarily interested in working single-band DX, these antennas can provide very good service at minimal cost. Fortunately, the advent of the toroid balun transformers with variable impedance ratios has also eliminated the feed problem once associated with these antennas. The purpose of this discussion is not to re-present every type of wire antenna array developed. A few examples are given and the use of a toroid balun illustrated. One can, however, glance back in some of the older antenna manuals and magazines and find any number of elaborate wire arrays to which the same feed techniques illustrated here can be applied.

Variable Impedance Toroid Balun

A toroid balun is usually thought of being a 1:1 or 1:4 ratio type unit. That is, going from 50Ω unbalanced to 50Ω balanced or from 75Ω unbalanced to 300Ω balanced. But any toroid

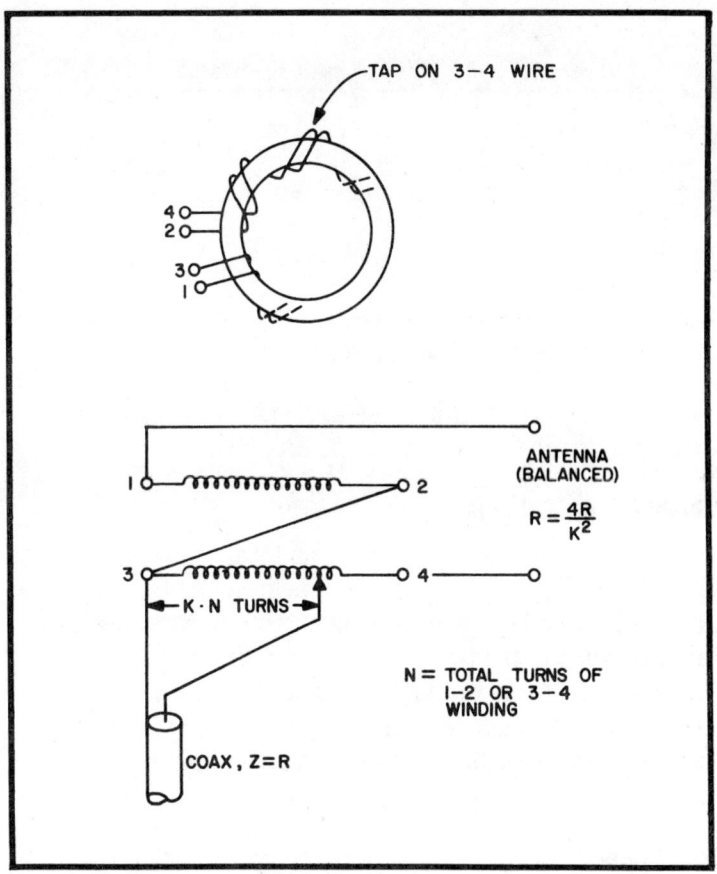

Fig. 2-52. Variable impedance transformation balun 1:4 to 1:10 or more.

balun kit can also be used as a variable impedance balun with transformation ratios greater than 1:4 possible up to about 1:10. Figure 2-52 shows a typical toroid balun winding. The instructions contained in any balun kit can be used to place the initial windings on the toroid core for a 1:4 balun. Note that if the coil tap on the 3-4 winding is placed at point 4, one has a normal 1:4 balun. If, however, this tap is moved closer to the 3 terminals, the transformation ratio of the balun increases according to the formula shown. For instance, if the tap were placed at the quarter way winding point between 3 and 4, that is one quarter of the turns from 3 to 4 away from 4, the transformation ratio would be approximately 1:10. A 50Ω unbalanced input would be transformed to a 500Ω balanced

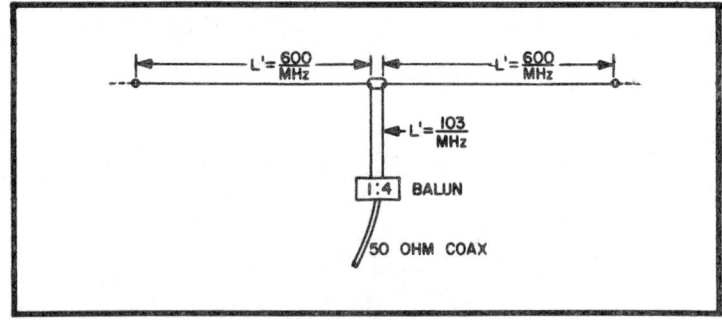

Fig. 2-53. Extended double Zepp with balun feed.

output. In a similar manner, the other tap points can be figured out for any impedance transformation ratio.

Double-Zepp Antenna

The Double-Zepp antenna is a form of extended dipole as shown in Fig. 2-53, where the dipole elements are made as long as they can be while still having the radiation pattern of the antenna not split up and remain at right angles to the line of the antenna (in and out of the page). The gain achieved is an easy 3 dB. A small phasing section is still required at the center of the antenna, as shown, before the connection of a 1:4 balun.

¾λ Dipole

The ¾λ dipole of Fig. 2-54, also has its main radiation at right angles to the line of the wire and produces 3-4 dB gain. This form of antenna may be somewhat easier to construct than the Double-Zepp since the balun (a 1:6 unit in this case) may be connected directly at the center of the antenna. The

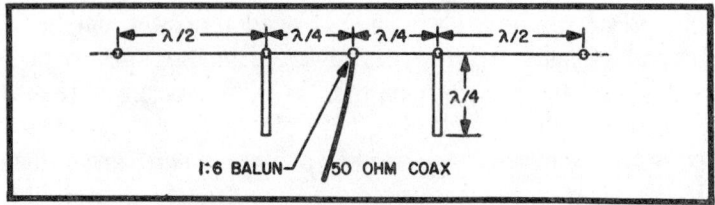

Fig. 2-54. Classic collinear array balun feed. See text for extending antenna to increase gain.

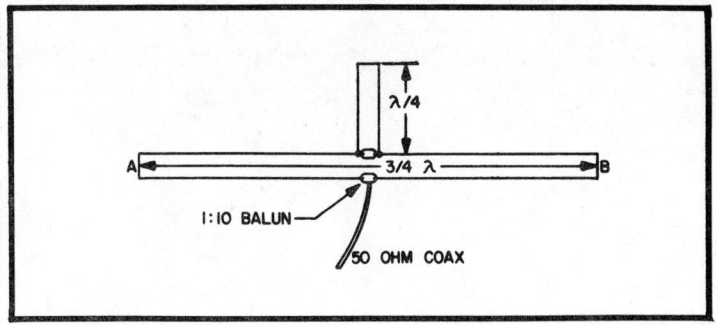

Fig. 2-55. ¾λ folded dipole may be operated on two bands if desired by using stub switch.

phase reversal stubs between the ½λ elements can be made of simple 300Ω twinlead shorted at the far end. The antenna can be extended with another ½λ element on each end (and a ¼λ stub to connect to the adjacent ½λ element) to raise the gain anothe dB or more. In this case, a 1:10 balun has to be used to feed the antenna.

Dual Band ¾λ Dipole

The ¾λ dipole shown in Fig. 2-55, can be used either as a single band or dual band antenna. Its total length is ¾λ long at the lowest frequency band used. If used as a single band antenna, the shorted ¼λ stub shown is not required. If it is to be used as a dual band antenna, the stub is made ¼λ long at the lowest frequency band. On the next higher harmonically related band, the stub will act as a short circuit, since it becomes ½λ long, and allow the antenna flat-top to properly resonate. Whether used as a single band or dual band antenna, it can be fed via a 1:10 balun.

Super Loop

The large loop antenna shown in Fig. 2-56, can be mounted from a tower or other support. Its radiation is horizontally polarized and broadside to the plane of the array (in and out of the page). The gain is about 4 dB in both directions. It can be fed directly from a coaxial line via a 1:10 balun at the base as shown.

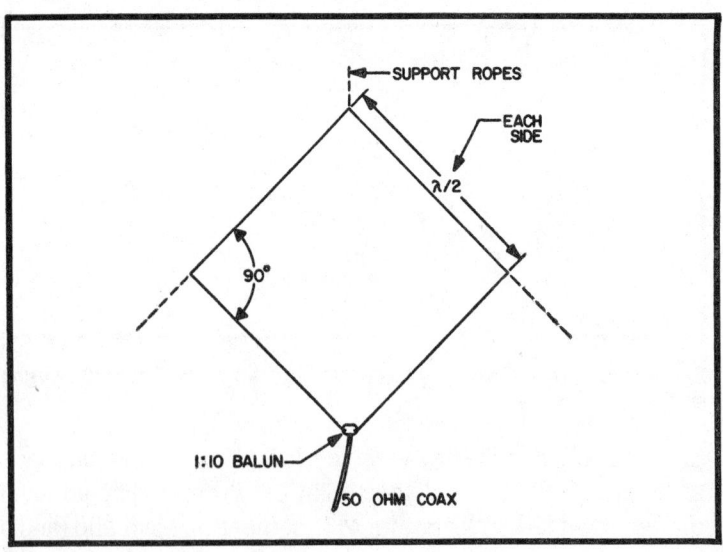

Fig. 2-56. Super loop or bi-square requires only single support and produces easy 4 dB gain.

The array of Fig. 2-57, is just one small example of a curtain array including such types as Sterba, Bruce Arrays, etc. The gain that such arrays can provide become quite significant if one has the space to extend them one to two wavelengths. In this case, the array will provide a broadside gain of 7.5 dB in both directions. The antenna can be fed at the point shown via a 1:6 balun. The phasing line between the

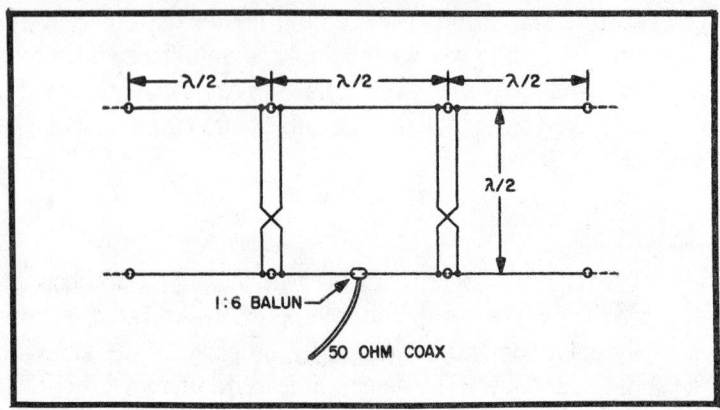

Fig. 2-57. The six-shooter array provides 7.5 dB of gain. Lower elements should be ¼λ high.

upper and lower set of elements can either be open wire line or 300Ω twinlead with a single twist.

Summary

Many other antenna forms which present a resistive impedance on a single band but of an awkward value can be fed via a properly constructed balun. Other antenna types which suggest themselves are V beams, rhombics, half rhombics and single tilted wire antennas.

A SIMPLE, SHORT 160 METER ANTENNA

A thirty-foot vertical antenna, top-loaded and using as good a ground as possible, can produce good results on 160. The antenna described here is designed for field day and portable operations, but may be made a permanent installation. It makes use of readily available materials and is inexpensive to build and easy to tune up.

Almost anything can be used for a mast, as long as it is capable of supporting the loading coil and can be made a solid, low-resistance structure. The mast used was made of telescoping 8-foot sections of aluminum tubing, using self-tapping screws at the joints for good bonding. Other possibilities might be copper water pipe, thin-wall tubing such as conduit, or even a wooden mast with a few heavy aluminum wires stapled to it to form a low-resistance conductor. Ninety lb nylon string was used for guys, but if a permanent installation is to be made, wire guys and suitable insulators should be used. Four layers of scotch electrical tape are wound over the bottom of the mast as an insulator, although a better insulator might be made by using a section of old bicycle innertube slipped over one end.

The adjustable end-section above the loading coil is a broadcast-band replacement automobile whip of the telescoping type, designed to fit over the broken-off stub of the old antenna and is available at most auto supply and electronics stores. This whip is used as a tuning device. Varying its length will allow tuning of resonance at the portion of the band desired, in the same manner that the commercial mobile whips are resonated.

The loading coil form is a piece of PVC water pipe, 1½ inches i.d., with a 1⅞ o.d., and makes a light, strong coil form of excellent Q. It is readily available at plumbing supply, hardware or home handyman centers. The coil itself consists of 10 in. of #20 closewound. The ends are secured with a drop of epoxy cement after passing through a hole in the form, and lugs are soldered to the ends for connections as shown in the sketch.

Fig. 2-58. A simple, short, loaded 160-meter antenna.

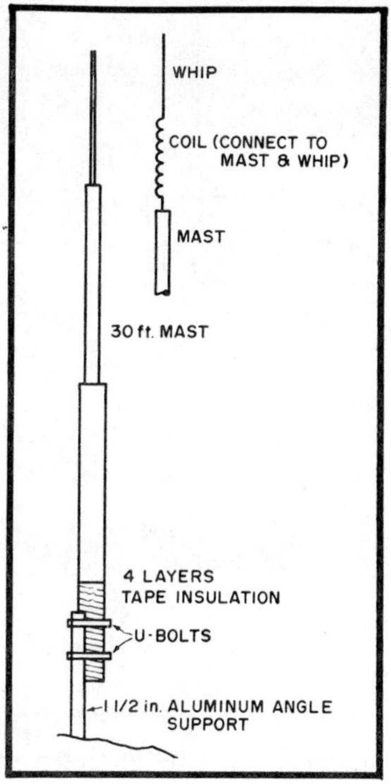

The end pieces, which are held in place with a screw passing through the coil form and end piece, are made of 1-inch thick lucite, 1½ inches in diameter to fit inside the coil form. They can easily be fashioned using a coping saw and a hand drill. The bottom end piece is drilled to fit over the end of the mast. The top stud is a 2 inch 5/16 bolt, passing through the center of the top end piece and allows the whip to be clamped solidly over this stud using the screws provided by the whip manufacturer.

After the entire loading coil assembly is completed, place a piece of shrink-tube over the coil and apply heat. This will form a good weather seal and offer protection to the coil windings.

Tuneup of the antenna is simplified by placing the coil on the top section of mast, supporting it temporarily, and adjusting the whip for resonance as indicated by lowest swr. Adjust for resonance 25 kHz HIGHER than the desired operating frequency, then when the rest of the mast is added, the

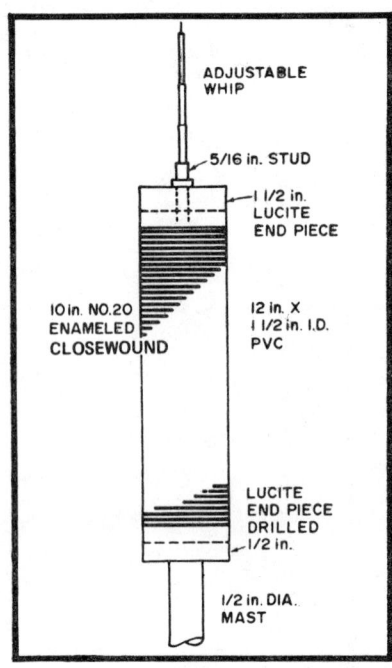

Fig. 2-59. Detail of the loading coil.

frequency will be at the correct point. If a grid dipper and impedance bridge are available, they will simplify the procedure. A direct connection to 50 Ohm coax is satisfactory, and no elaborate matching system is needed. The test antenna showed 1.2 to 1 at resonance. In use, the antenna will show a VERY narrow frequency range, on the order of 8 kHz plus and minus the selected frequency for an swr below 2 to 1. However, this is not too bad on 160, since the band is most active on the bottom 1.800-1.825. Any segment which is active in a specific area may be selected and the loading adjusted for that frequency.

The best available ground should be used. A rod driven a few feet into the ground is useless except for lightning protection. Use a connection to a cold water pipe, and at least two 130 foot wire radials, zig-zagged if necessary, to fit the space available. DON'T shorten the wires, even if the space is tiny. Run the radials straight out, and don't bury them unless necessary and then only a scant inch deep.

This antenna works well for local contacts, and does well on DX if the band is quiet.

NOVEL 160 METER ANTENNA

An antenna like this can be mounted almost anywhere. Limited space means almost nothing. It can be fitted to accommodate the contour of just about any size or shape of building or lot.

In very noisy locations, consider that an antenna of this type has probably the lowest noise factor of anything possible to erect. Static discharges are minimal. While lightning hits can ruin just about any equipment, the chances with an antenna of this type would be less than most conventional antennas if a small spark gap across each meter is provided.

This is a nominally *grounded* antenna—that is, one end is connected directly to ground—rather unconventional, but practical.

It is a random-length antenna. As long as there is *some* wire, somewhere, of reasonable length, it will radiate, and quite efficiently. The only drawback is that it takes two guys to properly adjust it. But when properly adjusted, the swr should be around 1.1 at the tuned frequency, and not more than 1.5 when 25 kHz away which can cover the whole 160 meter band nicely without subsequent adjustments, if tuned for the center frequency of the band. The test antenna has 64

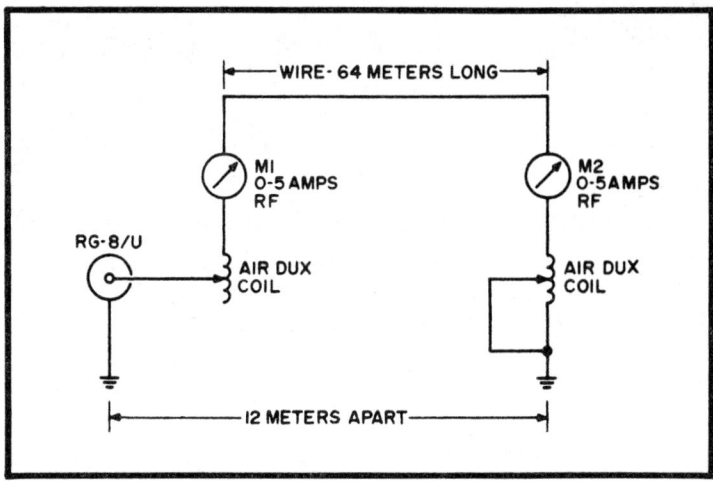

Fig. 2-60. A 160-meter test antenna.

meters of wire. It is strung from the first tuner in the basement out through a basement window, up alongside the house (with standoffs), and around the side of the house (near the top) at roof level. From there it runs up at an angle to a tower, back down to a pole near the front of the house, and from this pole down the front side of the house (on standoffs) to a window in the front of the basement, through it, and down to the second tuner and grounded.

Note from the diagrams that two tuners and two meters are required. The meters actually should be a matched pair if possible, i.e., of same make and range. Two 0-5A rf meters were used, 2% accuracy.

The two tuners were identical Air Dux coils 6.35 cm in diameter, 14-gage wire, spaced, with 40 turns per coil. For 160 meters about half the turns were used. The coils could have been half the number of turns. For any higher frequency bands, less turns would be satisfactory, depending of course on the amount of wire between the two. The two coils are about 60 feet from each other (the length of the basement), so no inductive effect was noticed.

The grounds at each end were positive. In addition to making use of the water pipes, three 250 cm ground rods were driven at the feed end, and one at the grounded end. The water pipes all through the house acted as a counterpoise-ground.

The tuning is a matter of cut-and-try, tapping down from the zero-turns end of each coil one turn at a time (equally) until each meter reads the same as the other, at which point the antenna will be tuned and the 50Ω feed point from the RG8/U cable matched.

One very good reason why the antenna was more efficient than most on 160 was that for about equal power QSO's the S-meter readings were better on reception than the readings on the other ends. On transmission, better reports were received than other stations like distances away with comparable powers and different antennas.

The principle of operation of an antenna of this type seems rather obscure. It might be compared to a radiating transmission line, with equal current at each end, balanced.

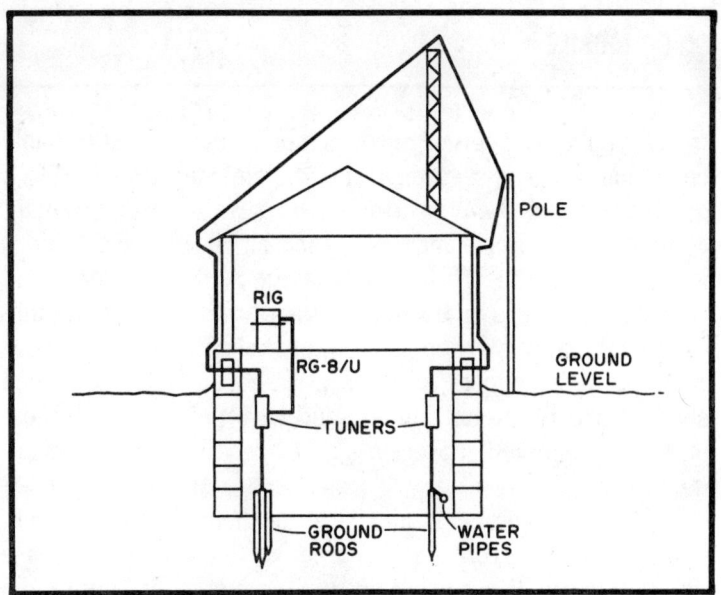

Fig. 2-61. This 160-meter antenna works well.

This antenna works well on 160. Further evaluation of its merits will necessarily have to wait until many hams try something similar on 80, 40 and 20.

GO-GO-MOBILE

If you've had the urge to go mobile at the least possible cost—particularly regarding the mobile antenna—here's a good, inexpensive way to do it. All the parts are easy to obtain, construction is simple and quick, and a highly effective antenna is the result. The "Hi-Q" coil arrangement has been found very effective, and on field tests, the performance of this unit exceeded that of two popular commercial antennas. Comparative S-meter reports at several hundred miles' distance showed one full S-unit higher, and local field strength measurements showed appreciable gain over the commercial units.

The base section is electrical-mechanical tubing (EMT, which is light, strong, and attractive). The top whip is a walkie-talkie, CB replacement unit, or a standard auto radio item—whichever is preferred. With their sliding sections, these units give smooth and rapid adjustment to your exact transmitting frequency.

The unit illustrated here covers 40, 20, 15 and 10 meters by tapping the coil and adjusting the height of the top whip section.

Construction is begun by fitting a plastic or maple rod into the EMT base section tubing and securing it with a self-tapping screw through the tubing into the plastic rod. The next step is to mount the whip by tapping into the top of the plastic rod with the same screw-thread size as provided on the bottom of the whip.

The loading coil is supported on the plastic rod by three wires, top and bottom, soldered 120° apart, on the coil turns at each end of the coil. These wires are then bent toward the plastic rod and clamped in place by the worm-drive hose clamps. One of the three wires on each end of the coil should be extended to provide electrical connection to the base section on the bottom and to the whip section on the top.

The pipe cap on the base section is drilled with a ⅜"-hole and a ⅜"-24 threaded cap screw is inserted and soldered in place. This is best done over a gas flame, flooding plenty of

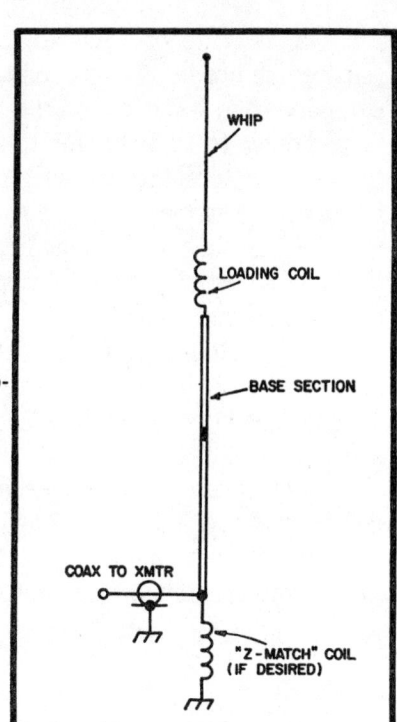

Fig. 2-61A. The loaded go-go-mobile antenna.

solder over and around the head of the previously well-cleaned cap screw. Then the EMT base section is fitted with an EMT compression fitting, which is then screwed tightly into the ¾" pipe cap.

The antenna is now ready for installation—run 52-ohm coax between the transmitter and the antenna base, making certain that the coax braid is well grounded. The antenna must now be tuned to your operating frequency; this is best done with an SWR meter and test clips, tapping down a turn at a time until the lowest SWR is obtained. Coarse adjustment may be made with a GDO, if available, followed up with fine adjustments obtained by changing the length of the top whip in increments of ½" or so. The overall length of the top whip should then be measured for future reference when making an appreciable frequency change within the band in use. After setting the top whip for the correct length for the operating frequency, the sliding section can be easily locked in place with a single wrapping of transparent Scotch tape.

On the antenna illustrated, the 40-meter phone band required all but two of the coil turns, and the top whip was extended to 47½ inches; on the 20-meter band, the coil was tapped down to 10 turns; for 15-meters, 5 turns. For operation on 10 meters, the coil is completely shunted and resonant frequency adjustments are made entirely by adjusting the top whip section. In order to minimize the amount of dielectric in the field of the coil, no cover was used, thereby retaining the highest "Q" and efficiency.

If a spring mount is used with this antenna, a fish line is used to stabilize it and to minimize swaying while under way. Tie the line between the car and the underside of the coil support.

If you haven't gone mobile before, you'll be surprised and delighted at the additional pleasure to be obtained from ham radio and the amazing DX that can be worked using an efficient, center-loading mobile antenna.

WHIP ANTENNA ADD-ON

Various attachments for vhf mobile whip antennas have been described which can be used to modify the directive pattern or even the radiation polarization of the basic antenna. Most of them provide a significant gain factor.

Often it is desired to mount only a simple whip antenna on an automobile for use on vhf bands. Such an antenna suffices in many cases for local contacts while "in motion," especially in large city areas. However, when outside the local area and especially when the automobile is stationary, a more effective antenna is necessary if any contacts are going to be made with reasonable signal levels.

Two approaches can be taken toward this type of antenna problem. The one approach used by many vhf mobile enthusiasts for stationary hill-top operating is to pack a full-size beam and mast. Unpacking, assembling, connecting, etc. of such an antenna can be quite a chore. Of course, the results are often worth the effort as far as vhf dx is concerned. The other approach is to utilize some form of auxiliary antenna structure which can simply and quickly be attached to the mounted whip whenever the automobile is stationary. A number of problems arise when one tries to devise an auxiliary antenna structure. It must provide a significant improvement in antenna performance, not upset the matching conditions to any great degree to the whip antenna, be relatively simple in construction and, yet, attach to the mounted whip in much the same fashion as a cap is put on a bottle.

Basic Considerations

Adding an auxiliary structure to a $\frac{1}{4}\lambda$ whip mounted in the middle of the roof of an automobile has been done before. Figure 2-62 shows the addition of a $\frac{1}{2}\lambda$ element to the basic $\frac{1}{4}\lambda$ whip. The phasing stub is necessary to insure the same direction of current flow in the $\frac{1}{4}\lambda$ and $\frac{1}{2}\lambda$ radiating elements and prevents the vertical radiation pattern from splitting and sending most of the radiation out at useless high angles. Otherwise the phasing stub need not be used as far as impe-

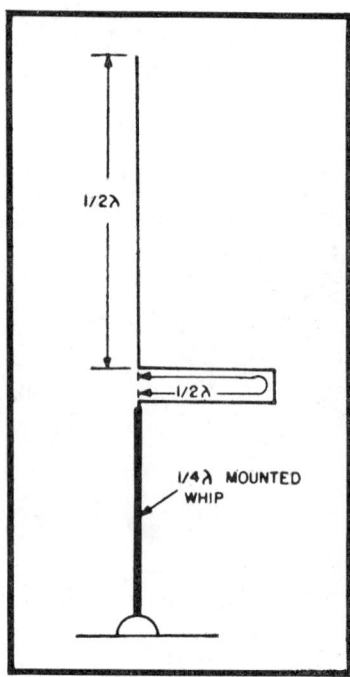

Fig. 2-62. Addition of phasing stub and ½λ radiator to basic ¼λ whip retains vertically polarized radiation, omnidirectional in the horizontal plane but with about 3 dB gain in the vertical plane.

dance matching is concerned. The phasing stub can be formed of stiff wire rod wound in a circle around the vertical elements. A gain of about 3 dB will result but the radiation will still be vertically polarized and omnidirectional in the horizontal plane. This design can be extended to as many elements as desired, but the basic radiation characteristics will not change, except for gain in the horizontal plane.

Another concept is necessary to modify the performance of the basic whip antenna. It's based upon the old Windom antenna idea Fig. 2-63. This was a form of off-center-fed dipole. It used a single wire feed line which represented a 600 ohm feed system using an earth return path for the "missing" half of the usual 2 wire feed line. Due to the earth return path, it performed best over earth of high conductivity.

The usual mobile mounting of a vhf antenna provides an excellent high conductivity ground path via the automobile body. So, ground conductivity is not a problem in adopting the "Windom" feed system concept to a mobile whip. What is different, however, is the impedance relationships. The Windom feed line was connected to the dipole flat-top at a point off center which provided a 600 ohm impedance. The transmitter end of the feed line was connected to an antenna coupler set to match the feed line impedance. If the ¼ λ mobile whip is visualized as the Windom feed line, it does not match the impedance of the coaxial feed line to which one end is attached. The only way the impedances will match is if the ¼ λ whip-turned "Windom" feed line also acts as a ¼ λ transmission line impedance transformer between the coaxial transmission line and the antenna structure which it feeds. This requires that the antenna structure (the whip "add-on") present a very high impedance at the point of connection.

Typical "Add-On" Designs

If the feed system concept explained in the preceding paragraphs is understood, find many designs that could be used as whip "add-ons." The few that are presented here are relatively simple and seem to work well.

Figure 2-64 shows two configurations. The Spider-Leg (A) provides horizontally polarized radiation with a bidirectional radiation pattern centered on the intersection of the legs. The horizontally polarized radiation could be further emphasized by keeping the entire length of the legs mounted horizontally. Some gain in the horizontal plane could be achieved by making the legs longer in multiples of ½ λ.

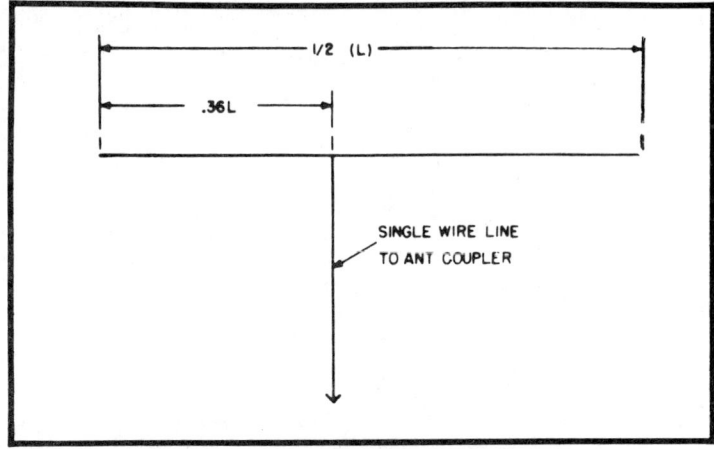

Fig. 2-63. Old style window antenna. Its single wire feed system idea is used as the basis for other antenna types shown.

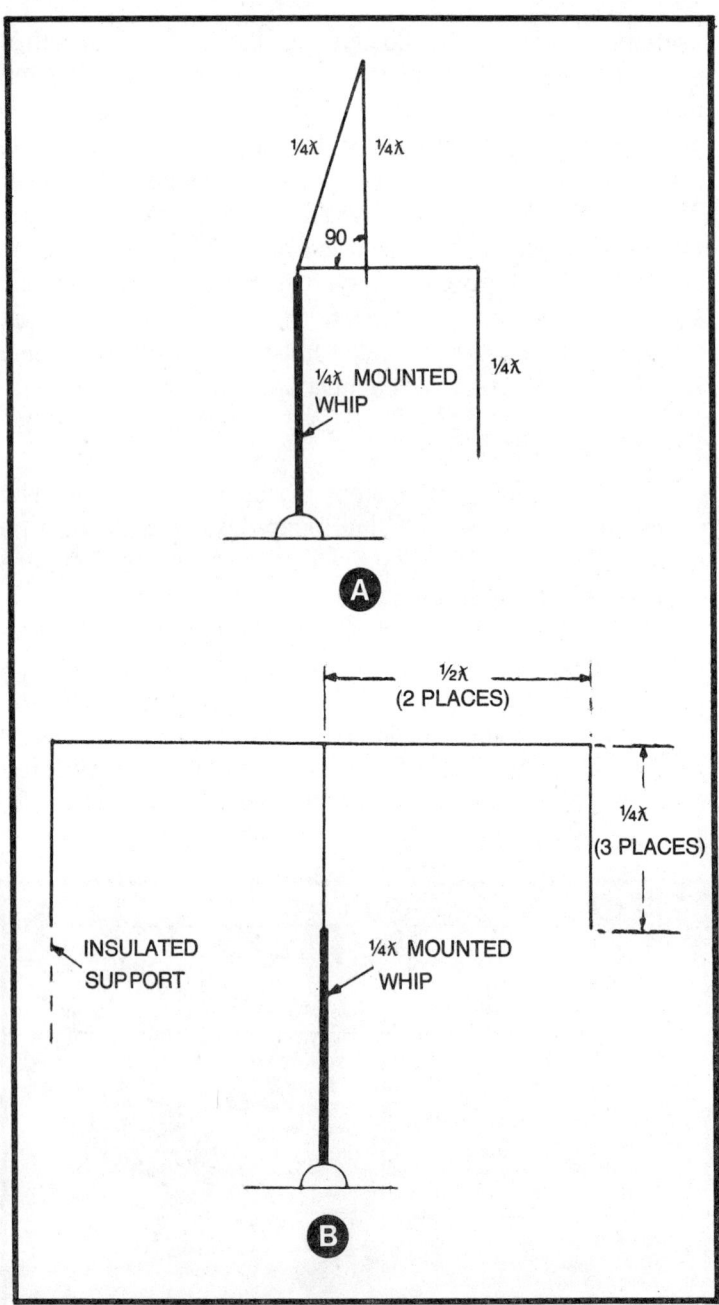

Fig. 2-64. "Spider-leg" configuration (A) provides combination horizontally and vertically polarized radiation. Modified "bobtail" configuration (B) produces quite directional vertically polarized signal.

The modified "Bobtail" antenna of Fig. 2-64 B produces a vertically polarized signal but one with a good amount of gain (about 7 dB). The radiation is bidirectional and broadside to the array.

Figure 2-65 shows a more elaborate "Zig-Zag" add-on. In spite of its appearance, the radiation produced is horizontally polarized with several dB of gain and a bidirectional radiation pattern. Due to the current reversal that takes place every ½ λ, the vertical components cancel while those in the horizontal plane enforce each other. If more convenient structurally, the antenna can be mounted with the first element in a horizontal plane to form a staircase outline, but this would result in the production of some vertically polarized radiation.

It would seem feasible to form beam antennas by placement of parasitic elements either in front of or in back of the main antenna. The spacing would have to be kept fairly wide

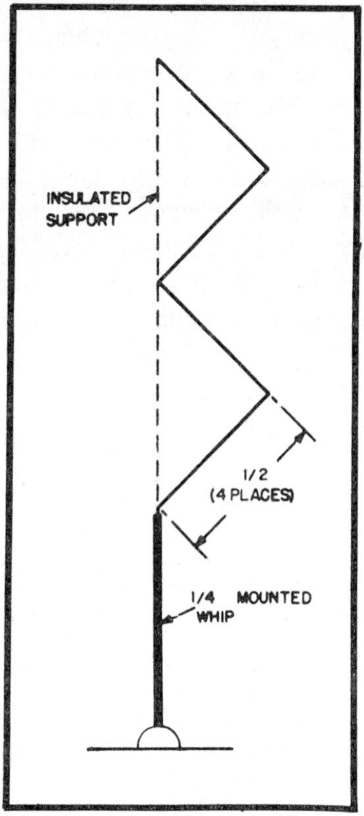

Fig. 2-65. "Zig-zag" configuration provides directive horizontally polarized radiation. Two sections are shown but only one may be used, if desired. The gain will be reduced but the polarization will remain horizontal.

181

($¼\ \lambda$ or more) in order not to appreciably lower the feed point impedance of the main antenna since there is no way to adjust the match to the main antenna.

Construction

Depending upon how durable an add-on must be made, it can be constructed from anything from No. 8 to No. 10 wire to 3/16" or ¼" duraluminum rod. It is suggested that for initial experimentation, the "add-on" be constructed from heavy gauge wire with wooden or plexiglas rods used for insulated support structures when necessary. The add-on can be secured to the whip by a variety of clamps.

An swr meter should be inserted in the coaxial transmission line to the whip to determine the effect upon swr when the add-on is used. Generally, the swr should not change to any great degree. It may, in fact, improve slightly, in some cases, depending upon how good the original match was, by canceling some reactive components.

The improvement in performance can be checked with another station by simple comparison. When making such checks, due consideration should be given to the fact that some add-ons will change the polarization characteristics of the radiated signal. Therefore, the station with which checks are made should have available an antenna that will match the polarization of the radiated signal. If this is not observed, very confusing results will occur as an antenna polarization difference between stations can easily produce a 20 db loss in the received signal.

QUICK BAND-CHANGE MOBILE ANTENNA

All antenna systems represent a compromise. One of the greatest is that of squeezing a 3 MHz quarter-wave onto the back of an automobile.

A look at the price tag affixed to commercial antennas should provide you (as it did me) with a strong incentive for "brewing your own."

And price, while a major factor, may not be the only consideration. For example, I needed a positive, quick-change antenna, to hit Air Force MARS frequencies. Namely, 7305, 4590, and 3311 kHz.

The antenna shown in the picture is the result of four brewing ventures. Prototype 1 was temporary, being wound on a bamboo fishing pole. Taps were brought out, with many trials and more errors; however, this temporary antenna worked so well it literally stayed on the car until the base "rotted" off. Had I coated the fishing pole with eopxy, I would probably still be using it. This antenna was wound with plastic-coated 20-gage stranded wire.

At the demise of the bamboo pole, I decided to be smart and refer to the handbook. I read and studied, and ended up with a very nice looking "outer-space" antenna, with a big-wire coil in the center. The transmitter loaded. The receiver worked swell. Only one small problem: Nobody could hear me.

Maybe it wasn't quite that bad. But the signal was down many dB. The center-loaded antenna did not deliver the signal that the top-loaded fishing pole punched out.

So I set to work to rebuild a better looking and more rugged "fishing pole."

The antenna now in use has two coils, each 2 ft long. The only reason for using two coils is that by loosening a clamp, the top section of the antenna may be easily removed. Also, it requires a long coil to hit 3311 kHz, and the necessary forms may be available in shorter length. Whether the coils are wound in one or two sections makes no difference in the operation or tuning of the antenna.

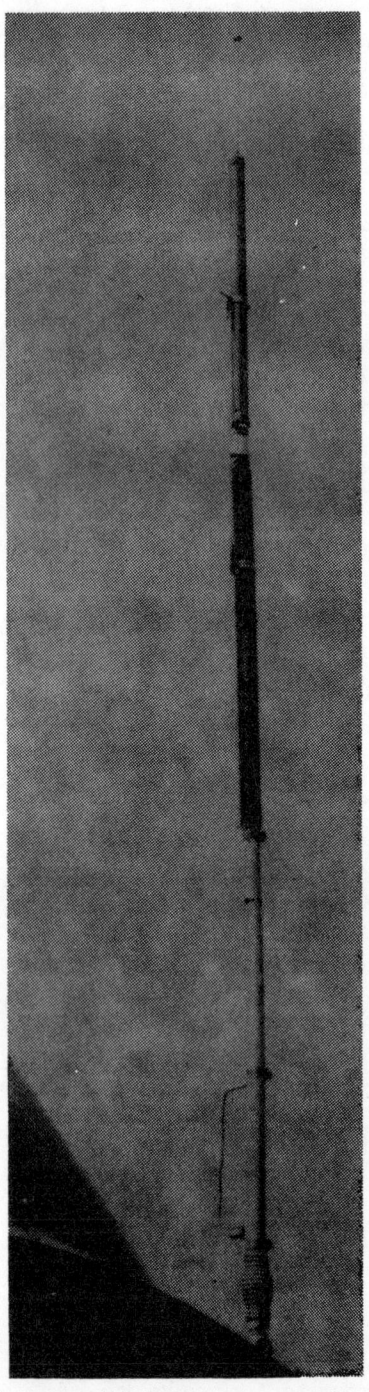

Fig. 2-66. Antenna installed on Chevrolet with Motorola SB3. The coil is wound in two sections but would be equally successful if wound in one section. Past models have been made in one section. Shorting bars are in place for 40 meters. Two sections are used here because of available coil forms.

Also, a shorter coil length may be used if only higher frequencies are to be covered. For 40—10m, a coil about 14 in. long is required. (Increase base length to 60 in.) For 75—10m, approximately 3 ft is needed.

Actual construction and tuning of the antenna is easy. No special tools or equipment is necessary.

The base section of the antenna is a piece of aluminum tubing 30 in. long. The bolt, with threads to match a heavy duty spring mount, is securely anchored in the base with epoxy. Three sheet-metal screws with the tips filed so that each barely touches the bolt, are useful in two ways. Helping to center the bolt, and keeping the epoxy plug from turning in the tubing after hardening, braid is soldered to the bolt and

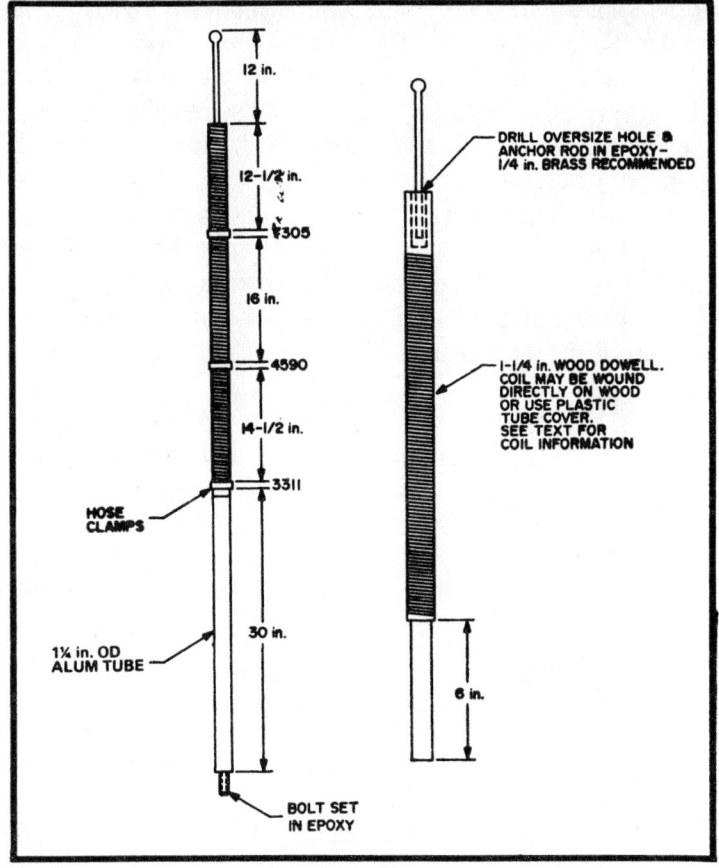

Fig. 2-67. Antenna materials and dimensions.

brought out through a small hole in the tubing. The braid is later clamped to the antenna with a hose clamp which reinforces the base and holds the 47 dial light indicator socket. Wadding keeps the epoxy where needed. Be careful. The bolt must be straight when the epoxy is poured.

One satisfactory way I have found for winding the coil is directly over a wood dowel. If you have access to a lathe the dowel should be grooved to space the wire about one diameter. Or you may wind a string or another wire between turns, which may be later removed. (The coil should never be closewound, unless plastic insulated wire is used. The plastic provides the spacing.)

At least 6 in. of the wood dowel should be sanded, whittled, or turned down to fit snugly into the aluminum tubing base. Slot the top of the tube 1 in. A hose clamp will hold the coil in place and provide a terminal for the wire.

The tip section of the antenna may be any conductor. However, since this tip will receive some hard bumps, I recommend a section of ¼ in. brass rod. Solder a wire to the rod. A ⅝ in. hole is drilled in the end of the wood dowel coil form. Wire and rod are anchored in the dowel with epoxy.

After the coil is wound and tip fitted, cover it with a good coat of resin to securely fasten the wire in place.

Another method of winding the coil is to use a piece of 1¼ in. plastic pipe or other coil form and slip this over the wood dowel, which may be an old broomhandle with a layer of tape applied. Fasten the ends of the plastic to the dowel with screws. Dowel should be long enough to extend 6 in. down into base. Tip is attached as previously outlined.

Or the plastic tube may be "splinted" to the base and temporarily "outside taped." Then the coil form and 6 in. of the base are filled with epoxy. The tip is placed in position and the whole thing hardens. This method is especially useful for the short coils, 40m and up.

A 36 in. plastic tubing form may be bought at a sporting goods store for about a dime. Just ask for the plastic covers designed to keep golf club handles from being scratched in the bag.

I have used coils constructed all of these ways. Using the wood is easier and works equally well. In any case the wire must be cemented in place.

Fig. 2-68. Closeup of slider tap and shorting bar connections.

After the coil is dry and both ends of the coil attached electrically to the base and tip, you are ready to start the tuneup.

For each frequency to be covered you will need a slider. Each slider is made of one hose clamp and one 6-32 brass bolt and nut. File the bolt head slightly flat. Drill a hole through the hose clamp. Insert the bolt from inside the clamp, securing with the nut. The bolt head will make contact with the wire coil.

By using the approximate measurements shown, estimate about where the tap should be for the frequency you wish to hit. With a coarse file cut through the resin and any insulation on the wire for 4 in. Your coil at this point will resemble a 100W slider resistor.

Now install the antenna on the car and place the receiver in operation. Work on the highest frequency first. For example I tuned 7305 kHz first. Then 4590, and finally 3311 kHz.

Attach a wire to the top of the base section long enough to reach the top contact area of the coil. Run the wire back and forth over the contact area. You will hear the receiver background noise peak sharply. If it fails to peak, but gets louder in one direction, you may have to file some more. At the point where the receiver peaks, the antenna will take a load, but will probably require a bit closer adjustment.

Next install your output indicator, a 47 light bulb. I leave this indicator in place at all times. It provides positive "on the air" indication, not to mention what happens at night when a car pulls up behind and sees the sideband modulation waving in the air. The indicator is a loop of wire about 16 in. long, with one side of the loop taped to the antenna at the base and the 47 bulb in series with the outside of the loop. If it is desired to leave it permanently in the circuit it may be tapped across about 8 in. of the base using the antenna itself as one side of the bulb loop.

Using the bulb as an indicator, finish tuning until the desired loading is obtained.

Fig. 2-69. The 47 dial light output indicator. This light bulb shows at all times positive indication of on-the-air and is all you need to tune and load the transmitter. The loop in this instance is completed by the antenna itself. Can also be used as a complete loop taped to the antenna.

Repeat the tuning procedure for each lower frequency desired. Then permanent shorting bars should be made. Brass welding rod and alligator clips are ideal.

Changing frequency is simply a matter of removing, or using the proper shorting bars. If several frequencies close together are required, several taps can be brought out every turn or every two turns, and a braid with attached alligator clip may be used to short out the unused portion.

This antenna has delivered plus performance. And when it comes to changing bands, it can be done in less than 30 seconds without hurrying, and this includes getting out of the car and back in. Every time the antenna will be in perfect resonance.

I scrounged everything. But if you have to lay out the cash, $5 should suffice.

Interested?

COVERING ADDITIONAL FREQUENCIES WITH DIPOLES

A half-wave dipole, center fed with a low-impedance transmission line is simple, inexpensive, is easy to erect, and it is an efficient radiator. Unfortunately, it is usually considered to be a singleband antenna. But this is not necessarily true.

First, at its resonant frequency, the center impedance of a ½λ dipole has a purely resistive value of approximately 70Ω and closely matches standard 50 and 72Ω transmission lines. But when a dipole is operated off its resonant frequency, its center impedance becomes complex and no longer matches low-impedance lines. At its second harmonic, for example, the center impedance increases to well over 1 kΩ and is obviously a very large mismatch to a low-impedance line—as trying to operate such a 3.5 MHz dipole on the 7 MHz band quickly proves.

At the third-harmonic frequency of the dipole—so the theory goes—the feedline again "looks into" the center of a ½λ dipole (that has a ½λ dipole connected to each end) and again sees a low-impedance resistive load. At progressively higher harmonic frequencies, the center impedance of the dipole swings between a very high value on even harmonics to a low value on odd harmonics.

These facts prompt many amateurs to attempt to use their 7 MHz, low-impedance, center-fed dipoles on the 21 MHz band as third-harmonic dipoles. The usual results are that it is impossible or difficult to load the transmitter properly on 21 MHz, and the measured swr on the transmission line is high. But all is not lost. Once it is learned why these antennas apparently defy basic antenna theory, the knowledge to persuade many other dipoles to work efficiently on two or more amateur bands can be used effectively.

The basic problem is that, although radio waves travel through space at the velocity of light—roughly 186,000 miles or 300,000,000 meters per second—their velocity of propagation slows down on a conductor. Consequently, a ½λ (the distance a radio wave travels in the period of a half cycle) is

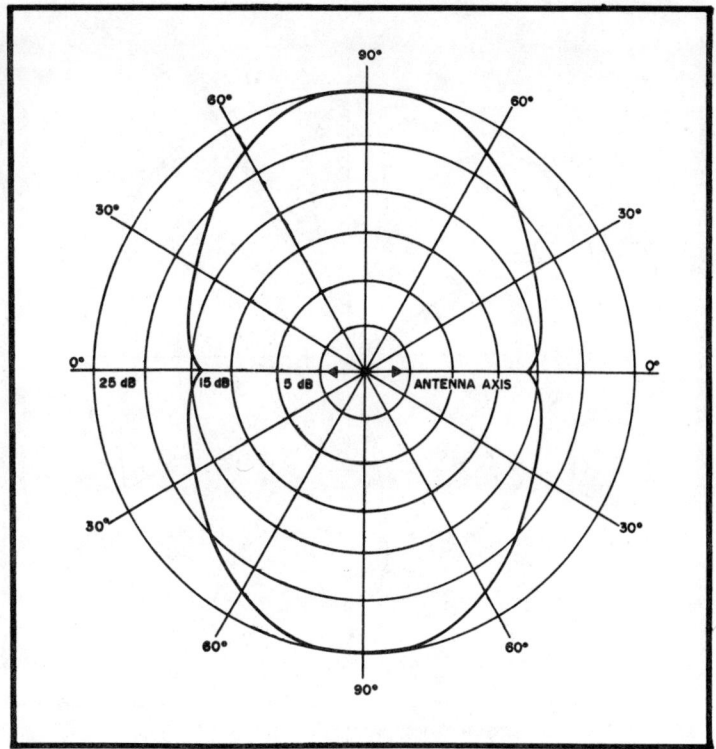

Fig. 2-70. ½ λ antenna pattern.

measurably shorter on a conductor than in space. Worse, the speed and shortening effect varies from conductor to conductor.

In an antenna, for example, the amount of shortening depends on the ratio of its length to its diameter,— and, especially, the capacitance between its end sections and surroundng space. As a conventional antenna has only two ends, no matter what its length, a short antenna (measured in fractions of a wavelength) is shortened a greater percentage than a longer one.

The standard formula for calculating antenna lengths compensates for this variable factor. The formula for the length in feet is:

$$L = \frac{(492)\ (n-0.05)}{f} \text{EQ-(1)}$$

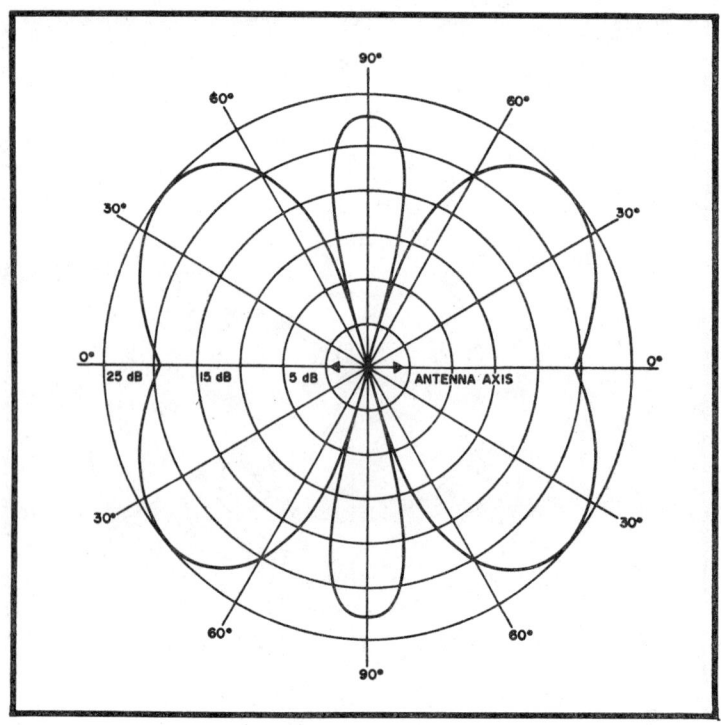

Fig. 2-71. 1½ λ antenna radiation pattern.

where n equals the number of half-wavelengths in the antenna, and f is the frequency in megahertz.

When n = 1, (a ½λ dipole), the formula simplifies to:

$$L = \frac{468}{f} \quad \text{EQ-(2)}$$

From either formula, the length of a ½λ antenna for 7.175 MHz (the center of the 7 MHz Novice band) is 65 ft, 2½ in. And Eq. 1 shows the length of a 1½λ antenna for 21.15 MHz in the 21 MHz Novice band to be 68 ft 7¼ in. Working the formulas backwards, the 7.175 MHz dipole has a third-harmonic resonant frequency of 22.25 MHz. Conversely, the 21.15 MHz, 1½λ antenna resonates at 6.822 MHz as a ½λ antenna. Forwards or backwards, there is enough difference in the lengths of the antennas for an oversize yardstick.

These figures indicate why the average 7 MHz dipole does so poorly on the 21 MHz band. Being resonant far outside the high-frequency edge of the band is of minor importance as far as radiating efficiency is concerned. If it can be fed power, the antenna will *radiate* the power. Rather, the problem is that the off-resonant condition of the antenna produces such a high swr on the transmission line that the output circuits of most transmitters cannot compensate for the mismatch.

But by increasing the length of the antenna to make it resonant at the desired 21 MHz frequency, these problems disappear. In fact, the antenna should outperform a ½λ dipole for the same frequency, simply because a harmonically operated antenna exhibits gain in its favored directions over a ½λ dipole. While the gain of a ½λ antenna over a ½λ dipole is only 0.8 dB, it is enough to refute the belief of many amateurs that operating a low-frequency antenna on a higher-frequency band is an inefficient substitute for separate dipoles on each band.

At first glance, it may seem that increasing the antenna length for improved 21 MHz results is simply trading one set

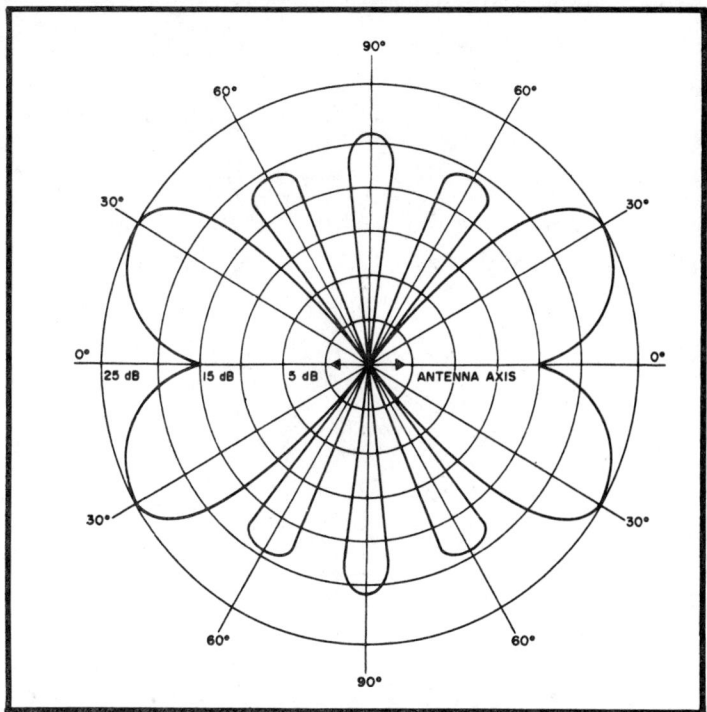

Fig. 2-72. 2½ λ antenna radiation pattern.

of problems for another. Certainly, the increased length resonates the antenna outside the low-frequency edge of the 7 MHz band. Fortunately, however, most transmitter output circuits have enough range on the 7 MHz band to compensate for the increased feedline swr at the operating frequency. And the mismatch is more acceptable on 7 MHz than on 21 MHz, because electrically a feedline of a given physical length is only a third as long as wavelengths on 7 MHz as it is on 21 MHz. Therefore, a particular value of swr increases the losses in the transmission line less on 7 MHz than on 21 MHz.

Moreover, if the antenna is an inverted V with the ends close enough to the ground to be reached without undue difficulty, its length can be adjusted for operation at the desired frequency on the 7 MHz band, and extensions approximately 20 in. long may be clipped to its ends for 21 MHz operation. The extensions may be allowed to drop vertically from the ends of the antenna.

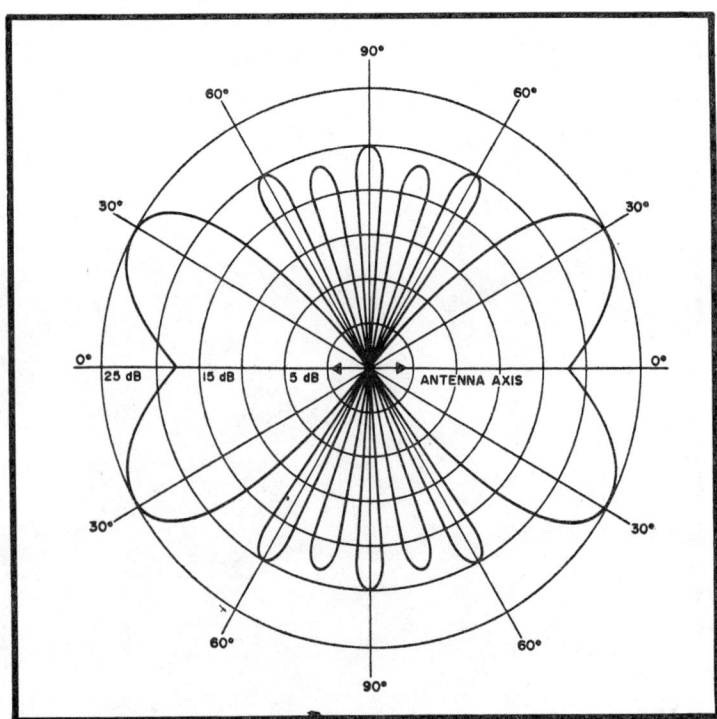

Fig. 2-73. 3½ λ antenna radiation pattern.

Other antennas and other bands. Plugging the appropriate figures into the formulas will show that a ½λ dipole for 3.925 MHz (119 ft, 3 in.) will also resonate as a 3½λ dipole near 28.73 MHz. Other figures show that a 116-footer resonates near 4 MHz as a ½ λ dipole and near 21 MHz as a 2½ λ antenna. This pair of frequencies will not be of interest to many amateurs, but it might interest some MARS members. Also, if the ends of the antenna can be reached from the ground, cutting it for 2½λ resonance in the 21 MHz band will permit clipping extensions to its ends for operation in the 3.5 and 28 MHz bands.

The possibilities: A Novice might cut his dipole for 2½λ resonance near 21.15 MHz and use 6½ ft extensions to reach 3.725 MHz. A General class phone operator, on the other hand, might select a length of 114 ft for 2½λ resonance near 21.36 MHz and 2½ ft extensions for operation in the 3.8 and 28 MHz bands. A third possibility is a 1½λ antenna for around 14.3 MHz (101 ft 6 in.). A couple of clip-on extensions at each end will permit operating on the 3.5, 21, and 28 MHz bands.

Chapter 3
In-Door and Limited-Space Antennas

MINIATURE ANTENNAS FOR 80 THROUGH 10 METERS

You can build *antennas* that are only 1/20th conventional size and, amazingly enough, have almost imperceptible losses, $-1\frac{1}{2}$ dB, compared with full size half waves.

Theory

The theory involved is quite simple: The *inductive* reactance of a short copper element and the capacitive reactance of a ¼ λ section of open wire line are combined to form a resonant antenna system. See Fig. 3-1.

Since the open-wire line carries currents 180° out of phase and the wires are separated by only .024λ, there is very little radiation from this section. All radiation, therefore, takes place from the short copper element.

Losses

Reducing the size of an element lowers the radiation resistance considerably. An element only .024λ long, the length used here, shows a radiation resistance of only 0.5 Ω. However, the efficiency of the radiator remains better than

Fig. 3-1. The inductive reactance of a short copper element and the capacitive reactance of a ¼ λ section of open-wire line are combined to form a resonant antenna system.

98% since the ohmic resistance of the short ¾ inch copper elements average less than 0.008 Ω. I^2R losses in the open-wire line section are shown graphically in Fig. 3-2. If the line is made of at least #16 wire, the losses are small averaging slightly over −1 dB.

In addition to the I^2R (heat) losses in the element and line outlined above, there is an additional loss, termed the directivity loss, that results from shortening the radiator. This is illustrated in Fig. 3-3. The radiation from a full size half-wave element forms the classic figure eight pattern. However, as the length of the element is made smaller and smaller the ovals of the eight become more nearly circular although the general radiation pattern remains the same. The result is a loss in the immediate forward direction of −0.39 dB, and some filling out along the sides.

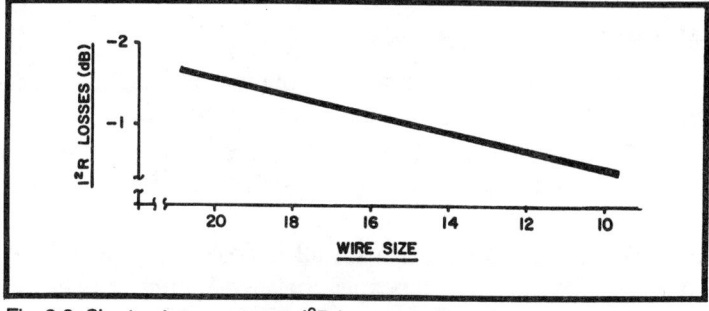

Fig. 3-2. Short antenna average I^2R losses vs. line wire size.

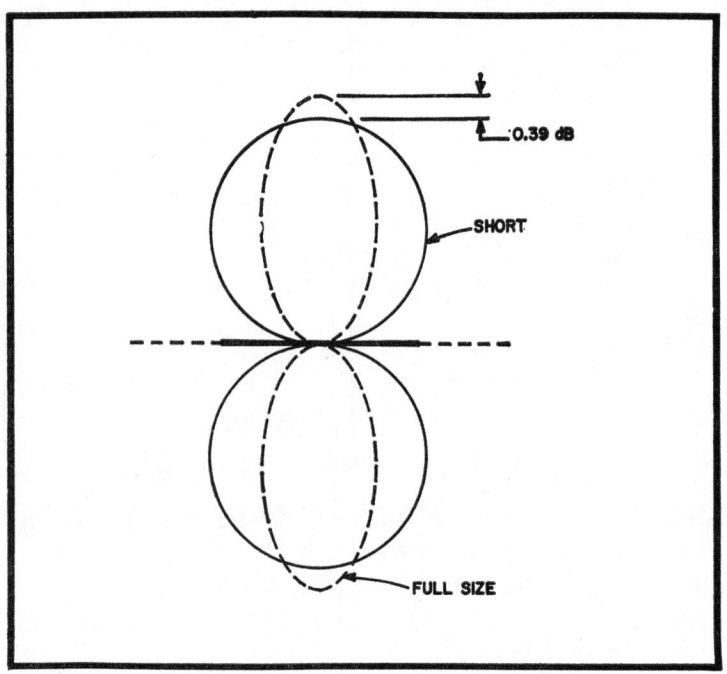

Fig. 3-3. Field pattern comparison short vs. full size element.

Summarizing: The directivity plus I^2R losses in these short element antennas average −1½ dB when compared with full size elements, a loss that could hardly be detected in the received signal.

Construction

Table 3-1 includes all the information required to size antennas for 10 through 80 meters. Figure 3-4 shows suggested construction details.

Additional Construction Notes

Horizontal arrangements are shown, although, vertical polarization could be used just as well.

Individual coaxial feeds are used on each band; however, one could design a parallel single feed that would function with very little additional loss.

Tap distances for use with 52 Ω coax are shown; however, feed lines of any impedance, balanced or unbalanced, can be used by merely tapping down on the line. Series capacitors

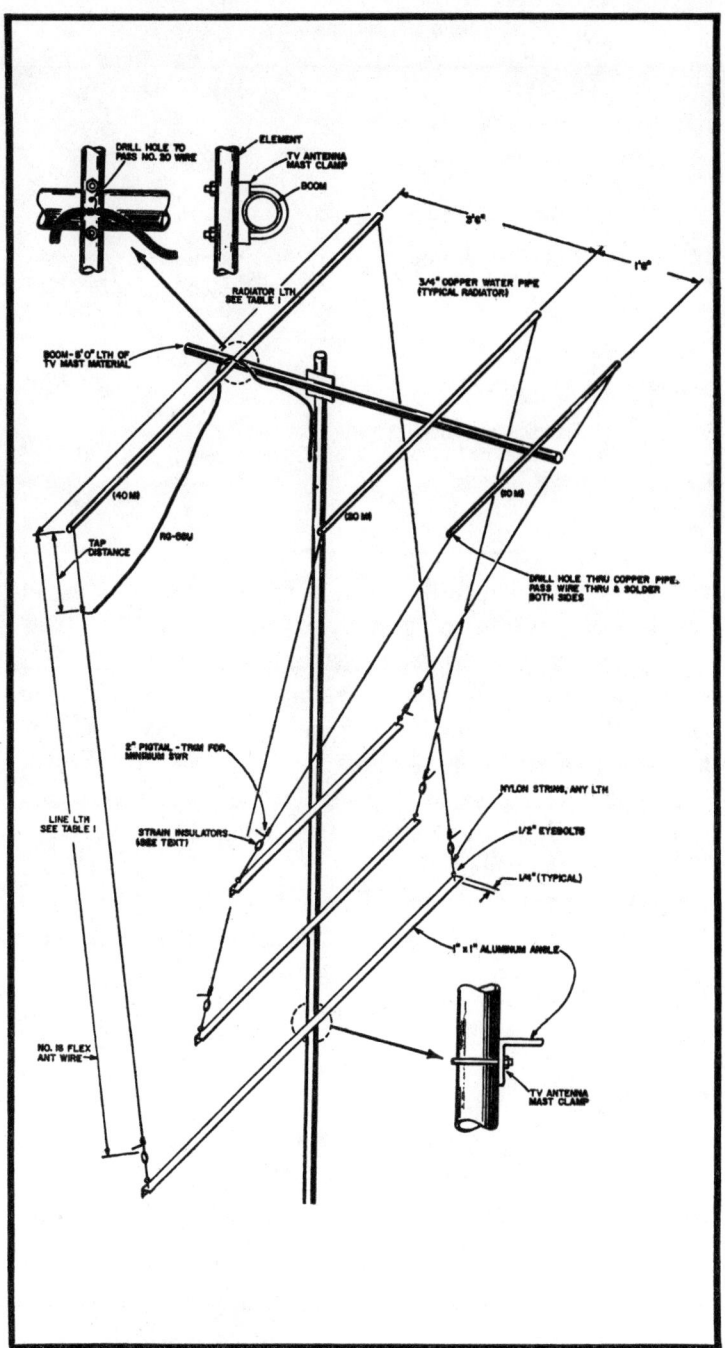

Fig. 3-4. Construction details.

Table 3-1. Antenna Sizes.

Band	Radiator	Length Line (1)	Tap (2)	B.W. (3)	Loss (4)
10M	10"	8'3½"	3"	245 Kc.	−1.3 dB
15M	15"	11'1½"	4"	215 Kc.	−1.4 dB
20M	20"	16'8½"	4½"	140 Kc.	−1.5 dB
40M	40"	33'1"	7"	80 Kc.	−1.7 dB
80M	80"	62'10"	12"	50 Kc.	−2.0 dB

Notes: (1) Adjust line length for SWR 1:1 (see text)
(2) 52 Ω tap
(3) Bandwidth at SWR 2:1
(4) Includes line I^2R losses and 0.39 dB directivity loss. #16 line wire size assumed.

are not required since the system is resonant and a purely resistive load is offered to the feed line.

Good quality, moisture resistant end insulators should be used since extremely high r.f. voltages appear at this point. High impedances and higher voltages are the effects of standing waves on the open-wire line.

Because the capacitive reactance (of the ¼ λ line section) changes rapidly with frequency, the tuning of the line is quite sharp. The dimensions given in the table, if followed closely, will place the resonant point of the antenna at the *lower* end of the respective band. The construction details show short pig-tails on each side of the line. These should be trimmed one-quarter inch at a time until the SWR is 1:1 at the operating frequency.

Design Details

The inductance, L_R, of short lengths of ¾ inch copper pipe is shown graphically in Fig. 3-5. The inductive reactance, X_R, may then be calculated from:

$$X_R = 2\pi f_{mc} L_R$$

The length of open-wire line, 1° (degrees), required to furnish the necessary resonant capacitive reactance can be determined from:

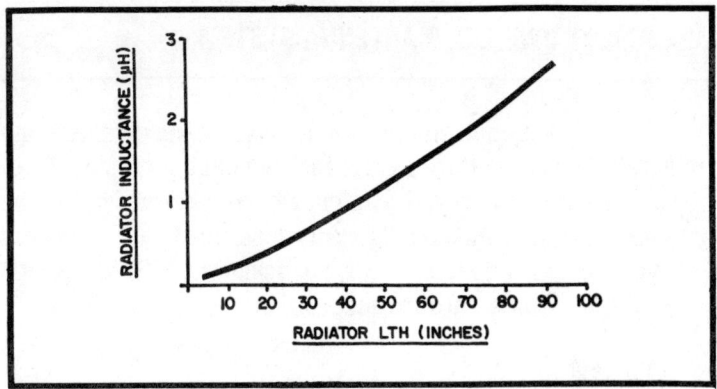

Fig. 3-5. Radiator length vs. inductance.

$$\text{Cot } 1° = \frac{X_R}{Z_0}$$

The line impedance, Z_0, for various spacings is shown in Fig. 3-6.

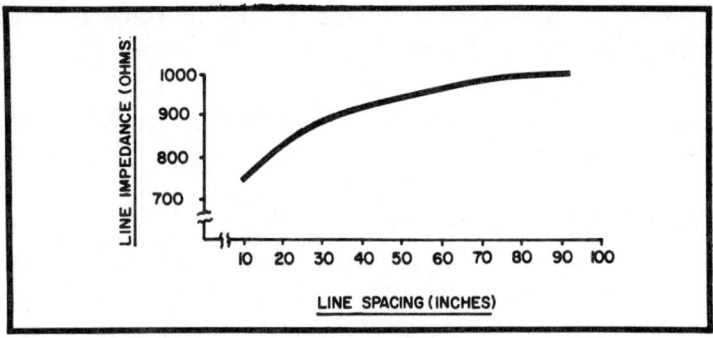

Fig. 3-6. Line spacing vs. impedance, Z_{01} wire size #16.

Results

Tap distances, antenna "Q", and bandwidths were first calculated and then substantiated by testing.

Antennas for each of the bands have been constructed and used with results comparable to any of the full half waves used at various times. Comparisons were run against a long wire (275 feet) antenna by switching between the two. In the direction of the maximum lobe of the long wire, the long wire outperformed the miniatures by ½ to 2 S units. In all other directions the miniatures were equal to or better than the longer wire.

APARTMENT DWELLER'S ANTENNA SYSTEM

The basic requirements for an apartment dweller's antenna system are 1. the antenna shall put out a good signal, i.e. if a station can be heard, it can usually be worked, and 2. the antenna shall be as inconspicuous as possible so that neighbors and your landlord do not object, particularly where the apartment lease forbids antenna installations.

System Testing

The antenna is tested during a period of high activity on the various bands—during a contest. The contest period is chosen so that reports from a large number of stations can be obtained in a relatively short period of time. Note that the failure of a station to respond to a call is a valid "negative" report.

The invisibility of the antenna is tested by usage. It is only visible during actual operating periods. If at those times it is inconspicuous, then no complaints will result. If after a month or so, no complaints have been received, the test is successful.

System Description

The antenna used is a Hustler mobile whip, operated out of an apartment window at about 25 ft above ground as shown in the sketch. This antenna system was chosen in preference to others because it was easily collapsible and did not fall down under the weight of ice or due to the effects of high winds.

Setting Up

Attach a hook or other fastening element to the wall or roof outside the building above the window. If a convenient tie point is available, it may of course be used. A piece of string is tied to the hook and a loop about 3 in. diameter is tied in the other end of the string. The mast is screwed onto the bumper mount and laid to one side. The resonator element is passed through the loop, then screwed to the mast. The antenna is then pushed out of the window and the bumper mount placed

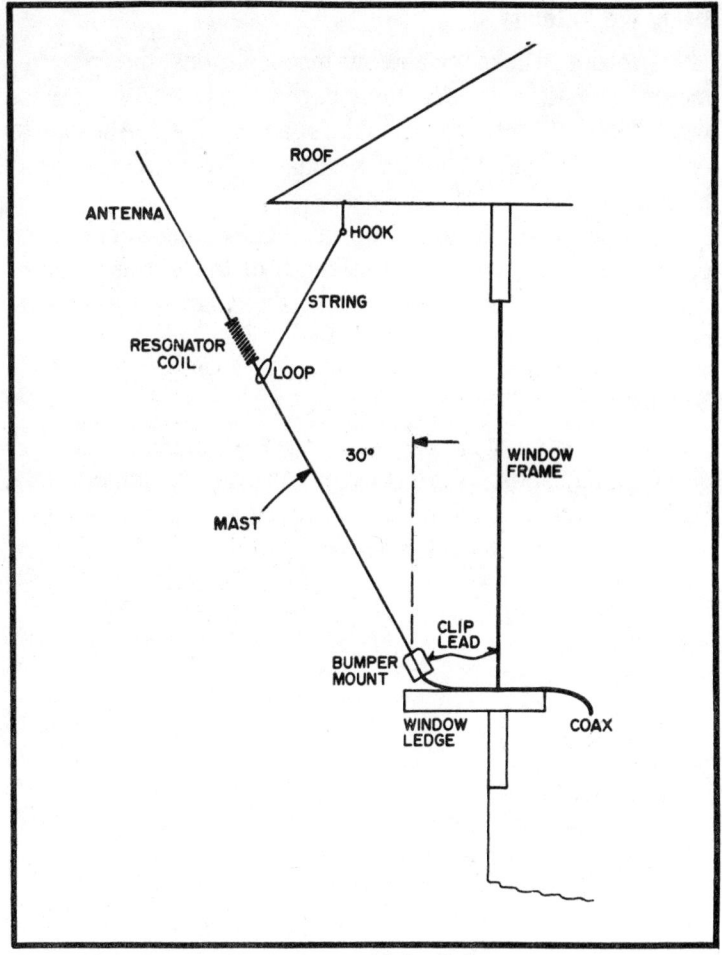

Fig. 3-7. Antenna set-up on apartment window ledge.

on the windowsill. The mast is supported by the string. The window is then closed, wedging the antenna in place.

The first time the antenna is used, the length of the whip is adjusted for minimum swr in the same way as is done in a mobile installation.

Dismantling the antenna only requires opening the window and pulling it in. The string may be left hanging or tied down to the window frame by catching it between the window and frame when the window is closed. The whole operation of erecting and dismantling the antenna takes only a matter of seconds.

Using the Antenna

In order to keep the antenna inconspicuous, the 75 and 40 meter resonators are used only during the hours of darkness when the whole antenna cannot be seen. By day, the small size of the 15, 20 and 10 meter resonators keeps the antenna inconspicuous.

In use it is possible to alter the effective angle of radiation of the system by varying the length of the string so as to change the angle of the antenna with respect to the vertical position.

Summary

The apartment dweller's antenna system puts out a good signal on all bands. While it will not replace or outperform a beam or a quad, it does allow transmitting operation under many unfavorable conditions.

BASE-TUNED CENTER-LOADED ANTENNA

There is nothing new about base inductive loading of a vertical antenna. The method is convenient and often allows easy construction. The disadvantage, of course, is low efficiency when the antenna is shorter than ¼λ long at the operating frequency. As shown in Fig. 3-8, the high current flow at the base of the antenna means that the greatest current flow takes place through the coil's relatively small conductor. Various solutions have been tried to get around the problem of distributing the current flow in an antenna such that the reactive elements necessary to bring the antenna system into resonance do not also produce the greatest losses. The helical antenna was one solution. By distributing the inductive loading along essentially the entire length of the antenna, the average current flowing through the base section was reduced. Center lumped inductive loading Fig. 3-8 has become the most popular method, however, since it allows the heavy base current to flow through the bottom section of the antenna, which is not loaded, and because of its constructional advantages, particularly for mobile antennas.

The method is not always practical for fixed station situations, however, where the antenna center may not be acces-

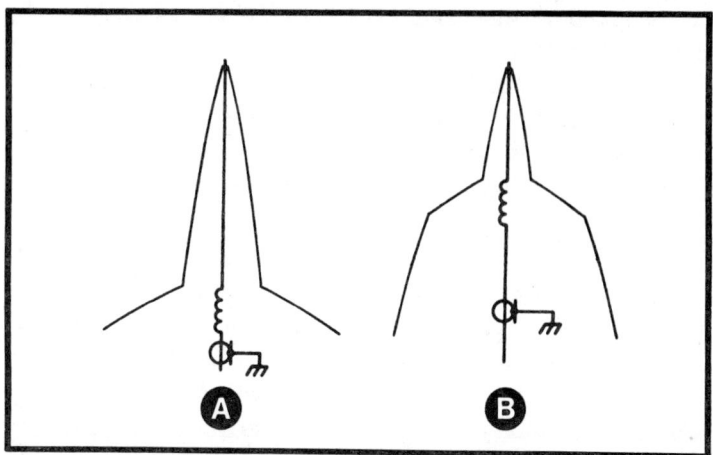

Fig. 3-8. Improved current distribution of center-loaded antenna (B) over base-loaded antenna (A) allows highest currents to flow where ohmic losses are low.

sible to change coils for various bands. Usually, the fixed station operator who is constrained to using a vertical antenna on several bands has had to settle for base loading and could simply take the approach of making the base loading inductor as hefty as possible in an effort to reduce its losses. This discussion explores a method by which the efficiency of such antenna can often be considerably improved by a form of pseudo-center loading using a transmission line to transfer the reactance of a loading element located physically at the base of the antenna electrically to its center.

A note should first be made concerning the function of the loading inductor in a center loaded antenna. The inductive reactance is necessary to make the electrical length of the antenna equal a ¼λ so that the base impedance will match a coaxial transmission line and the antenna will accept a transfer of power from this line. Unless the loading inductor is extremely long or wide, it does not radiate any appreciable signal. The radiating is done by the rest of the antenna. If the copper losses in the inductor are not kept low, the inductor will act in a manner similar to a dummy load and simply produce heat. The inductors used for the center loading of short mobile antennas are usually not shielded because weather protection can be obtained by epoxy coatings and a shield will reduce the inductor's Q or efficiency. The coil, however, could be shielded with a large enough enclosure so that the effect upon the Q would be small and antenna performance would not be affected thereby.

So, if the antenna saw the proper value of loading inductance at its center, regardless of where the physical inductor was placed, it would operate the same as it did with the physical loading inductor at its center.

Reactance Transfer

One method that can be used to transfer a remotely located reactance to the center of an antenna is via a transmission line. Transmission lines can actually be used not only to transfer reactances but also to simulate them. The latter feature is commonly used when stub matching is employed to simulate the necessary inductive or capacitive reactance to correct a mismatched transmission line condition. By a combi-

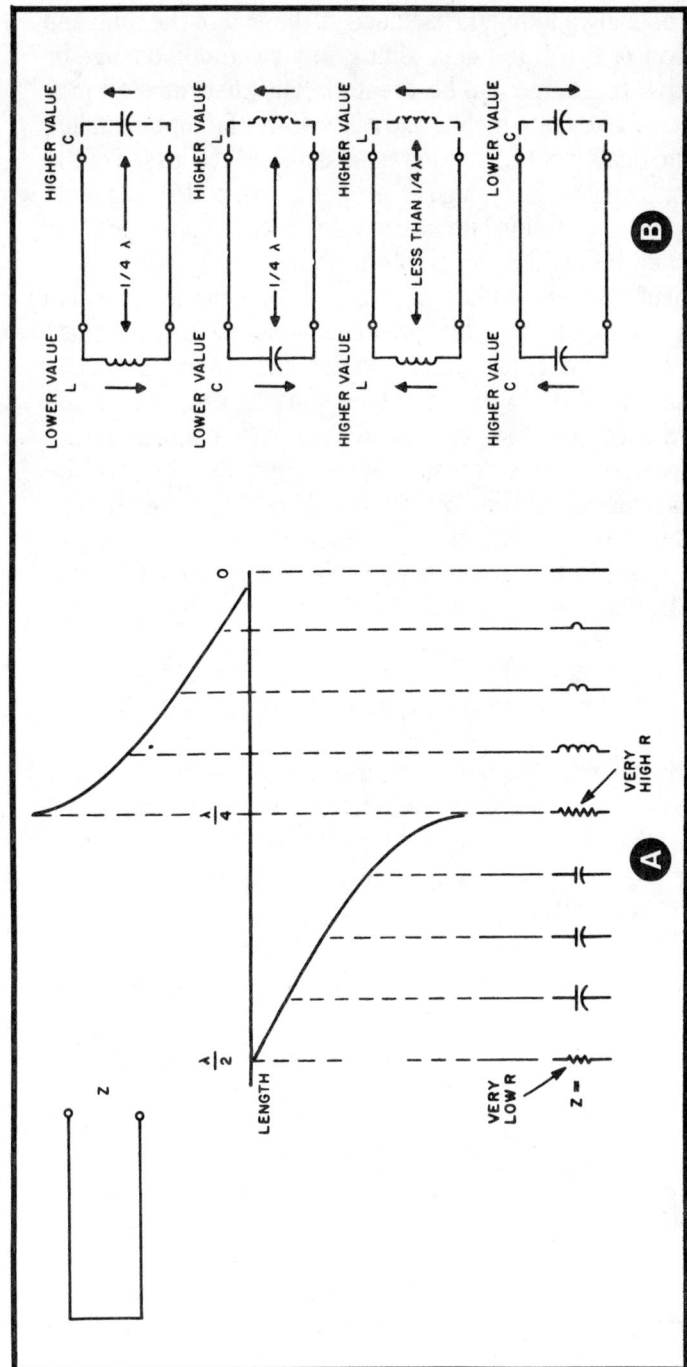

Fig. 3-9. (A) illustrates how length of transmission line can simulate a series of reactances depending upon its length. (B) shows some of the transfer reactances possible when a lumped reactance is used on the end of a stub. A line exactly ½λ long will transfer exactly the same value as the lumped reactance.

nation of using a lumped reactance at the end of the stub and the effect of the stub itself, almost any value of inductive or capacitive reactance can be created. The reactance so produced can have quite high Q and a low power factor depending upon the quality of the lumped reactance and the losses of the transmission line. By using the proper length and/or termination for a transmission line, it can also reflect an open or short circuit and be used as a remote switch.

Figure 3-9 shows how a length of transmission line also can act as either a capacitive or inductive reactance. The chart shows how a short circuited stub can increasingly act as an inductor until it is ¼λ long and then start simulating a capacitive reactance beyond ¼λ. If the stub were open-circuited, the opposite reactive condition would apply. That is, capacitive reactance would be produced until the stub were ¼λ long and then at exactly ¼λ length it would act as a short circuit.

By performing some simple math, the reactance produced by the stub when up to a ¼λ long is:

$$X_L \text{ (short circuited stub)} = Z \tan (2\pi L/\lambda)$$
$$X_C \text{ (open circuited stub)} = Z/\tan (2\pi L/\lambda)$$

Z is the impedance of the transmission line used to make up the stub and L/λ is the line length in wavelengths. The physical line length is affected by the velocity factor (usually .66 for coaxial lines) and shorter by that factor than a free-space wavelength.

Although the reactance values repeat as the stub is made odd multiples of a ¼λ long, such long lengths should be avoided if possible. The highest Q will be obtained when the stub length is less than ¼λ. Once the reactance produced by the stub is known, one can calculate or look up a reactance graph to determine how many micro-henrys or picofarads are produced at any particular frequency.

Also, to obtain the best Q, a cable of the lowest loss possible should be used and also one with an inner conductor radius to inner radius of outer conductor ratio of 3 to 5. Not all cables can satisfy the latter criterion, RG/59 cannot, for example. RG/58 and RG/14 both are usable and the latter cable is particularly useful, although it is a bit large (.55" diameter) to satisfy the requirement for low loss and the proper geometry for high Q as a stub.

Short or open-ended stubs alone are limited in ability to produce simulated values of capacitance or reactance simply because of the fixed impedance of the cable used. Replacing the short or open-ended termination with a lumped reactance will, however, extend the range to any desired value. As a general indication of what effect lumped constant termination will have in order to experimentally determine the required reactance, Fig. 3-9 can be used as a guide.

Practical Application

Figure 3-10 shows how the principle outlined for transfer of the center-loading reactance can be applied. By using the proper value of loading reactance, the overall antenna is resonated either as a ¼λ or ¾λ vertical, and so its feed point impedance will match a 50-ohm transmission line directly. The coaxial line used to transfer the base reactance is placed inside the lower element of the antenna. The reactive components

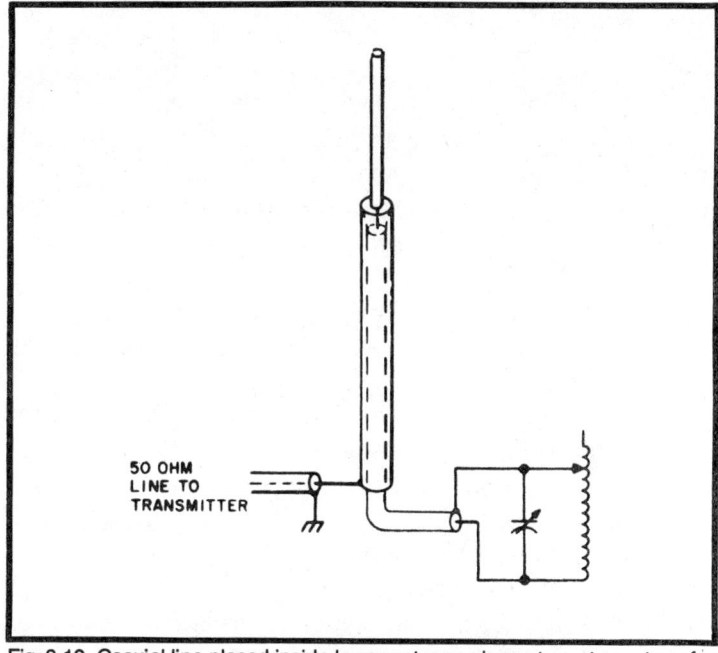

Fig. 3-10. Coaxial line placed inside lower antenna element reactance transfer from base to junction of upper and lower antenna elements.

209

themselves can be placed in a shielded enclosure, if desired, but note that both sides of the coaxial transfer line must remain insulated from ground. There will be some radiation from the coaxial transfer line although the portion enclosed in the lower antenna section will not cause any problem. The exposed portion can cause various problems, including pattern distortion of the vertical if it is made too long. Basically, the method of reactance transfer was meant to function with the physical reactance at the base of the antenna and not 100 feet away. Therefore, the dimensioning of the antenna sections and the length of the coaxial transfer line must all match to achieve proper operation of the antenna on various bands and positioning of the base reactive elements.

With some care, many combinations can be found that will work. One combination that was experimented with used a standard 10 foot TV mast as the antenna lower element and a smaller diameter 5 foot tube, insulated from the lower element, as the top element inside the ten foot section. Since ½λ of this cable on 10 meters was 10.8 feet, it provided a very convenient length to the base of the antenna. Either a B & W stock 3905-1 coil (6 TPI, 2½" diameter) or a 500 pf variable capacitor was used to resonate the antenna, depending upon the band and the antenna's mode of operation.

On 10 meters, some slight inductance was used at the base to make the antenna operate in a ¾λ mode in order to match the transmission line to the transmitter. On 15 meters, capacitive reactance was used to make the antenna operate in a ¼λ mode. On 20 meters some very slight capacitive reactance was also required. On 40 and 80 meters, the transfer line length is quite short in terms of wavelength and additional inductive reactance at the base was used.

Tuning of such an antenna can initially be done using an SWR meter in the transmission line to the transmitter. Try both the inductive and capacitive element at the base on each band, varying their value until the SWR is brought down as close to 1:1 as possible. Generally, the minimum amount of reactance that will achieve resonance should be used since additional reactance means additional losses. Some confusion will occur on the higher frequency bands because the length of the transfer line and antenna may allow various combinations of base reactances to resonate the system. The only solution in such a case is to use a field strength meter or, perferably, check signals with a distant station to determine which reactance value produces the best radiated signal. Once the proper reactance values are found, they can be bandswitched, if desired, although the tuning does become critical on some bands; a better solution might be to switch the inductor and capacitor (with various taps on the former) and use a simple tuning chart on each band to locate the correct settings.

With a vertical antenna, the transfer of the loading to the base does not affect the fact that either a radial system or good ground connection must be provided if the antenna is always operated in a ¼ or ¾λ mode for direct match to a 50 ohm transmission line.

Summary

The method presented for transfer of the center loading in an antenna to a more convenient location presents various interesting antenna construction possibilities aside from the one described.

If a location is used where no adequate ground connection is possible, it would seem possible to use the reactance transfer idea to resonate the antenna in a ½λ mode. The base of the antenna would then be at a high impedance point and an antenna coupler would be necessary to couple to a coaxial transmission line. Another approach to make an antenna with less ground dependence would be to form a dipole antenna, center loading each element via a reactance transfer cable to the center of the dipole.

Removing the physical reactance a distance away from the base of the antenna presents the problem of line radiation which was mentioned. A possible solution to this problem may be the use of triaxial cable with the outer shield grounded and the inner shield and inner conductor used for the reactance transfer. How long the cable could be while still retaining good loading Q and without having the tuning become too critical are questions that only trial and error experimentation can answer.

CITY DWELLER'S MULTIBAND ANTENNA

The antenna, being the most important single component of the station, poses an especially critical problem for the city-bound amateur whose space may be limited.

One satisfactory solution is described. The characteristics of several simple antenna forms are reviewed briefly, the goal being to incorporate as many desirable features as possible from each, into one efficient, compact multiband system.

The Horizontal Linear Dipole

The center half of a half-wave dipole contributes most to the radiation process. Increasing amounts of the radiated energy, which proceed directly upward, are wasted as the operating frequency is raised, never returning to receivers on earth. This energy may be channeled to lower, useful angles by forming the radiator into the popular inverted dipole. A more desirable omnidirectional radiation pattern then results, becoming slightly more directional off the radiator ends in some cases. This arrangement reduces the amount of ground space required and elevates the important center half of the radiator by the use of only one tall support. Efficient multiband operation requires open-line feeder and match box arrangements.

The Folded Dipole

Broadband characteristics and simplicity of construction are represented by the folded dipole. However, operation is limited to fundamental and third harmonic frequencies, such as 7 and 21 MHz. This antenna also may be installed as an inverted dipole to enhance low angle radiation.

The Trap Antenna

For fast band changing, the trap antenna is probably the ultimate choice. However, this convenience is accomplished at the expense of some losses in the isolation and matching networks. Also, in many designs, only a small portion of the structure is actively radiating on the higher frequency bands.

The extensive guying necessary to some vertical trap antennas has been known to elevate eyebrows, both within and outside the household.

The Windom Antenna

In the true Windom the single wire feeder offers the lowest loss feed line system, because no insulation is used except air. Multiband operation is available on the even har-

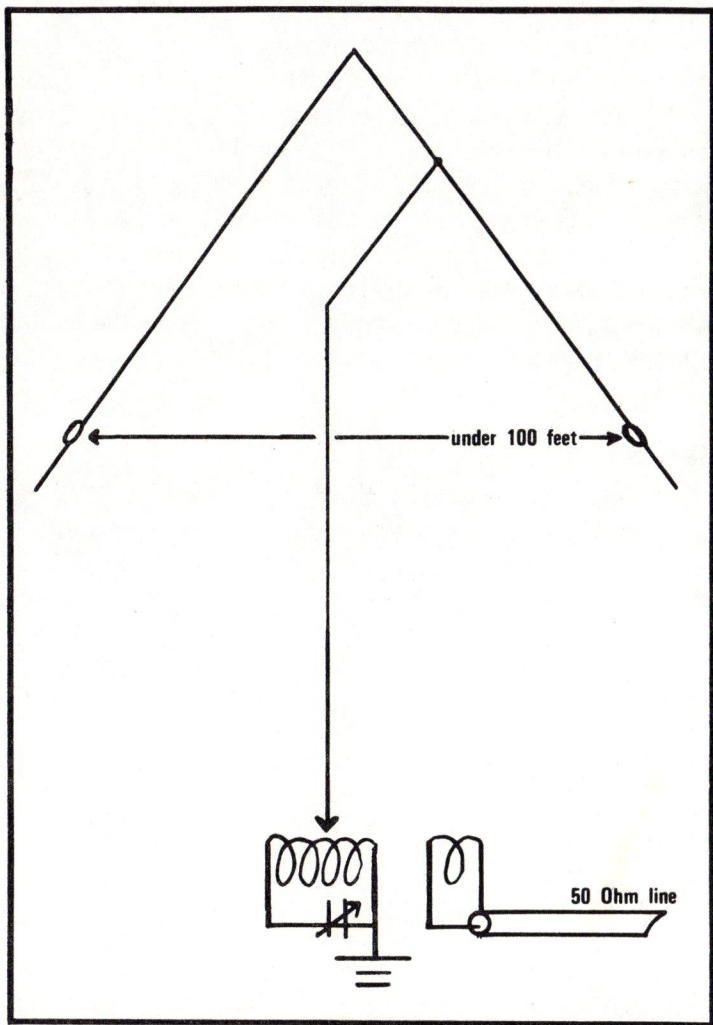

Fig. 3-11. Inverted Windom. The tank component ratings are the same as in the transmitter final.

monically related 160,80,40,20 and 10-meter bands, using a simpler tuner. The entire radiator actively contributes to its operation on every band.

Application

All of these simple antenna forms have been tested in the above order, mounted in the same position, over the past 5 years. The inverted Windom surpassed all others in signal reports, simplicity and space requirements.

The length, 140 feet, remains un-pruned for extra-class CW use on 80 through 10 meters. The single feeder is tapped at 20 feet off-center. The inverted configuration was borrowed from the inverted dipole antenna design and for the same reasons. The radiator can be supported by the family TV mast and will appear to be an unobtrusive pair of guy wires. The simple tuner in Fig. 3-11 is mounted at a window to avoid dielectric losses which would result if the feeder were routed inside a building. Tuneup is conventional, tapping the coil for the best transfer of energy, while maintaining resonance of the tank.

Results

Signal reports and resceived signal strengths favor the multiband antenna described over the discussed systems.

THE MINI VEE-BEAM

Many of the current crop of hams do not have even a nodding acquaintance with the Vee beam—yet it is an antenna with some very positive advantages: it will provide excellent gain on three or more bands; is easy to build and tune up for real efficiency over a wide range of frequencies; and it is so neat and inconspicuous that it will be a lot more popular with the neighbors than the usual quad or tri-bander.

Probably one of the reasons that the Vee beam has been so neglected is that those hams who do know something about the configuration usually feel that the antenna is not practical unless located on an acre of ground.

Happily, this point of view is not quite true. Size does help, of course, but it is possible to build a Vee beam which provides very worthwhile gain on 20-15-10 meters, yet can be put up on the average city lot.

Such an antenna does perform effectively. The one illustrated was first put on the air on 28 megacycles. In actual use, a flattering number of CQ's resulted in three and four station pile-ups on the frequency—from stations 1500 to 2500 miles away.

The theoretical gain of a Vee beam of the dimensions shown is approximately 6 dB on 10 meters, or about that of a typical 3 element beam. On 15 meters, gain is somewhat less, approximately 5 dB, and the gain is in the neighborhood of 4 dB on 20 meters. A long wire antenna of any type will deliver more in actual communication gain than a parasitic array which is rated at the same level.

This is *probably* true for at least two reasons. First, long wire antennas do not seem to require as much height, to deliver low angle radiation, as is needed for the smaller beams. Second, receiving "efficiency"—if there is such a thing—seems to be markedly better. At any rate, a Vee beam or other long wire type of antenna, will often dig out DX which other local stations are having difficulty receiving, and pull in that DX with less QSB. This could be accounted for by the fact that the long wires develop both vertical and horizontal

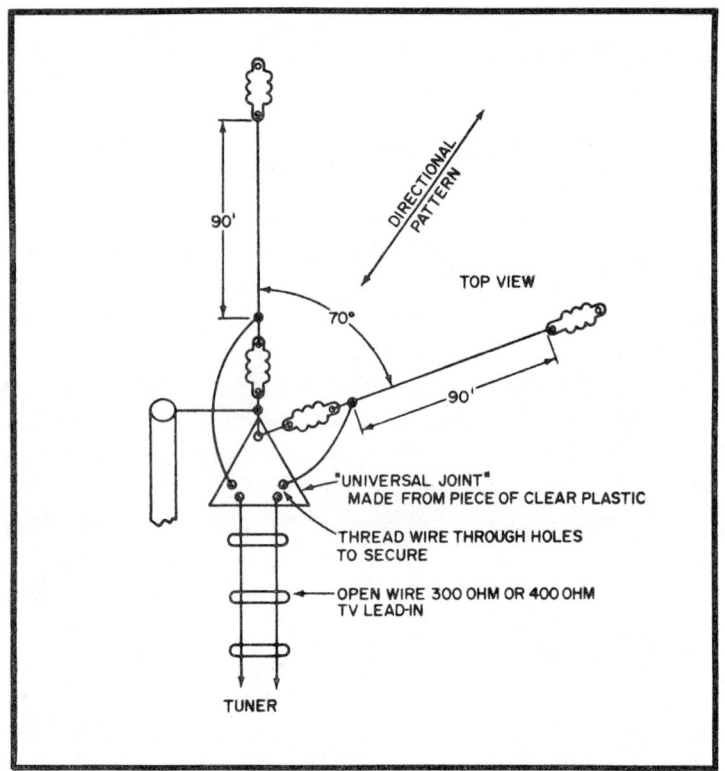

Fig. 3-12. Small Vee beam layout.

polarization—and in addition, the larger capture area provides for a kind of diversity reception.

Perhaps *none* of these theories is entirely correct—but there is no question about the results; long wire antennas, particularly Vee beam and rhombics, have justly earned a reputation for generating outstanding signals.

A Practical Vee Beam

The drawing Fig. 3-12, shows the layout of a Vee beam which is small enough to be practical for many hams. It is fed by a tuner, the circuit of which is shown in Fig. 3-13. Despite its simplicity, the tuner will bring the antenna to exact resonance at any frequency in the three bands, and will provide a virtually perfect load for the transmitter, allowing standing wave ratios as low as 1.1 to 1, or beyond the accuracy of the usual swr meter.

The antenna is likewise very simple, consisting of two wires, mounted parallel with the ground, and as high as local conditions permit. The antenna shown in the photograph is approximately 25 feet above the ground, higher, of course, would be better.

Ideally, the feed line would drop down from the antenna at right angles, however in practical application this frequently is not possible. Figure 3-12 shows a kind of "universal joint" made up from a piece of clear plastic, which serves as the termination point for the feed line, and allows the feed line to leave the antenna at almost any convenient angle.

The feed line itself is the open wire "ladder" type of TV line, either 300 ohm or 450 ohm (preferably the latter). This line is adequate for any SSB rig which does not have a linear amplifier—or for a CW rig up to 200 or 300 watts. For powers above that, the line should be made up with number 14 wire, and either ceramic or clear spreaders.

Remember that even a 90 foot wire, loaded up with ice, can become pretty heavy, so be certain to use strong turnbuckles, screws, etc.

The 70-degree angle is about optimum for an antenna to be used on three bands. However, if this angle isn't possible, narrow it to 50 degrees, or, preferably, use a larger angle, up

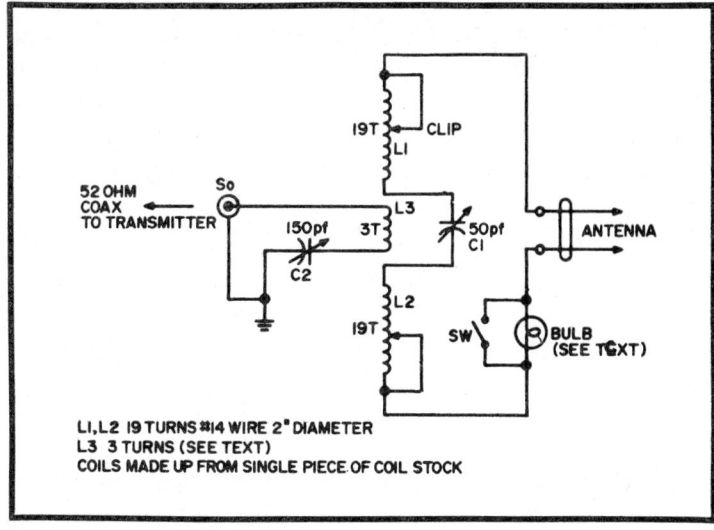

Fig. 3-13. Antenna tuner circuit.

to as much as 90 degrees. The last-named angle will reduce gain somewhat on 10 meters, but will give better results on 20 meters.

The Vee beam provides a "figure 8" bidirectional pattern. With the dimensions given, the pattern is quite broad.

Building the Tuner

The tuner for the Vee beam is made up on a 5 × 7 × 2 inch chassis. Building the tuner requires a minimum of parts: two tuning condensers, one length of coil stock, a switch, a dial light bulb, some insulators, and miscellaneous hardware.

The smaller of the two variable condensers, (50 pf.), should be double spaced. Since it is used at a low voltage point, even a midget variable like that shown is adequate for a fair amount of power. A single spaced variable condenser, 150 pf or larger, likewise should do the job even for the more powerful SSB rigs run barefooted. For full, legal power use condensers with more spacing at both points.

The coil can be of standard coil stock (B + W 3900).

Notice that the coil is actually *three* coils. This is accomplished by clipping the coil stock at the proper point with sidecutting pliers. Clipping the coil is easy, but following the turns by eye to make certain which winding is which can be confusing. The best answer is to use an ohmmeter as a continuity meter—or a flashlight bulb and a battery. It sounds ridiculous, but the tuner illustrated failed to work properly, and it took all afternoon to discover that the connections to the clipped-apart-coil were wrong!

The tuner is so simple that the actual wiring is at most an hour's job. Use some scrap lengths of No. 14 wire, if possible, for the leads. Note that the variable condenser above the chassis is mounted on an insulator so that both the rotor and stator are "above ground." The variable condenser below the chassis can be mounted directly to the chassis, since the rotor is grounded.

Tuning Up The Antenna

The Mini Vee beam, like any center fed antenna fed with a tuned line, can utilize a simple resonance indicator which shows when there is actually power in the antenna. And this is what counts—*not* the swr ratio. A low swr is a fine idea, especially with an antenna fed directly with co-ax, but contrary to what appears to be an unfortunate popular opinion, a low swr reading *may* mean simply that you have lucked into a critical length for the coax you are using. Low swr does *not* guarantee your antenna is working efficiently.

With a tuner like that shown you have the best of both worlds: the tuner resonates the antenna and its feeder—*and* allows tuning out the reactance on the coax which runs between the tuner and the transmitter. An swr meter inserted in this line *will* show low swr, and furthermore the reading is honest.

The resonance indicator on the tuner is simply a dial light bulb in series with one feeder. (The system is balanced, so in theory, at least, the same current will appear on both wires of the feed line.)

A number 44 bulb will handle up to 35 watts or so—and if two bulbs are put in parallel, they will handle considerably more power. Higher power will require use of a Christmas tree light bulb, or, if *it* blows out, simply clipping a flashlight bulb over a portion of the feeder, as shown in Fig. 3-14.

Step One

In tuning up any antenna system, start out by tuning up the transmitter to a suitable dummy load. As the first step it is always nice to know that the transmitter is putting out power.

Next, tune up on 28 MHz. Remove the dummy load from the transmitter and hook up the coax to the tuner. Place the clips on the tuner coil at approximately 8 turns from each end.

Set condenser C2 at approximately one-half capacity (assuming 150 pf).

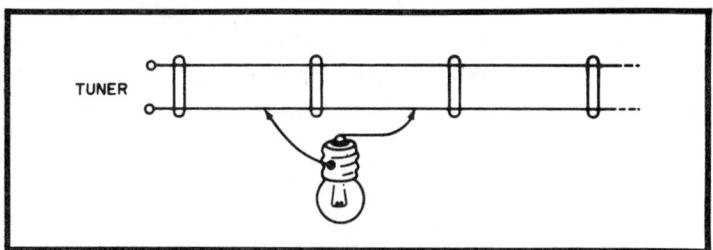

Fig. 3-14. Resonance indicator on feedline.

Adjust the load controls on the transmitter so that meter readings indicate the antenna appears to be taking some power. Now rotate C1, and watch for the bulb to light. If it does not, throw the switch, and try again. If there is still no indication, move the clips either closer to the center, or further out. Try to have the same number of turns on both sides.

In a couple of minutes, a point where tuning condenser C1 through resonance will cause the bulb to light should be reached. Now, adjust C2 for maximum brightness of the indicator bulb, and "touch up" the tuning of the transmitter. The name of the game is to achieve maximum brightness on the indicator bulb without, of course, overloading the transmitter—which probably has a plate milliameter to indicate proper input.

After discovering the proper tap points for 10 meters, indicate them with a dab of paint (or finger nail polish) alongside the clips so that it will be easy to return to the same spot.

Tune up on 15 meters and 20 meters follows exactly the same procedure. When maximum brilliance on the bulb is obtained the system is in resonance.

Additional Bands?

Yes, the Mini Vee will tune up on the other bands as well. It will tune up on 6 meters and it may tune up on 2 meters as well, although most of the coil will have to be shorted out and it would be better to use a smaller coil. The radiation pattern will not be a bi-directional "figure 8" (as will be true on the lower frequency bands) but rather a figure 8 plus a number of other lobes, many of which will be highly useful, and tend to give coverage in all directions.

In addition, the antenna can be used as a random length long wire (fed with an L network) on 80 meters and 40 meters so the Mini Vee is actually an all-band antenna. All of this can be done with a somewhat more complicated tuner plus a simple switching arrangement.

AN EFFICIENT INDOOR ANTENNA SYSTEM

Having to work with an indoor antenna system inside an apartment or house is, of course, a major handicap. No indoor antenna system will ever work as well as an outdoor antenna system constructed of the same materials, at the same height, etc. However, rather than take the defeated approach to the indoor antenna problem, it is very worthwhile to examine the possibilities concerning what things can be done better with an indoor antenna system than with an outdoor antenna system. After all, with an indoor antenna system, the materials used are not subject to the same wear or stress requirements as those on an outdoor system, and the antenna is usually more accessible to make adjustments.

One of the chief factors that is desired to achieve in any antenna system is low-loss. That is, regardless of how good the matching is to an antenna system to transfer power to the system, one still wants to keep the basic Ω loss of the system as low as possible. Such a condition insures at least that each delivered watt of power really radiates and also leads the way to the development of a broad-band or multiple-band antenna system which does not require critical tuning.

Thin copper sheeting is an ideal antenna material, and most amateurs would have been using such material if it were as readily available as common household aluminum foil. The advantages are numerous as compared to the aluminum foil material various amateurs have used for the indoor construction of loop or dipole antenna systems. The losses of copper are far lower and the copper can be directly soldered with ordinary soldering materials. Unfortunately, one can't walk down to the nearest hardware store and obtain a roll of thin copper. But, it can be found by searching out the various wholesale metal product outlets.

The total cost will depend upon the length of sheeting purchased.

But, why try to obtain such copper sheeting or foil when, if copper is so desirable, copper tubing is readily available from plumbing supply houses? The advantage of the sheeting is that it covers far greater surface area for less cost and it is far easier to handle and form in different antenna shapes.

A Practical Antenna

One of the simplest but most effective indoor antennas which can be constructed if space is available is an ordinary dipole. In one case, a dipole antenna which is constructed is shown in Fig. 3-15. The 12' wide copper sheeting is cut with a pair of heavy shears to two strips of 6" width and each strip used as the arm of a dipole. Little loops of wire are soldered to the top edge of each sheet at intervals and these loops used to attach plastic cord which in turn is used to suspend the antenna from a roof beam, at about a 12" spacing from the beam. At the center of the antenna, the copper strip is folded together towards the center where the coaxial feedline is attached. The folded over edge of the strips is soldered along each edge to the body of the copper strip. This is done to insure absolutely minimum resistance at this high current portion of the antenna. The copper strips are first cut to "formula" length for a regular dipole on the band being used. However, there is no way to predict exactly how much longer the antenna will be than required. One has to use an swr meter in the feedline and carefully trim the antenna length down until proper resonance is found. This procedure is easily done with a pair of shears, trimming the copper stripping down equally at both ends of the dipole until a 1:1, or as close as possible to 1:1, swr ratio is achieved in the center of the band for which the antenna is cut. This procedure requires patience but it is absolutely essential. One of the greatest faults made with indoor antenna systems of the self-resonant type is that many operators forget that the capacitance of the building structure surrounding the antenna completely changes its resonant frequency. The antenna must be cut for resonance where it is mounted or one will end up blaming the indoor location for poor performance results which are not really justified.

A Multi-Band Antenna

The use of the copper stripping to construct an indoor antenna demonstrates its versatility when constructing a multiband parallel dipole type of antenna system. The multi-band

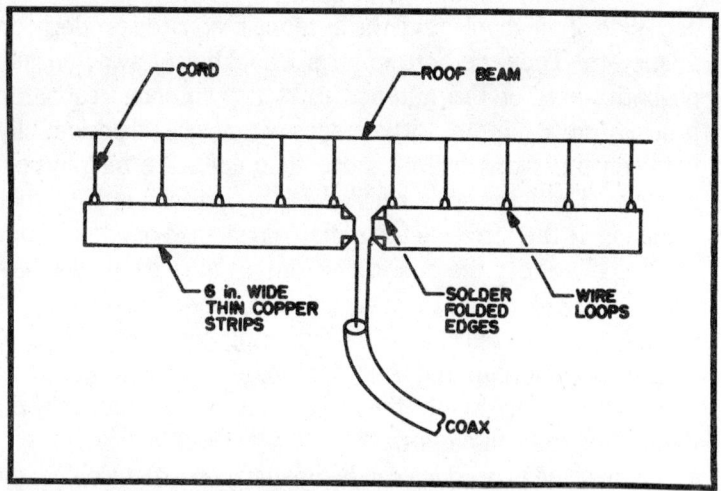

Fig. 3-15. Basic dipole constructed from thin copper strips.

type of antenna can be made for any combinations of bands, although the space available in most indoor situations will allow it to be constructed for only 20, 15 and 10 meters or some two band combinations of these frequencies. The basic multi-band antenna is constructed for the lowest frequency to be used like the antenna shown in Fig. 3-15 and tuned up for operation on this band. Then each side of the basic dipole is cut using shears to form either two or three strips out of each dipole side as shown in Fig. 3-16. Try to cut the copper so there is about a ¼″ gap between the strips. Now, if the basic dipole were cut for 20 meters, the center strip would be cut back equally on

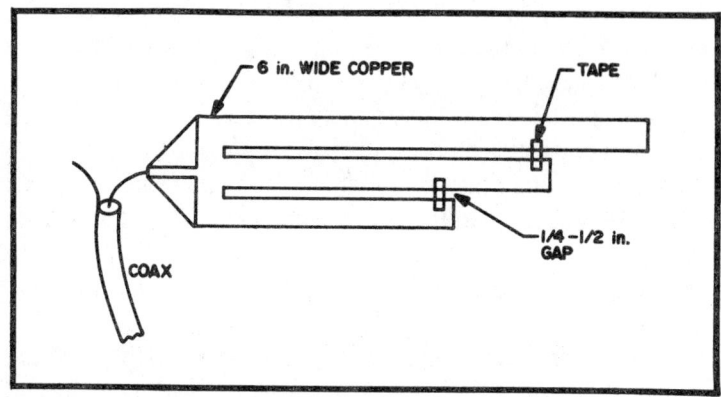

Fig. 3-16. One half of a tri-band dipole.

each side of the dipole until the antenna resonated properly on 15 meters. Then the bottom strip would be cut away equally on both sides of the antenna until the antenna resonated properly on 10 meters. Little pieces of tape placed periodically between the strips will be more than adequate for physical support. The large area surface of the antenna is such that trimming of the strips to form dipoles on each band does not appreciably affect the resonance on any one band, but one should recheck the resonance on each band. Corrections, if necessary, are easily done by soldering on a few inches of copper stripping cut off during the tuning process on each end of the dipole strip. Solder these correction strips vertically on the end of each dipole strip.

The same procedure can be used to construct almost any form of dual or triband antenna when there is sufficient space to run a dipole on the lowest frequency band being used.

Variations

The ease with which the copper stripping can be bent and, particularly, soldered makes it possible to vary the construction of an indoor antenna to suit almost any situation. For instance, as shown in Fig. 3-17, if not enough space is available to run out a full length dipole, the dipole strips can be bent to hang vertically at the end of the antenna to make up the necessary length. Inductively loaded or trap antennas are also easily constructed by soldering the necessary components between sections of the copper stripping. A 80-10 meter loop antenna can be formed as shown in Fig. 3-18 by constructing as large a loop in the attic as space will permit to be hung and using a trans-match type of tuner to resonate the system. Don't hang such a loop horizontally unless it is relatively small and operation is desired only on the 80 or 40 meter bands. The reason for this is that the dominant radiation from a loop will either be broadside to the plane of the loop or along the plane of the loop or a combination thereof, depending on the relationship of the loop size in wavelengths to the frequency being used. A horizontally placed loop operated on the higher frequency bands might well operate in a mode such that the dominant radiation is wasted because it is straight up and down.

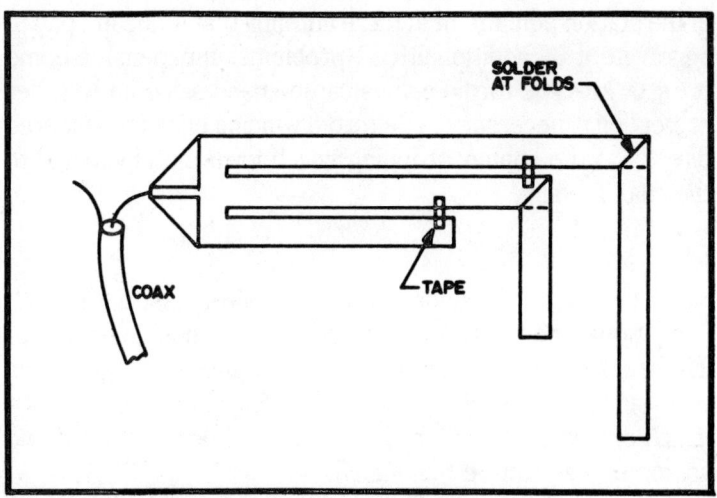

Fig. 3-17. A space-saving version of the multi-band dipole.

Baluns

The use of a balun for an indoor antenna system is recommended. There are usually enough problems with rf fields with indoor installations because of the close proximity

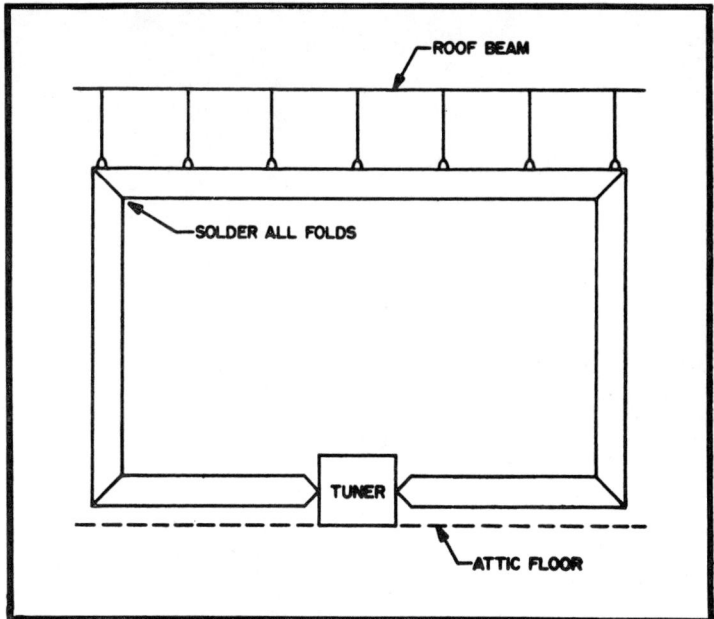

Fig. 3-18. A multi-band loop for suspension in an attic.

225

of the station equipment and the antenna that it doesn't pay to aggravate it by additional R.F. problems. Inexpensive home brew or kit-type toroid baluns can be used since no weather protection is necessary. The toroid winding ends are soldered directly to the copper stripping which form the dipole and to the coax feedline.

Conclusions

The usage of indoor antennas is often associated with lower power operation so that one suffers a double handicap. The use of copper stripping as described goes just about as far as economically possible in keeping antenna losses low. So, if a match of whatever power is available is obtained at least one source of loss can be minimized.

A word of caution when handling copper stripping, especially the hard-drawn type. With normal care, there is no problem in the handling of the material. However, the edges, when cut with shears can become like knife blades.

THE SMALL LOOP ANTENNA

When attempting to operate an amateur radio station while living in an apartment it is often desirable to use the smallest antenna possible. Since in the interests of social harmony it is often mandatory that the antenna be concealed, the antenna may be restricted to indoor operation. In the course of trying to match a number of different shapes of wire to a coaxial transmission line it was noticed that when the loop forming the gamma match in the center of an 'S' shaped antenna was made large enough, a close match to the transmission line could be obtained. While the first surprise was that the loop forming the gamma match was a large as it was, the second was that the match to the line was little affected when the arms of the 'S' were shortened and even removed. That the loop which remained was of a useful size for an antenna was evident since at 7.15 MHz the length of wire in the loop was about 6.7 meters (22 feet) and the total height when erected vertically was about 2.1 meters (7 feet). An antenna this size was easily manageable while there was simply not enough room to put up a half wavelength antenna 20 meters long.

At first it seemed to be a bit strange that a loop antenna this small would have as high a radiation resistance as it did. In a number of books the radiation resistance of a loop antenna with a uniform current distribution is calculated and a loop with a circumference of about .7 wavelength is needed to obtain a radiation resistance of 50 ohms. Since this would mean the calculated loop circumference would be some 4.5 times the actually measured size a qualitative check was made of the current distribution in the loop. As can be seen in Fig. 3-19 the current distribution was certainly not uniform. Indeed it was not even symmetric about the two points where the antenna was fed; a much greater current flowed in the side connected to the capacitor. This held true with the connections to the coax braid and center conductor switched. Once it is established the current is not uniform it is to be expected that the radiation resistance will be higher than the uniform current

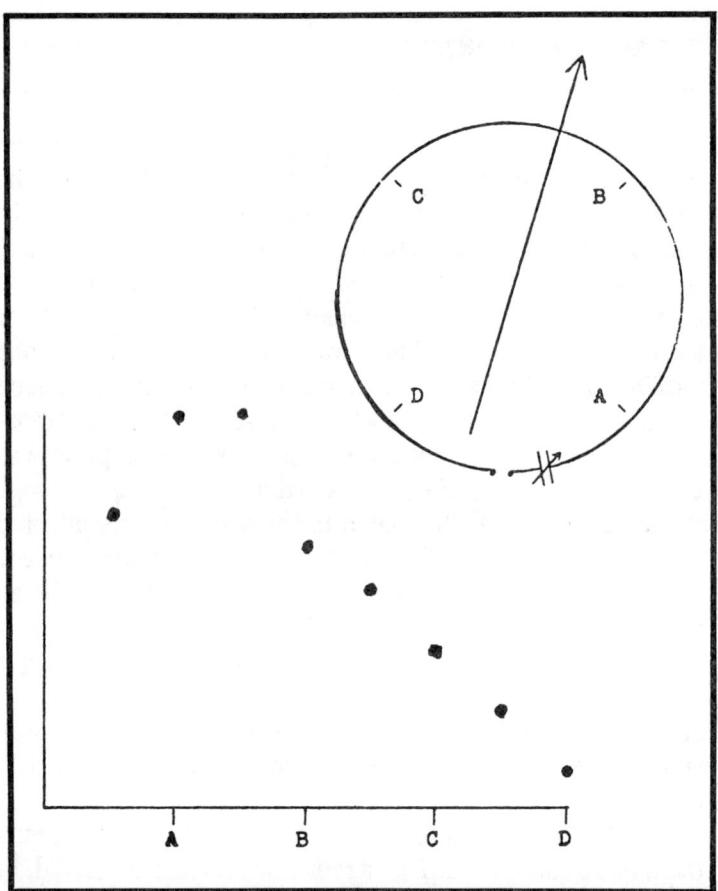

Fig. 3-19. A plot of the approximate current flowing in the antenna. Current is plotted on the vertical axis in arbitrary units. The slanted arrow indicates the general radiation polarization of the antenna positioned as shown.

model would predict. Thus it appears that a more accurate physical model would have to be used to explain why the current flow assumes the form it does. (These measurements are not precise but are probably accurate enough to ascertain the antenna current. They were made with a loop of wire a couple inches in diameter held several inches from the antenna. This test loop and a four germanium diode bridge were mounted on the end of a four foot piece of plastic pipe and the DC output was fed through a coax line to a galvanometer.)

It needs to be said that the above mentioned and all the following measurements were made with the antenna indoors on the ground floor of a two story frame apartment building. The concrete floor, which is essentially at ground level, was covered with at least two layers of regular aluminum foil. This gave a solid ground plane roughly 6.7 by 4 meters and .0032 cm thick (which is about the skin depth of a 7 MHz rf current in aluminum). The antenna was made of 19 strands of aluminum wire in a loose bundle. Each strand was slightly less than 1/16 inch in diameter. The antenna was erected in a vertical plane and fed with RG-58/U at its lowest point which was 20 cm (8 inches) above the ground plane. The ground plane was not directly connected to the antenna, feed line or signal source. A transceiver was used as the rf source and the swr meter gave a 1.0 to 1.0 reading for a 51 ohm load. The antenna was about 3.5 meters from the transceiver.

The small loop antenna, as might be expected, is inductive and from Fig. 3-21 it can be seen that the inductive reactance increases as the length increases. This can be contrasted with the short linear dipole which looks capacitive and whose reactance goes to and crosses through zero as its

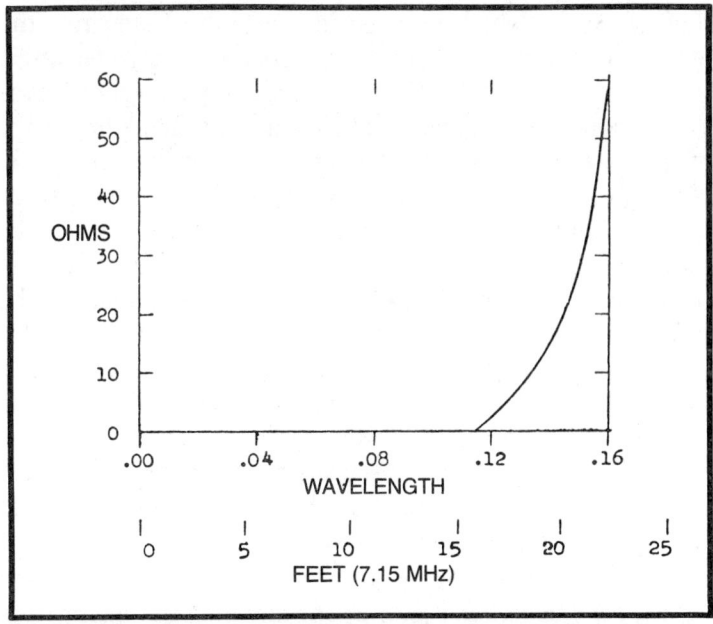

Fig. 3-20. The radiation resistance of the loop antenna versus the length of the circumference measured in fractions of a wavelength. The bottom scale gives the circumference in feet for a signal frequency of 7.15 MHz.

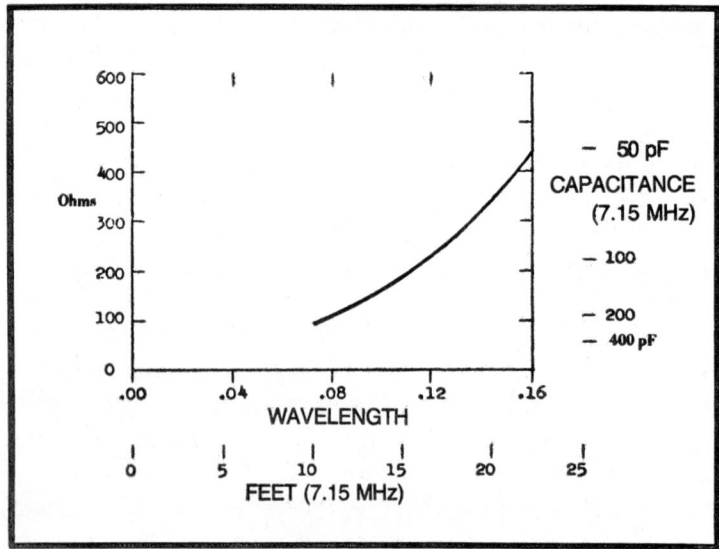

Fig. 3-21. The inductive reactance of the loop antenna versus the length of the circumference. The scale on the right indicates the value of the series capacitor needed to tune out the antenna's reactance for a signal frequency of 7.15 MHz.

length is increased. The radiation resistance of the loop as a function of its length is given in Fig. 3-20. Comparing the points on the graphs where the radiation resistance equals 51 ohms it can be seen that the reactance is almost eight times as large as the resistance which indicates that the setting of the capacitor in series with the antenna will in practice be critical. Adding to this problem is the very rapid change in resistance as the length increases, which indicates that the length of the antenna will also be a critical factor. Experience confirms that only small variations in the length and in the capacitance can be tolerated if a close match to a transmission line is sought. These readings were taken at low power with a calibrated 100 ohm carbon potentiometer inserted in series between the coax inner conductor and the variable capacitor.

The swr of the loop antenna across the entire 40 meter band is shown in Fig. 3-22. Here the length of the antenna and the setting of the capacitor were chosen to give the best match at 7.15 MHz. As can be seen the swr is less than 1.6 to 1 even at the band edges. In Fig. 3-23 the length of the antenna was not changed but the capacitor was adjusted to give the lowest swr at each frequency. One can see that for a fixed length the

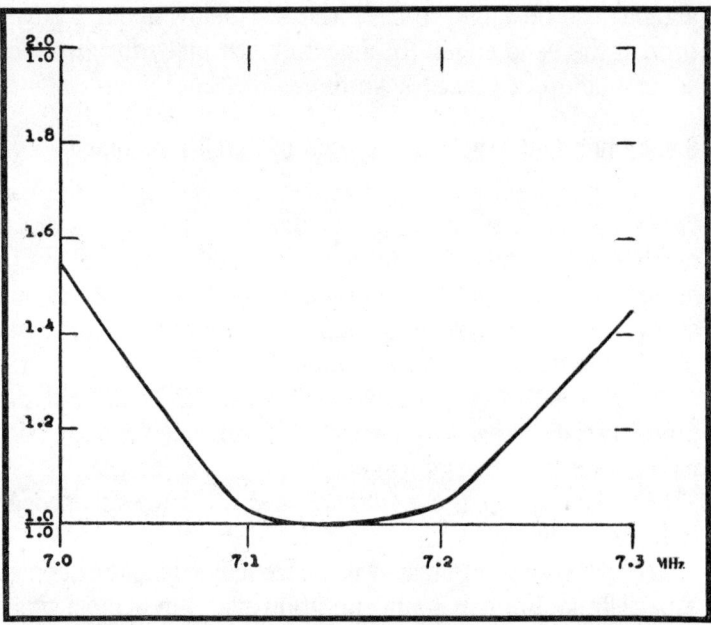

Fig. 3-22. The swr of the small loop antenna across the 40 meter band. The antenna was matched to a 51 coaxial transmission line at the band center.

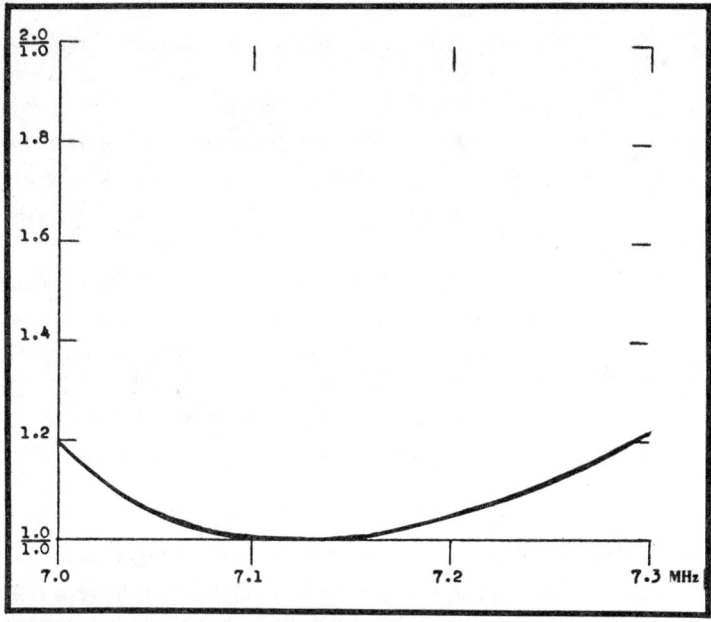

Fig. 3-23. This is the same plot as Fig. 3-22, except that the series capacitor was adjusted to give the lowest swr at each frequency.

antenna can be tuned to keep the swr below about 1.2 to 1 even at the band edges. In Fig. 3-24 swr plots are given for several different capacitor settings.

Some Important Practical Aspects of Small Antennas

While a loop antenna that is .05 wavelength high can radiate as well as any simple dipole antenna, there are a number of very important practical considerations that have to be kept in mind. Indeed, these considerations apply to any short antenna whether it is a dipole, loop or whatever, even though the discussion here will focus on the loop. The first concern is with tuning out the reactance and obtaining the proper radiation resistance while the second involves minimizing nonradiative energy losses.

1. Tuning out the reactance and obtaining the proper radiation resistance.

Most short antennas will be reactive and the loop is especially so. For maximum operating efficiency in most situations, it is desirable to have the feedline see a purely resistive load. It is also usually desirable that this resistive load be of a particular value. Since in most short antennas a small change in the length will have a profound effect on both the resistance and the reactance, the length has to be determined rather accurately. One difficulty here is that the "right" length is going to depend on such things as the proximity of conducting bodies, antennas, feedlines, towers, etc., as well as the height above ground and the nature of the ground itself. The reactance needed to tune the antenna will be similarly affected. What this means in practice is that the length of the antenna and the value of the tuning reactance will have to be adjusted in each situation to make sure a reasonable match is being obtained. Thus it is essential that an swr meter, antenna bridge or other accurate device be used to determine that the antenna does indeed provide a close match to the feedline being used.

In connection with tuning the antenna there is another detail to consider. Since the loop antenna has much inductive reactance, the capacitor used to tune it will also have a high reactance which means that the voltage appearing across the capacitor will be large. For the loop antenna under considera-

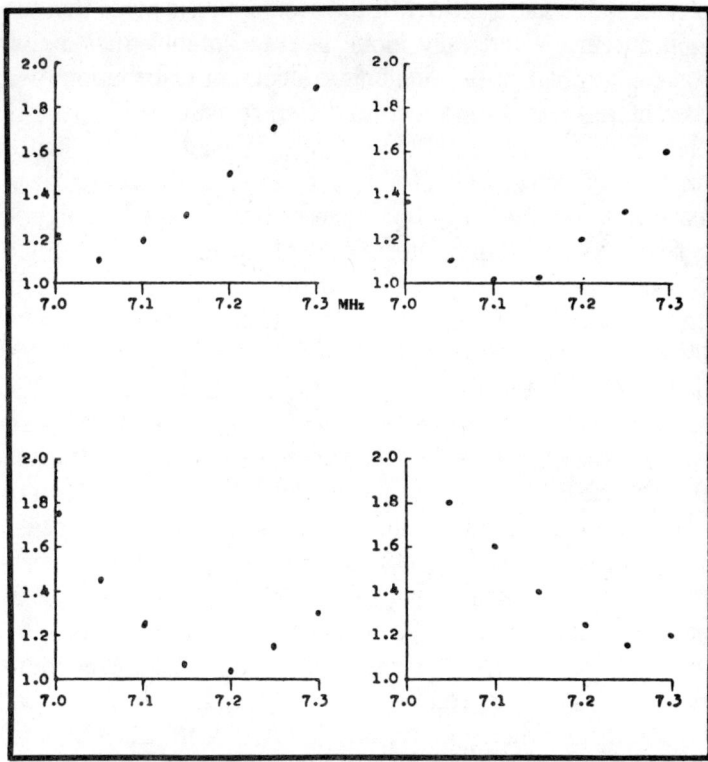

Fig. 3-24. These plots are the same as Fig. 3-22 with the exception that each one was made with a different setting of the series capacitor. The antenna length remained fixed.

tion here it means that 100W of rf fed into a 50 ohm loop will produce about 800V peak across the capacitor, which indicates that even at this power level there are few capacitors other than air or vacuum dielectric ones which can be used that will not arc through or burn out. For an rf power level of 2 kw PEP into the loop it means that something like 3600V appears across the plates, which indicates that the capacitor will have to be chosen with some care.

2. Minimizing nonradiative energy losses.

In principle the ability of an antenna to radiate does not change as the antenna is made smaller since the current goes up as the length is reduced. The problem is that as the current rises the energy lost too heating the antenna wire increases as the current squared so that what may have been built to be a small antenna may in fact be a big resistor that generates a lot of heat and radiates little rf. If the assumption is made that the loop antenna electrically looks like an antenna that is .05 wavelength high with a nonuniform current distribution then the antenna has an intrinsic radiation resistance of about ½ ohm. This low value should not be confused with the much higher resistance presented to the feedline. In order that at least 90% of the energy be radiated, this means the antenna wire has to have a circumference of about 4cm or, for a single conductor, a diameter of about 1.3cm (½") at 7MHz. All that is important is the effective 'surface area' of the conductor. The thickness can be quite small since a flat strip of aluminum foil that is four layers thick and 2cm (¾") wide is sufficient. If a small antenna is constructed by joining sections of tubing or wire or whatever together than care must be taken to insure that the joints do not constrict the diameter enough to create points of high resistance and accompanying high heat loses.

The second source of nonradiative energy loss that affects every antenna, regardless of its size or type and in almost all locations, is the heating of the ground near the antenna. This subject is somewhat involved and will be mentioned only briefly here even though ground losses are probably the major limiting factor on the lower ham bands in most amateur radio stations. The less the height of an antenna the greater the ground losses are. Indeed, for antennas close to the ground surface, the ground loss is horrendous and only a small fraction of the rf fed to the antenna is radiated. Ground losses can, however, be greatly reduced and such low antennas can be highly efficient radiators if an adequate ground system is used. While the size of the ground system needed will depend on such factors as the ground conductivity, the height and size of the antenna and the frquency, a minimal system might consist of 100 wires each a quarter wavelength long shallowly buried in a radial pattern.

The aluminum ground plane used with the 7 MHz loop antenna should be about as efficient as an extensive wire ground system of the same dimensions (6.7 by 4 meters). Assuming a local ground conductivity of 15 millimhos per meter, a rough calculation indicates that, neglecting wire losses, the efficiency of the antenna system is reduced by ground losses to about 35%. Thus about two thirds of the rf energy fed to the antenna merely heats up the ground near the antenna and is wasted; 35% is radiated to the atmosphere. Wire losses in the loop antenna itself may reduce this to near 30%. To compound the losses a bit further, if the final amplifier in the transceiver used here is assumed to be about two thirds efficient, then a dc input power of 200W to the final amplifier finally results in about 40W of rf being radiated to the atmosphere—this is an overall efficiency of about 20%. Without the ground plane the losses would be expected to be much greater than they already are.

Some Possible Uses of a Short Antenna

The small loop has two disadvantages relative to a conventional full size dipole. First it has to be constructed of a much larger wire size to minimize resistive wire losses. Second it has to be tuned fairly accurately—both the length and the series capacitance have to be within narrow tolerance limits. In addition, the capacitor has to have a hefty voltage rating which for many hams means using a suitable air or vacuum capacitor or a section of coax trimmed to the proper length. In practice, once they have been recognized, the disadvantages can be readily overcome, giving the user a small antenna that can be expected to work in much the same manner as a regular dipole with the same orientation at the same height.

Since the 7 MHz loop is about 7 feet in diameter it is possible to construct it of a self supporting conductor such as copper or aluminum tubing or fairly stiff ¾" diameter coax and to suspend it from a single support. It could, for example, be mounted instead of a center supported horizontal dipole or inverted V. The loop has the advantages that no end supports are needed and that it will usually be possible to orient it as desired instead of orienting it to fit space limitations. The loop also possesses the unique advantage that in such an installation it can be orientated to radiate vertically; it should work much as would a vertical dipole whose center was as high as the center of the loop. In choosing a particular orientation it should be noted that in Fig. 3-19, with the antenna fed at the bottom, the orientation of the radiated rf will be mostly vertical with a sizeable horizontal component; it looks like a tilted vertical. For true vertical radiation it will be necessary to rotate the loop, capacitor and feedline counterclockwise. A suitable clockwise rotation will make it radiate like a horizontal dipole. It will also radiate horizontally if the plane of the loop is parallel to the ground. In many installations it will be possible to erect two or even three loops, so the operator has a choice of polarization.

In adddition to being usable where space limitations rule out a full size dipole, the small loop antenna lends itself to emergency and portable operations. If erected close to the ground without an extensive ground system, it can be expected to work as badly as any other antenna at the same height, although there are many situations in which such performance is adequate. The loop is a lot smaller than a full size dipole and it can be oriented to radiate vertically or horizontally. As indicated in Fig. 3-22, it can be constructed to cover a fairly wide range of frequencies and still maintain an acceptable swr. It might be noted that while the effect depends on the conductivity of the ground, the radiation resistance of an antenna that is not above an extensive ground system will be affected by the height above ground. As a rough figure, the effect can become important below a quarter wavelength and drastic below an eighth wavelength.

Table 3-2 gives rough length, capacitance and minimum single conductor diameter (for aluminum) for the lower ham

Table 3-2. Approximate Length, Capacitance and Minimum Conductor Diameter for the HF Ham Bands.

Amateur Band	Loop Circumference	Series Capacitance	Conductor Diameter
160 meters	88 feet	200 pf	1.0 inches
80	44	100	.7
40	22	50	.5
20	11	25	.4
15	7.3	12	.4
10	5.5	6	.3

Approximate loop circumferences, tuning capacitances and minimum conductor diameters (if a single round conductor is used) for the lower ham bands.

bands. With aluminum conductors, either solid or hollow of the diameters given, the wire losses should be less than 10%. Losses in a copper wire of a given size should be about .8 times the aluminum losses. To obtain a suitable conductor it is possible to use two conductors with half the diameter, four conductors with a quarter the diameter, twenty conductors with a twentieth the diameter, etc., provided the wires are well separated from each other and not tightly bundled together. These numbers were obtained by scaling the 7 MHz results. If "f" is the frequency, the length and the capacitance are proportional to "1/f," while the conductor "surface area per unit length" is proportional to "$1 \div f^{1/2}$." The exact length and capacitance values required may well vary with different installations and it is urged that, as with beams, quads, and many other antennas, the length of the loop and the series capacitance be adjusted for an optimum match to the feedline with the antenna at the intended height and orientation.

THE 18 INCH ALL-BAND ANTENNA

A simplistic explanation of this antenna is that it is not an antenna in the common meaning at all. Rather it might be thought of as a capacitor coupled to all the rest of the universe. In this universe are quite a few other antennas connected to transmitters.

The size of this capacitance must be very, very small as its elements are very far apart. Any capacitance existing must represent an extremely high impedance. Thus it is necessary to build a transformer which will convert this high impedance down to something manageable, like 52 Ohms.

Today, this enterprise can be managed with real components by the use of an FET source follower, as shown in the diagram. An input capacitor is used to prevent inadvertent DC coupling to the FET gate and its subsequent destruction. At the input of the FET is also the only special component, high quality low noise metal or carbon film resistor of about one meghohm resistance.

This is required to reduce resistor noise at the input which will soon develop if carbon composition resistors are used. Of course, a choke would solve all the problems if only one could find one with a high enough impedance and no strange characteristics like resonances at undesired frequencies or low Q figures to eat up the signal.

The "antenna" portion of the unit also is a bit critical in that the capacitance to the rest of the world often needs to be adjusted to fit a particular location in the real world. For example, a nearby broadcast station or a bad fluorescent light could cause cross modulation. The "antenna" used is a replacement receiver whip which could extend from about 8 inches to about 4 feet.

But 18 inches seems best for most locations. Remember again this is a capacitive device, and any stray capacitance to ground from the antenna or circuitry leading to the gate of the FET causes the signal to be divided into an unwanted capacitive branch. So use a big insulator at the base of the whip and a short lead to the FET base. The units built use "free form"

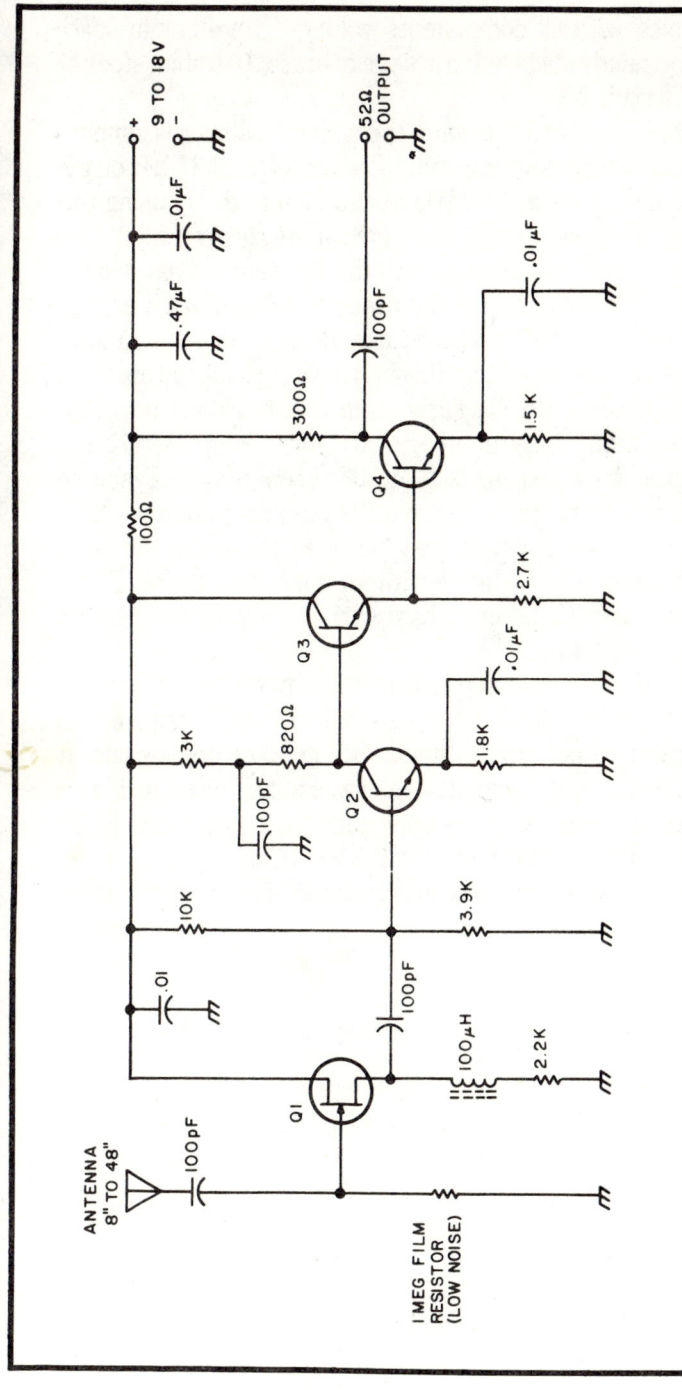

Fig. 3-25. Q1 = 2N3819 or equal. Q2, 3, 4 = any 200 MHz, 20V, NPN transistor, i.e., 2N918, 2N6008, etc. R = Carbon composition, except where noted, all ¼-watt.

electronics with all components soldered together in space and the grounds soldered to a piece of brass .010 shim stock or printed circuit board.

All of the circuit following the source follower is simply a 30 dB gain broad-band amplifier. The roll-off is at 3 MHz on the low end and at about 35 MHz at the high end. By using this amplifier, the result is a lot more gain at the receiver input (and possibly some cross modulation if the antenna is extended too far). The extra gain makes the receiver think it has a quarter wave whip connected to it at any frequency from 3 to 35 MHz.

It should be noted that this particular broadband amplifier has been designed to be fairly foolproof, but don't bring the input and output close together. Hard core cases of local cross modulation may require a filter between the FET source follower and amplifier to remove the offending station. Or an FET broad-band amplifier may be used to handle the large signal levels of a local station without cross modulation. So far the capacitor-FET antenna has worked so well that neither has been necessary.

In use the capacitive antenna has been remarkable. It draws only about 6 mA from a 9V battery. Any frequency in the working range can be tuned with good strong signals. It doesn't seem to be frequency selective at any place within its range and once the whip has been adjusted for the location, 10 meters, WWV, all major shortwave bands, etc., can be tuned with nothing but a small box sitting on top of the receiver and a twist of the dial.

Chapter 4
Beams and Irrational Antennas

A WIDESPACED BEAM

This three-element, widespaced beam is plenty sturdy; it is of all-aluminum construction and can withstand winds of up to 80 mph.

When building this beam, *do not* alter any of the physical dimensions, as this will decrease the efficiency of the antenna. The frequency of the array is set in the middle of the band to allow its use on the CW and phone portions of 20m.

The elements are constructed of thin-wall aluminum tubing, of the diameter and length stated in Table 4-1.

Each element is constructed of seven pieces. The center portion is 1 in. inside-diameter aluminum conduit to give strength to the remaining portions of the elements. The conduit is slotted at each end on both sides for about 3 in.

The remaining portions are fitted together to the values shown in Table 4-2. About 4 in. from each individual piece of tubing is placed a self-tapping screw to insure that the elements do not move or rattle. At the end of each element is placed a drip hole about 1 in. from the end and a cork is press-fit in the end of the tubing to prevent the elements from whistling in the wind.

Table 4-1. Thin Wall Lengths.

ELEMENT SECTION	ELEMENT			TUBING DIAMETER
	DIRECTOR	DRIVEN EL.	REFLECTOR	
A	2 ft	2 ft	2 ft	3/4 in. O.D.
B	5-1/2 ft	6 ft	6-1/2 ft	7/8 in. O.D.
C	5-1/2 ft	6 ft	6-1/2 ft	1 in. O.D.
D	10 ft AL. CON.	10 ft AL. CON.	10 ft AL. CON.	1 in. I.D.

All three elements are constructed in the same manner, the only difference being their physical lengths. To obtain the proper length on each side of the center of the boom, the thin-wall aluminum portions of the elements are adjusted in or out of the aluminum conduit. Once the proper distance has been attained, drill a hole about 8 in. from the edge of the conduit and drop a self-tapping screw in, and also place a hose clamp about 1 in. from the end of the conduit.

Just a small note here on cutting the thin-wall aluminum. Most of the tubing comes in 12 ft lengths, so on the antenna portion cut the tubing in half and to obtain the proper lengths for the director and reflector for their overlap cut the tubing 6 in. off center.

The boom is a 27 ft 4 in. piece of 3 in. aluminum irrigation tubing. It is the most expensive single portion of the antenna, but is well worth the money spent, from at least the standpoints of the strength it gives and its light weight.

At each end of the boom there is a circular block of wood, the diameter of the pipe, which is fitted in the end of the pipe and then nailed. This precaution is necessary unless you are a bird lover.

Table 4-2. Lengths of Remaining Pieces.

ELEMENT SECTION	DIRECTOR	DRIVEN EL.	REFLECTOR
A	1 ft – 3 in.	1 ft – 3 in.	1 ft – 3 in.
B	4 ft – 9 in.	5 ft – 3 in.	5 ft – 9 in.
C	4 ft – 9 in.	5 ft – 3 in.	5 ft – 9 in.
D	10 ft	10 ft	10 ft

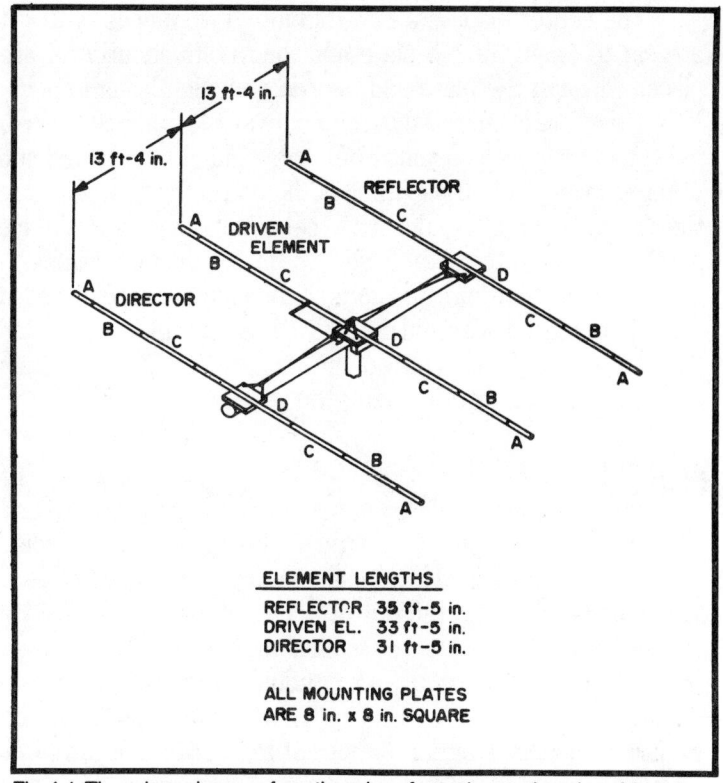

Fig. 4-1. The values given are from the edge of one piece to the edge of the other.

The main feature of this antenna is the method used to mount the elements to the boom and keep them there. Aluminum plates (0.25 in. thick) are used in this deal. The plate is held to the boom by two 3 in. muffler clamps. The plate in turn holds the element with two smaller muffler clamps as shown in Fig. 4-2. The plate is first mounted on the boom with the 3 in. clamps and tightened slightly. It should be mentioned that all hardware used was galvanized heavily and then lead plate was used on all the nuts and bolts to prevent seizing and rusting.

The aluminum conduit portion of the elements is mounted and tightened on each plate. There are two pieces of grappling iron, about 6 in. long, which are placed on each side of the boom, one each under the nuts which are the furthest away from the ends of the boom. Make sure that the distance from the center of one parasitic element to the other is 26 ft 8 in.

The center plate is a two-fold job. Two pieces of 2 × 4 are cut to length of the plate and long bolts about 7 in. are placed through the plate and boards and slightly tightened.

This plate is mounted to the boom as the other two were, using muffler clamps again. Then the conduit is mounted and clamped in its place. By sighting at the end of the boom, look at the three pieces of conduit and make sure that all three are parallel to each other, then tighten all the muffler clamps.

Now the remaining portions of the elements are placed in their respective places and finally the hose clamps are installed and tightened. When you do this make sure that the drain holes are on the bottom facing the ground.

Matching

On this particular antenna, a gamma match was tried and when adjusted properly, proved to be a very wise choice because the swr was flat across the band and did not exeed 1.2:1. It was constructed out of a TV antenna element. The shorting bar was constructed from aluminum and was made so that the center of the aluminum conduit to the center of the 48 in. piece of TV element was 6 in. This is very important. The capacitor was made out of a length of RG-8; 41 in. of the outer covering was taken off and then 40 in. of the copper shield had the same treatment. On the remaining 1 in. of braid, there was soldered a brass or copper bracket which will later be used to mount it to the beam. The remaining portion of the stripped end of the coax was placed inside of the gamma tubing.

The bracket must be mounted onto the boom and it must keep the 6 in. from center to center constant.

The bracket is held in place by two self-tapping screws placed on either side of the hump in the bracket. At the other end of the coax a coaxial connector was placed to provide easy connection to the feedline. The inside of the shorting bar is 44 in. from the center of the boom.

Installation

The beam is installed on the mast by means of another plate. This plate is made of ¼ in. steel and is drilled to accommodate the four long bolts which are on the center plate

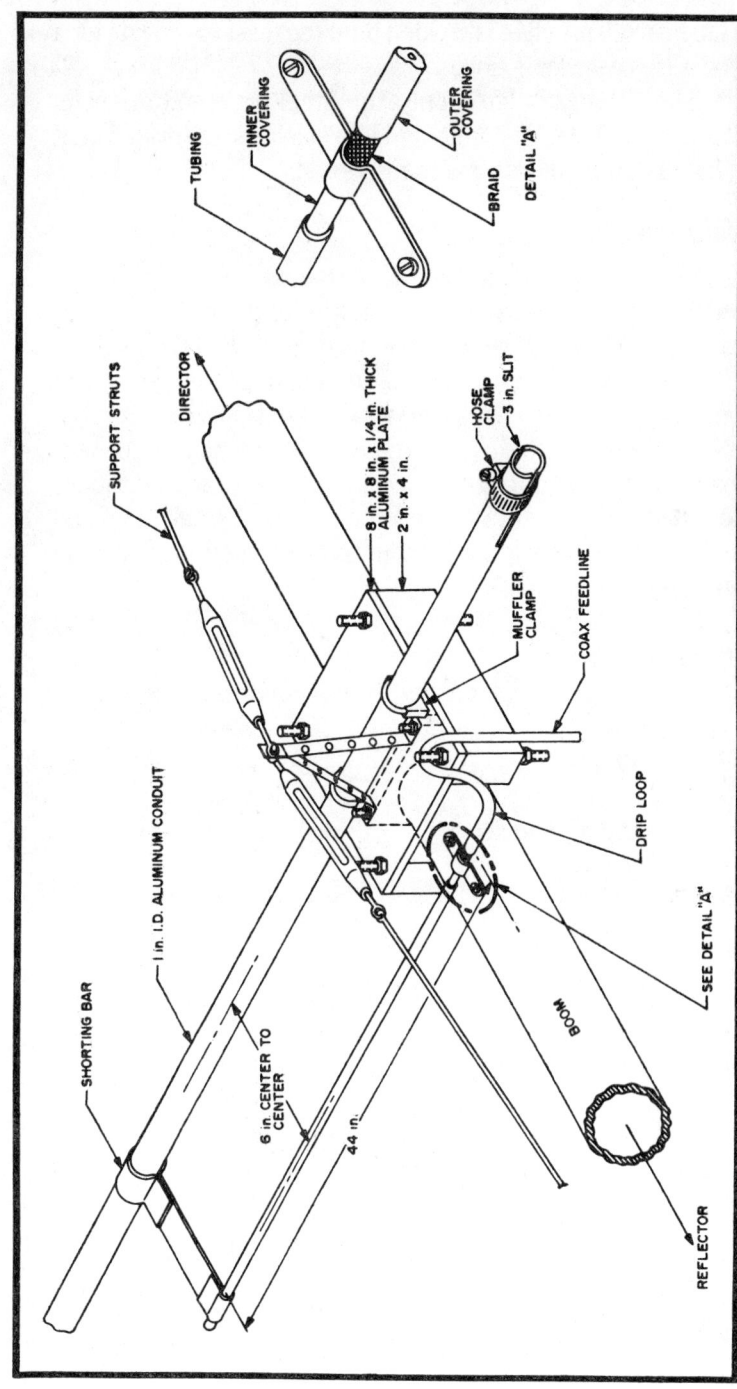

Fig. 4-2. This sketch shows the construction of the center plate of the beam and also the position of the gamma match.

of the beam. This plate is welded onto the mast to provide for a good slip-free connection.

The antenna is lifted onto the steel plate and the bolts placed in the hole. The nuts and lockwashers are placed on it and all mounting hardware is tightened.

Adjustment

There need not be any adjustments necessary to the elements if precautions were taken in acquiring the proper sized tubing and the measurements followed to the inch.

There may be, however, some adjustment needed on the gamma match. The values given were used on three previous antennas identical to this one and no adjustment was needed. However, should the need arise that it does need attention, then the bracket on the gamma match must be taken off and the end of the coax trimmed, about half to a whole inch until the swr is down to at least 1.5:1.

Specifications

Gain	8.5 dB over a dipole
FBR	25—30 dB
Side ATT	50 dB
Boom	27 ft 4 in. × 3 in.
Turning Radius	22.5 ft
Weight	45 lb

2 ELEMENT BEAM SPACED A QUARTER WAVELENGTH

This simple beam antenna for 15 meters can be made from wire elements strung between some trees and still provides various directional patterns.

These requirements are satisfied quite easily by a driven, two-element array with quarter-wavelength spacing. Quarter-wavelength spacing of two driven elements represents a very interesting case because of the variety of directional patterns which can be obtained without any complicated impedance-matching problems. This is due to the fact that at quarter-wavelength spacing the impedance of each element is almost the same as its free-space impedance, while at closer spacings the presence of each element severely affects the impedance of the other element.

The three directional patterns which can be obtained from such an antenna are shown in Fig. 4-3. The cardioid

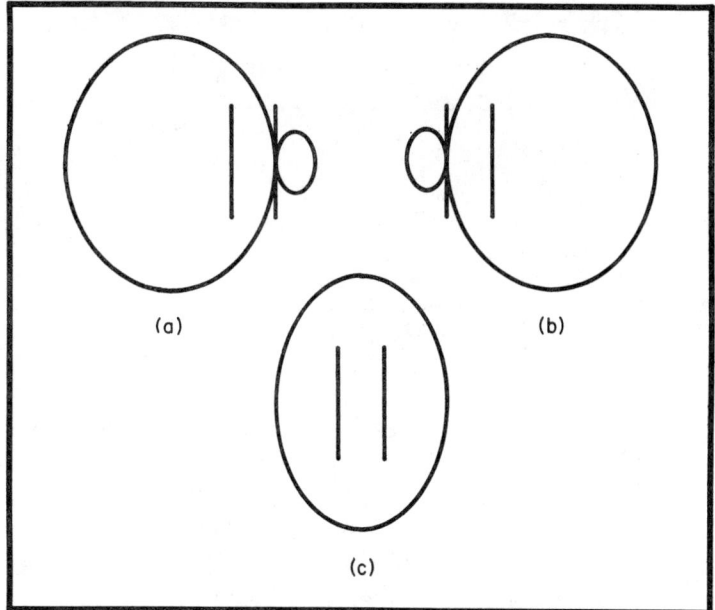

Fig. 4-3. A representation of the three directional patterns possible. (a) is the cardioid pattern obtained with a 90-degree phase difference between elements, (b) is the same pattern switched in the opposite direction, and (c) is the pattern of zero-degree difference between elements.

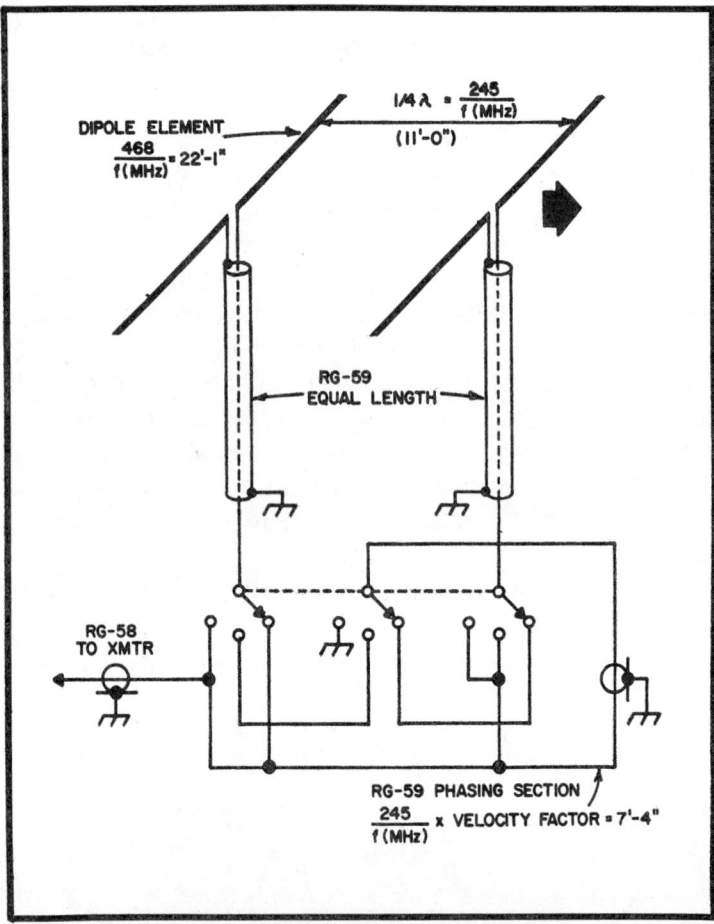

Fig. 4-4. The pattern selector switch. Maximum radiation is in the direction shown by the arrow for the switch position show. When the switch is in the bidirectional position, the antenna is also grounded through the quarter-wave phasing section as a lightning protection feature. The dimensions shown are for 15 meters.

patterns will provide gain of 4-5 dB while the bi-directional pattern in (c) of Fig. 4-3 will provide about 3-dB gain.

The antenna which is constructed for 15 meters is shown in Fig. 4-4. RG-59/U is used to feed each antenna as well as for the quarter-wavelength phasing section. RG-59/U was chosen because when the two feedlines are effectively paralleled by the pattern selector switch, an impedance of 36 ohms will result. When RG-58/U is used to the transmitter an SWR of about 1.5 to 1 should result.

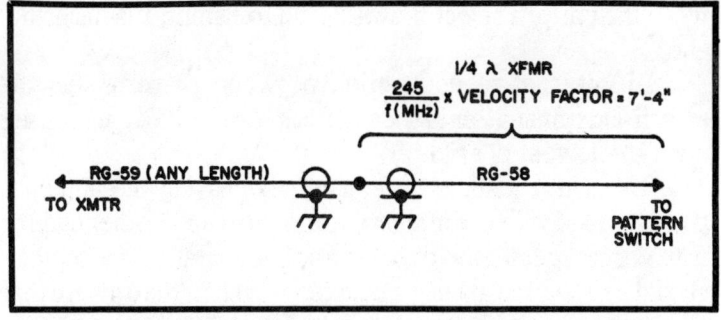

Fig. 4-5. An additional matching circuit which can be placed between the pattern switch and transmitter to improve the SWR. It replaces the RG-58/U phasing line shown in Fig. 4-4.

Actual practice will show an SWR of closer than 2.0 to 1, probably because of some slight mismatch between the RG-59/U and the dipoles. The 2.0 to 1 SWR should cause no difficulty as far as transmitter loading is concerned and the actual power loss in the short length of RG-58/U used bet-

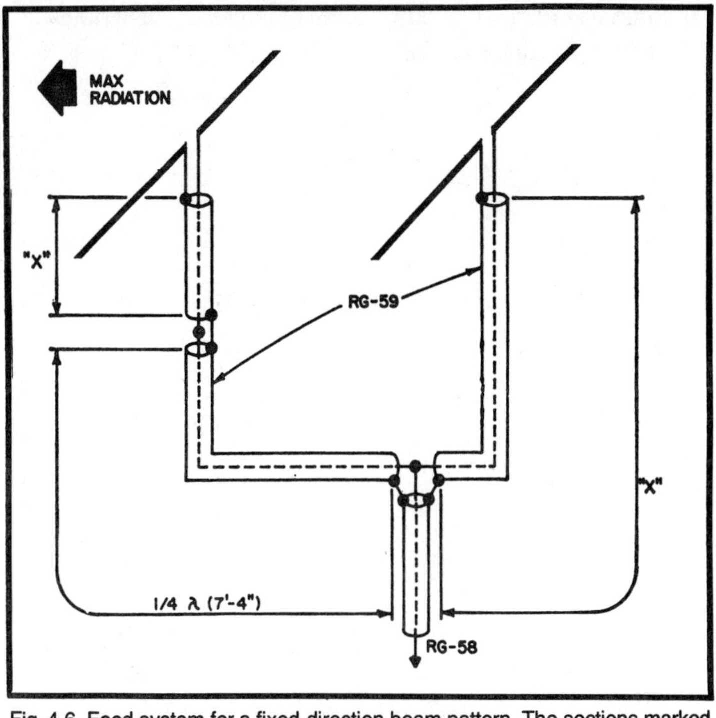

Fig. 4-6. Feed system for a fixed-direction beam pattern. The sections marked with an "X" are made up from RG-59/U and may be any convenient, but equal, lengths.

ween the pattern selector switch and transmitter is insignificant.

An alternative connection between pattern selector switch and transmitter is shown in Fig. 4-5 for those who insist upon the lowest possible SWR.

The same scheme of feeding and phasing the antennas could be used with an antenna dimensioned for another band or with vertically oriented dipoles. For horizontal antennas, they should be elevated at least a quarter-wavelength to insure that the impedance of the dipoles is 60-70 ohms.

For someone who is just interested in a beam pattern in one direction, the simple feed system shown in Fig. 4-6 can be used. The RG-58/U feedline should be limited to about 100 feet, however, because it may operate at an SWR of up to 2:1.

This type of antenna is certainly not new but the type of feed system considerably simplifies construction. The directivity is not as sharp as a two-element, parasitic beam but it provides almost the same gain in several directions at a minimum installation cost.

TAKE-APART BEAM

This portable take-apart 2-meter beam can be disassembled and stuffed into a long umbrella case for ease in carrying.

It features simple construction using junkbox parts and is ideal for portable operation.

The main boom and the take-apart extension is made from ½ in. aluminum tubing. The six elements are made from ⅛ in. brass rod with the ends threaded 6/32 to fit into the banana plugs. They could also be soldered in place. Millen #37222 binding posts are used for the jacks, and fitted in the holes in the boom and secured with 8/32 nuts. The slight offset of these jacks at each far end of the boom does not materially affect the performance. The ceramic center insulator is a surplus unit with 8/32 threaded holes to accommodate the binding posts. The threaded ends of the posts are shortened a bit so they will screw down tight to the center insulator, securing solder lugs or clips for the coax feed.

It works well and the compactness is a bonus for people who travel.

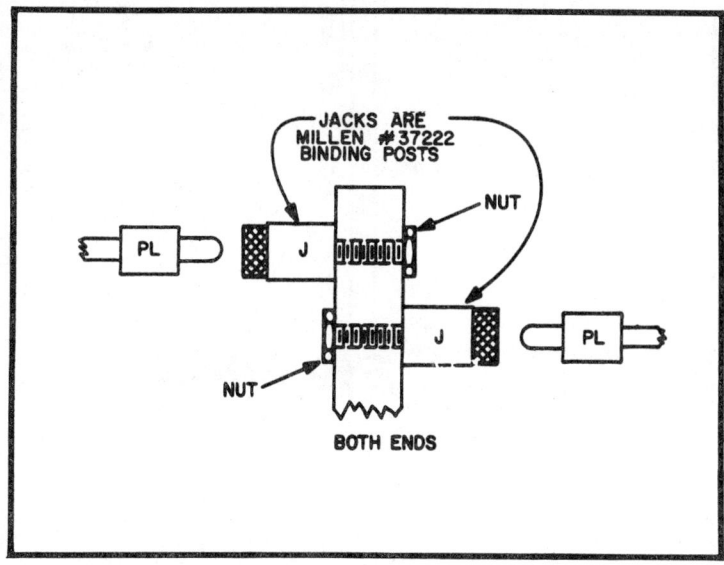

Fig. 4-7. Details of end elements.

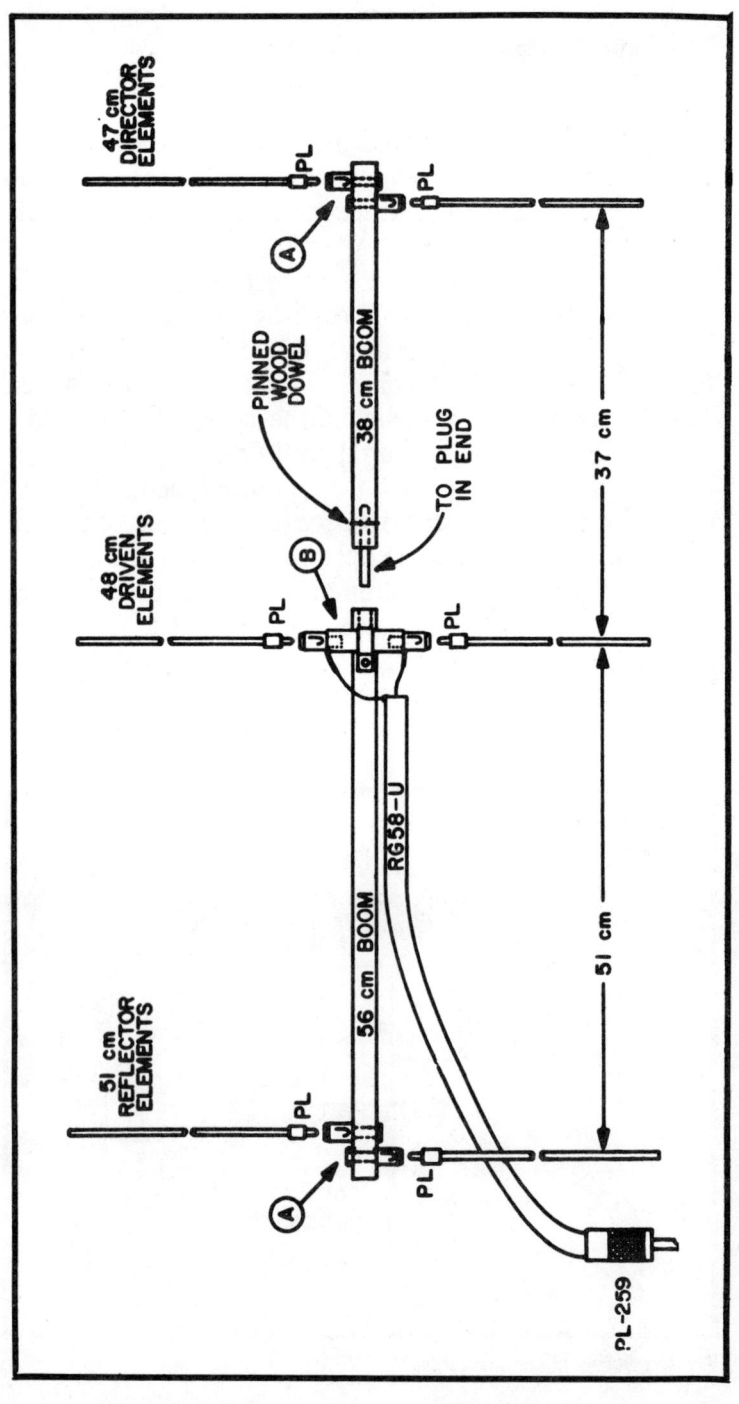

Fig. 4-8. The 2 meter beam. All elements are measured from center of boom.

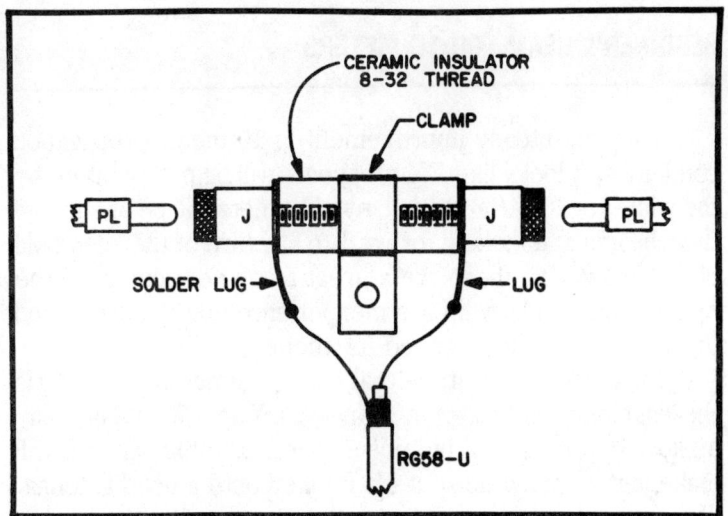

Fig. 4-9. Driven element details.

BEGINNER'S BEAM FOR 10 METERS

With the steady improvement in 10 meter propagation conditions, it looks like DX prospects will be pretty bright by the winter of 1967/68, and many old-timers will be dusting off their beams and looking forward to a return of the "good old days." However, there are a great many newcomers to the ranks of ham radio who are inexperienced on this band, and this article is really intended for them.

Most people will argue that the power output of the rig is the least important factor in 10 meter DX operation. Naturally a kilowatt will make a big noise, but a 100-200 watt rig will make just as much noise if it's hooked onto a good antenna. The size and weight of 10 meter beams are well within reason for even the most crowded back yard or roof-top.

The beam described here is ideal, especially for the newcomer, as it combines light weight, standard components, very simple construction, and of course, low cost. Despite the simplicity, the gain will be 7 to 8 dB for the three-element version, and around 9 or 10 dB for the four element one. For the small extra cost and work involved, the four element version is much to be preferred. The front-to-back ratio will also be better, and the extra gain is worthwhile.

Depending upon your operating preferences, the length of the elements should be decided by reference to the standard formulae:

$$\text{Driven element length} = \frac{473}{\text{Freq MHz}}$$

$$\text{Reflector length} = \frac{501}{\text{Freq. MHz}}$$

$$\text{Director length (both)} = \frac{450}{\text{Freq. MHz}}$$

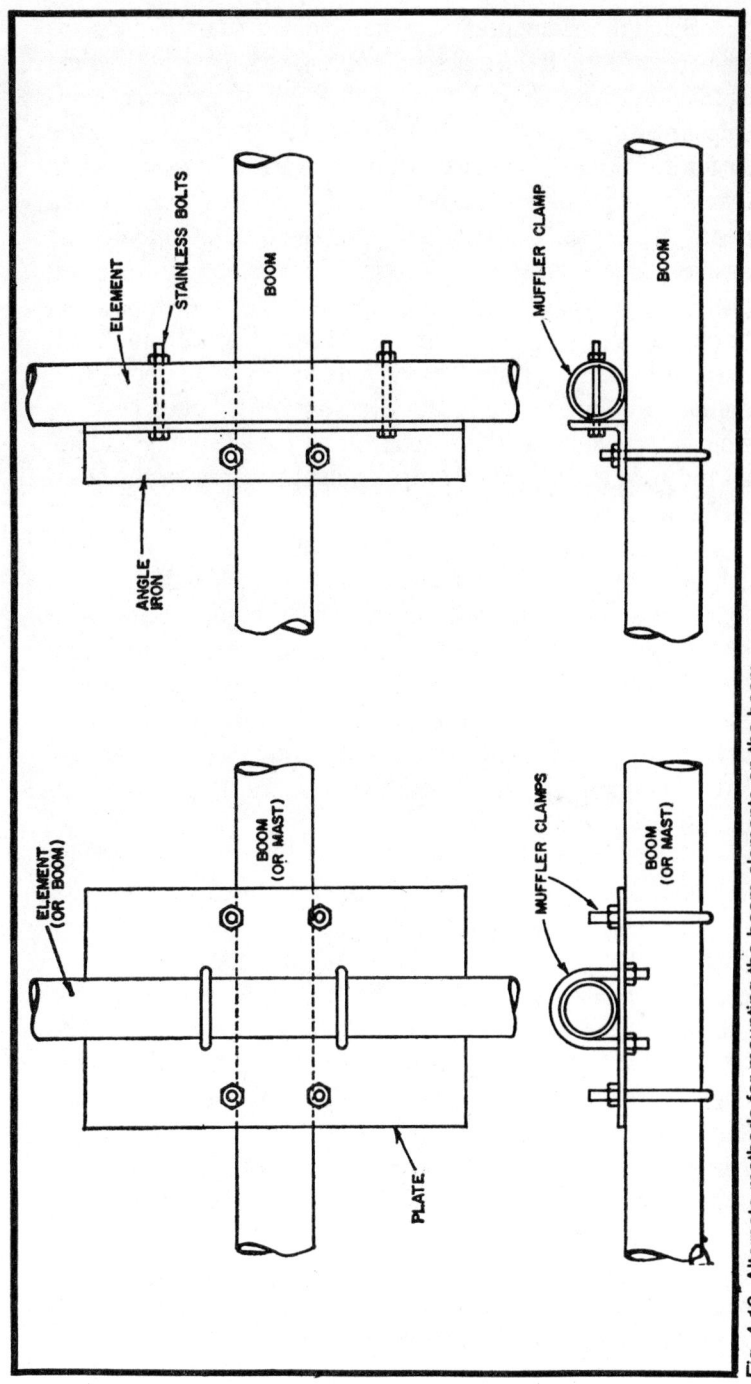

Fig. 4-10. Alternate methods for mounting the beam elements on the boom.

Since the elements are adjustable, the exact lengths are easy to come by. If the material is purchased new, choose rigid aluminum tubing 1″ and ⅞″ in diameter (or similar relationships in size) so that the center sections can be made of the larger tubing with the smaller tubing inserted into the ends to form the adjustable sections. For a 4-element beam, you'll need four lengths of the larger size, and three lengths of the smaller. This is assuming you get 12-13 foot lengths which are pretty well standard. The three lengths of smaller tubing can be cut into four foot pieces for the end sections, with a little left over. If cost is a prime factor, you can use old booms from defunct TV antennae as I did. I scrounged a bunch of these from a local service shop, took off all the elements and assorted junk, and ended up with excellent material for the beam elements.

Without doubt, the best material for the boom is old reliable irrigation tubing. The 2″ diameter stuff is fine, in a 20 foot length. This gives reasonably wide element spacing. As a matter of fact, a 5-element beam can be mounted on such a boom if you wish, but I happen to prefer the wider spacing. Steel TV masting is another common material which can be used for the boom, but it is quite a bit heavier and you may have to couple sections together to make up the required length.

Several methods can be used to mount the elements on the boom, as shown in Fig. 4-10. In both cases, standard automobile muffler clamps are used to fasten the element support plates to the boom. Make sure the clamps are given a couple of coats of rust-proofing first. By using the flat plates, the elements can be laid across the long dimension of the plate and fastened with small U-bolts. On my own model I used angle iron instead and fastened the elements onto the iron with *stainless* steel bolts. Be sure you use nothing but rust-proof hardware on the beam. There isn't much required, and the small cost is well worthwhile if and when you try to take it apart again.

Figure 4-11 shows the arrangement of the elements on the boom and the spacings used. Antenna handbooks give all sorts of opinions on which spacing is best, and why, but as a general rule the optimum spacing should be 0.2, 0.2 and 0.25

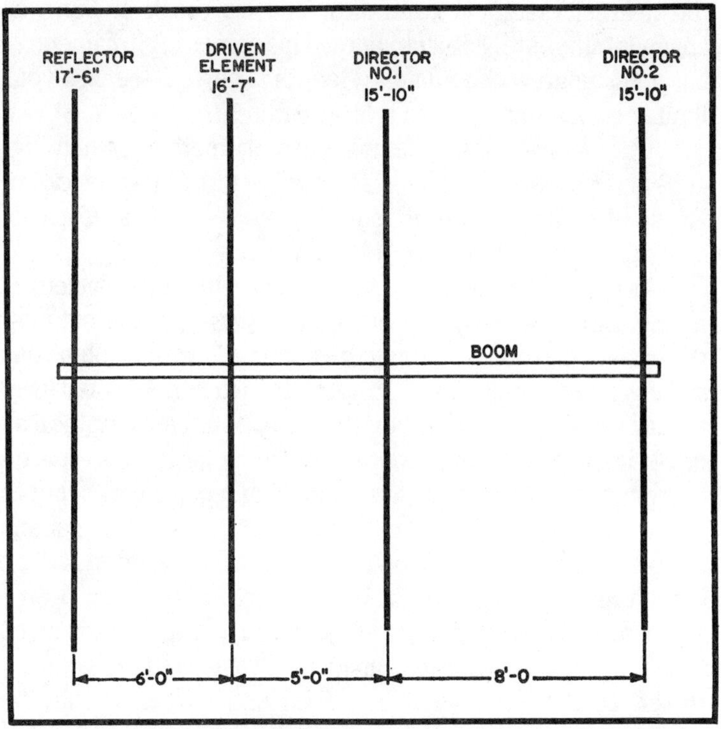

Fig. 4-11. Physical layout of the four-element ten-meter beam. Dimensions given are for approximately 28.4 MHz.

wavelength, reading from the reflector to the second director. I modified this a bit in an effort to get a higher front-to-back ratio, so feel free to change the spacing if you wish.

Figure 4-12 shows a typical element and how it is put together. Simple. All you need is a hacksaw, a screwdriver and two hose clamps per element. Depending upon how the tubing fits, you may need small shims to tighten up the joints. Incidentally, if aluminum tubing is not readily available, look up

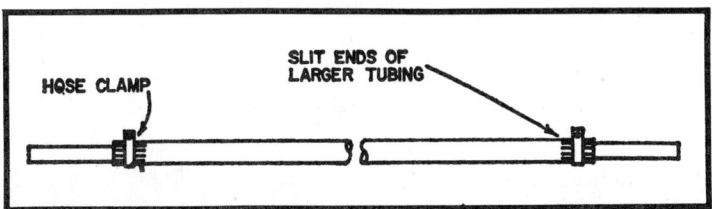

Fig. 4-12. Construction of a typical element, showing the adjustable end sections.

257

the nearest electrical contractor and his stock of thin-wall conduit, either steel or aluminum. This comes in all diameters, but unfortunately the standard length is only 10 feet, so your total requirements will be a little different.

The boom-to-mast clamping arrangement shown in Fig. 4-13 is probably in its simplest possible form. Two pieces of flat steel or iron and four muffler clamps—with a couple of coats of paint—will do the job very nicely.

With the whole beam assembled, the last problem is tuning. Since the majority of rigs today use coaxial outputs, the easiest method of feeding the antenna is with 52 ohm coax and a gamma match. This is diagrammed in Fig. 4-13 and uses a small variable capacitor mounted in a plastic refrigerator dish or similar weatherproof container. Use a fairly wide-spaced capacitor, not because of power handling requirements, but to prevent oxidation from shorting out the plates. The gamma rod is tapped onto the driven element at a trial position and the SWR is measured on the transmission line. Use as little power as possible for this adjustment procedure in order to reduce QRM. Carefully rotate the capacitor through its range and try to reduce the SWR as close to 1:1 as possible. It may be necessary to change the position of the tap several times, but usually the capacitor will do the trick after one or two trials. For this procedure the beam should be mounted reasonably well off the ground and away from trees, guy wires, etc. The ideal place for it is on top of your tower, but this may not be possible. The procedure will be infinitely easier if you can persuade someone to turn the capacitor while you watch the SWR meter *and* the resonance of the final in the rig. It changes considerably while all this is going on, so make sure you check it often. Actually, if you get the SWR down under 1.5 you can be pretty happy with it. It is debatable whether or not the extra effort of getting down to 1:1 is worthwhile.

The last problem is tuning the elements for best forward gain-or best F/B ratio. The two factors don't go hand-in-hand. Several methods can be used, all of which involve test dipoles, field strength meters, signals which stay steady enough to made adjustments and of course, the "friendly amateur a few miles away" who will dutifully do just what you want him to—baloney! If you figure out your dimensions properly by

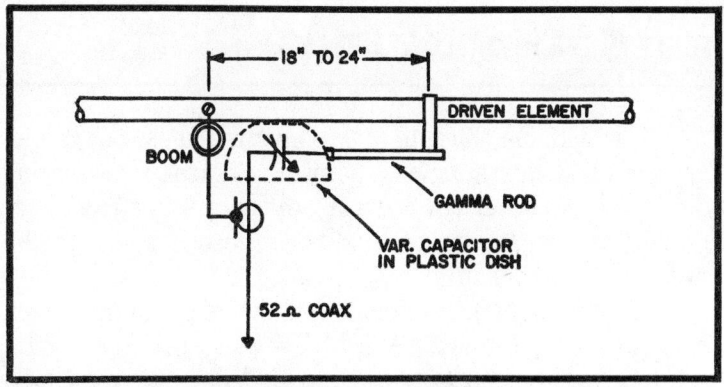

Fig. 4-13. Gamma match details for the four-element ten-meter beam.

formulae, measure the lengths exactly, and get the gamma match adjusted, you are very likely going to get just as much out of the antenna as if you spend a month fooling with it. It's your choice—the methods are detailed in the various handbooks. Personally I don't think it's worth the effort.

The tower and rotating system are up to the individual. However, the light weight construction should allow the use of a TV type tower and rotator. This beam will give the low or medium power operator many hours of fine contacts and provides a *kilowatt* type signal at a small fraction of the cost. Welcome to 10 meters.

FORTY METER INVERTED VEE BEAM

This antenna consists of two inverted vees, one driven and the other operating as a parasitic element. The spacing depends on whether you are using the parasitic as a reflector or director. It can be both with clip leads to change the direction of the beam.

When building the antenna, find the length of the driven element by the formula 492/FMC = Length. For making a reflector, add 5 percent to the length. A director would be 4 percent shorter. Space the elements at .15 λ, or about 18 feet. One important thing to remember when putting up the antenna is to keep the apexes of the inverted vees at the same height. This is done by using a boom or by using two equal length poles.

Results should be very satisfying, and this type of antenna can also be constructed for 80 meters.

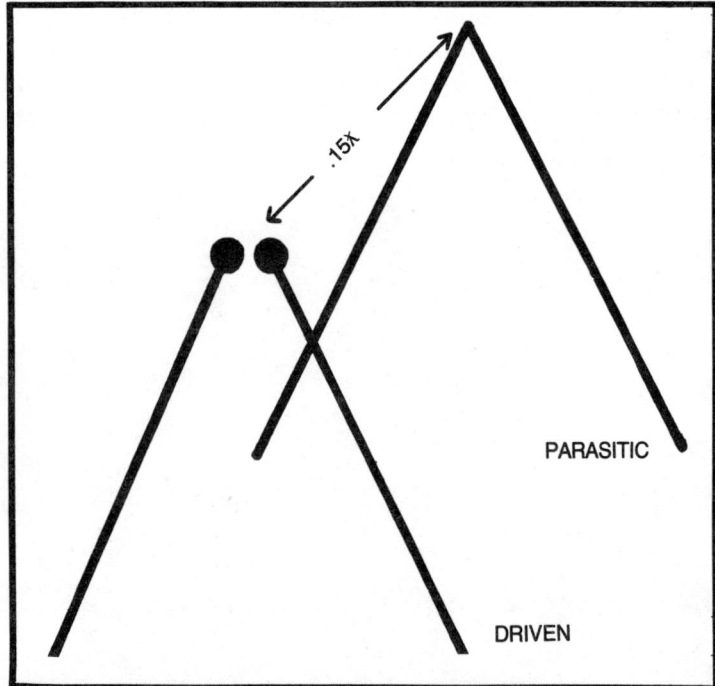

Fig. 4-14. Two inverted vees make a 40-meter beam.

AN 80 METER PHASED ARRAY

This 80 meter antenna will offer advantages over a simple dipole and will also include sufficient flexibility to permit direct experimental comparison of a number of antenna configurations which are of interest.

Approach

Consideration of space limitations (2/3 acre) and other practical constraints led to the choice of two parallel dipoles as the basic elements of the array. Since it was desired to switch to a unidirectional pattern and also to control the angle of maximum radiation, direct feed, rather than a parasitic array, was chosen.

Reference to the radiation patterns in the handbooks shows that a uni-directional cardioid (heart-shaped) pattern can be obtained in an end-fire array of two parallel elements, with a spacing of $\lambda/4$ and fed with a 90° phase difference. The radiation pattern in this case is a reversible cardioidal pattern with maximum gain in the direction of the lagging dipole element. This cardioidal arrangement was chosen as the basic horizontal directional array with other related options available by switching.

It is of interest to provide, in addition to the reversible cardioid, a 45° lag (higher radiation angle), a 0° lag (highest radiation angle—90°), and 180° lag (8 JK configuration—low angle, bi-directional) and, for comparison purposes, each of the two dipoles separately, This is a total of eight different pattern options.

Circuitry

The entire circuitry for the horizontal and vertical phased arrays is shown in the schematic diagram, Fig. 4-15. Instantaneous switching from one pattern to another is achieved by only three switches: a main selector switch S; the reversing switch X, which permits 180° phase reversal; and the 4PDT switch for changing between the horizontal and the vertical arrays.

For purposes of description, the system will be treated under the following headings: The Horizontal Dipoles; the Verticals; Impedance Matching; and the Switching Manifold.

Horizontal Dipoles

The original installation utilized two dipoles. The centers wire 46 feet above the ground with a horizontal spacing of 61 feet. The RG-8/U feedlines, one wavelength long, were inside the masts with the balun action and lightning protection as previously described. This original arrangement gave very good operation.

However, since it was desirable to have the lowest possible angle of radiation the centers of the two dipoles were raised to 61 feet ($\lambda/4$). This was accomplished by lengthening each steel mast by the addition of a thirty foot length of three inch diameter aluminum irrigation pipe at the bottom end of the mast. The steel mast is inside this pipe and the overlapping portion is bolted securely by use of ¼ inch plated bolts through the pipe and mast in perpendicular pairs. (No. 8 self-tapping screws in the steel mast served to space the mast within the pipe radially before the bolts were put in place.)

No data could be taken for comparison of these two heights but it is assumed that the 61 foot height yields a somewhat lower angle of radiation for each pattern option.

Verticals

The sixty-one foot masts are fed as top-loaded verticals. The horizontal dipoles are connected to the top of each mast and the two halves of each dipole are connected together by shorting the opposite end of the 1λ feedline.

Referring to the diagram, all of these connections are switched by means of the 4 PDT switch. This permits the selection of all of the vertical phasing options by the selector switch S just as for the horizontal system.

The resonating and impedance matching of the verticals is accomplished by the capacitors C and the inductors L. A noise bridge was used to insure adjustment to 52Ω resistive input at 3.955 MHz.

The 220 pF fixed mylar capacitor connected across the feedlines of the verticals during use of the horizontals serve

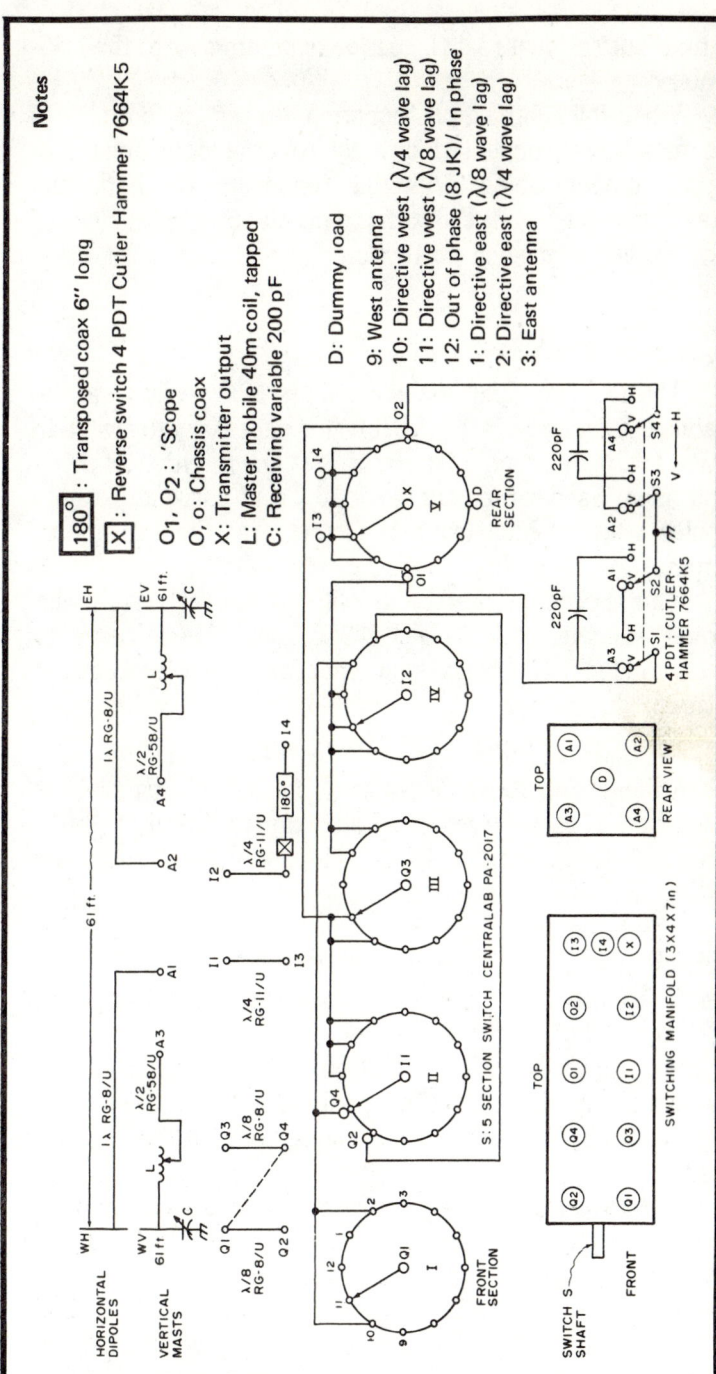

Fig. 4-15. Schematic diagram—80 meter phased array.

to tune out residual reactances for optimum swr to the horizontals.

When full lightning protection is desired the bottoms of the masts are connected directly to ground by means of copper jumper cables. With this connection the horizontal array can be used with DC paths to ground from both sides of each dipole, giving full protection against build-up of static charge.

Impedance Matching

The feedline input impedances are 52Ω resistive at the resonant frequency (3.955 MHz). It is necessary to switch-in phase lag by inserting a length of 52Ω line in either of these feed lines, as desired, and to feed equal currents to both dipoles while maintaining a 52Ω match at the transmitter output.

This is accomplished by use of two quarter wave transformer sections of RG-11/U (75Ω) coax. These serve to transform the 52Ω antenna input impedance up to 108Ω by the relation:

$$Zinput = Z^2 \text{ Line}/Zoutput$$

When these two 108Ω inputs are connected in parallel the resulting 54Ω value is well matched to the transmitter output.

Switching Manifold

The heart of the switching manifold is the 5-section 12-position switch, S. The current rating of this switch is sufficient to handle the full power as long as the transmitter power is removed before the switch position is changed. As the diagram shows, the system can be switched from the west antenna alone, at the 9 o'clock position, through the various angles of radiation to the east antenna alone, at the 3 o'clock position.

The reversing switch permits instantaneous switching of patterns, for example, from east to west, without having to turn the selector through the intermediate positions.

Only four of the twelve switch positions are not used: 4, 5, 7 and 8 o'clock. The six o'clock position is used for a dummy load.

The switches are mounted in the 3 × 4 × 7 inch aluminum chassis box with the sixteen coax sockets as shown. The box is mounted under a projecting top of the operating desk. The four lengths of coax used for matching and delay lines are wound on a wooden reel and placed inconspicuously behind the desk.

The connectors 01 and 02 provide inputs to the vertical and horizontal plates of an oscilloscope for a Lissajous display of the inputs to the two antennas. Thus, the phasing and the amplitude of the rf voltages can be continuously monitored, allowing any change in either antenna to be immediately noticed.

The scope shows a circle for the cardioidal patterns, diagonal lines for in-phase or out-of-phase, and a flattened ellipse for either antenna alone. (This pattern is elliptical rather than a straight line due to the rf energy picked up by the non-energized antenna.)

Performance

The performance of the array is excellent both for transmission and for reception.

The swr is consistently low (under 1½:1) for all configurations. The array shows a broadband behavior typical of coupled resonant circuits. The swr remains low throughout a bandwidth of some 400 kHz—only the phasing varies.

The measured front-to-back ratio is of the order of 15 dB and the gain is about 4 dB, for both transmission and reception. The improved operation for low angles of radiation is sometimes spectacular.

One of the most pronounced characteristics noted has been the great reduction of QRM for reception. The combination of the high front-to-back ratio and the low angle of radiation serves to reduce the level of some signal strengths while increasing the level of others. Thus, there is often, at nighttime, a sort of single-signal performance which is very gratifying. This single-signal selectivity of the antenna system is particularly impressive when the station being worked also has a low-angle directive antenna system. In this case the directivities complement each other with spectacularly strong signals at either end.

This array will be a great boon to stations located on the coasts as the 3 dB of power wasted out over the water could be largely utilized.

The only disappointment thus far has been the consistent weakness of signals from the vertical antennas. The separate vertical antennas are typically down about 10 dB compared with the horizontals and this inferiority carries over to the vertical array, regardless of direction or distance. The poor performance of the verticals is attributed to ground losses, with attendant high radiation angles, in spite of the fact that a parallel grid of about 3500 feet of ground wire is used.

Conclusions

A two-element horizontal phased array for 80 meters has been constructed with a total of eight pattern options available by direct switching. Operating results have confirmed the expected gains and front-to-back ratios. The performance of the unidirectional cardioidal patterns has been particularly effective, especially when the station being worked also has a directive antenna system.

TWINLEAD PHASED ARRAY

The ideal answer to many antenna problems would be directive, rotatable, have some front to back ratio, and probably most important, would be inexpensive.

The phased array to be described here is both simple and effective. The use of commonly available materials and "electric rotation" beats the high cost of rotors and the unavailability of some antenna components. While the antenna described was designed for use on forty and fifteen meters, there is no reason why this general design cannot be used on any of the other bands.

Theory

The array consists of a pair of folded dipoles fed 90° out of phase to provide end-fire directivity. This phase-shift is caused by an electrical quarter wavelength of 300Ω twinlead. By using a DPDT relay to switch the phasing line, bidirectional operation can be obtained. Each folded dipole has a characteristic impedance of 300Ω. Ideally, the two elements are fed at the center with electrical half-wavelength lines of twinlead. This brings the 300Ω resistive load present to the switching relay without inducing any reactive components. Of course if the length specified is too long for a particular installation, try whatever length is needed as long as both feedlines are of equal length. If more length is needed, use any integral multiple of the lengths given. At the relay the two 300Ω impedances are presented in parallel, transforming the impedance to 150Ω. This is fed through a 4 to 1 balun to bring the final impedance down to 35Ω. When fed with 50Ω coax (RG-8, RG-58) it will present an excellent standing wave ratio (SWR) across the entire band. This broadness is due to the inherent wide bandwidth of the folded dipole itself, plus the use of the 4 to 1 balun, which, besides bringing down the impedance, also reduces the change in impedance as the frequency is varied.

Construction

The elements, feedlines, and phasing line are all made of 300Ω twinlead. It is best to weatherproof the relay by mount-

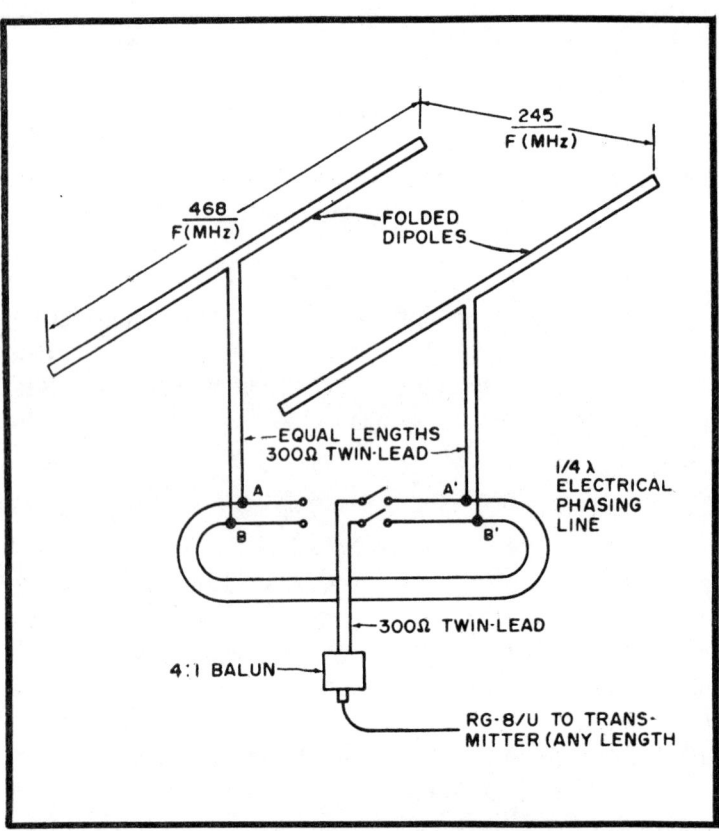

Fig. 4-16. Two element phased array. If SWR is unacceptable, interchange lead A with B or A' with B'. With the relay in this position directivity is to the left.

ing it in a plastic refrigerator box and then mounting the box on a tree or building. From the relay, zipcord can be run to a control console in the shack. It will be best to wire the relay so that in its unenergized positon the directivity will be in the most often used direction. This will cut down on wear and tear. If the SWR of the finished array is unacceptable, possible remedies include transposing the A and B leads (see Fig. 4-16) of the feedline, or changing the length of the line from the relay to the balun. The fact that changing the line length varies the SWR, shows that the line is not "flat," but this is really inconsequential as the loss from high SWR in the twinlead is less than that of an equal length of perfectly matched RG-58.

The antenna can be supported by towers, buildings or four conveniently placed trees.

Frequency (kHz)	Element Length		Spacing		Phasing Line		Elec. ½ Wavelength	
	Ft	In.	Ft	In.	Ft	In.	Ft	In.
80 Meter Band								
3550	131.	9.97	69.	0.17	56.	9.87	113.	7.74
3650	128.	2.63	67.	1.48	55.	3.19	110.	6.38
3850	121.	6.70	63.	7.64	52.	4.74	104.	9.47
3950	118.	5.77	62.	0.30	51.	0.82	102.	1.64
40 Meter Band								
7100	65.	10.99	34.	6.08	28.	4.94	56.	9.87
7250	64.	6.62	33.	9.52	27.	9.88	55.	7.76
20 Meter Band								
14100	33.	2.30	17.	4.51	14.	3.68	28.	7.35
14275	32.	9.41	17.	1.95	14.	1.57		3.14
15 Meter Band								
21100	22.	2.16	11.	7.34	9.	6.72	19.	1.44
21350	21.	11.0	11	5.70	9.	5.38	18.	10.76
10 Meter Band								
18050	16.	8.21	8.	8.81	7.	2.30	14.	4.59
18600	16.	4.36	8.	6.80	7.	0.64	14.	1.28

Fig. 4-17. A simple directive phased array chart.

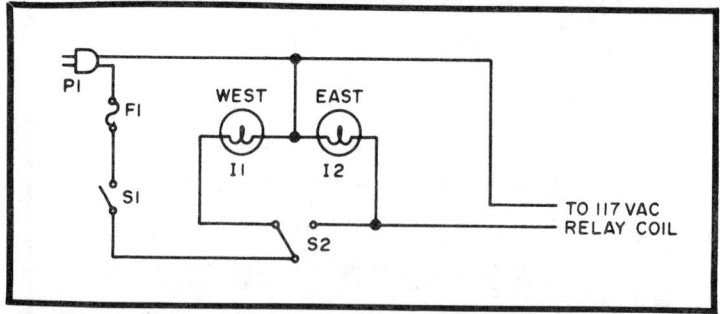

Fig. 4-18. This is a schematic of a control console that should be used where the most used direction is to the west. To use a lower voltage relay just put a transformer on the 117V AC output. F1—2 amp fuse; I1, I2—117V AC light; P1—standard power plus; S1—SPST switch (turns unit on or off); S2—SPDT switch (selects direciton).

Performance

This type of antenna cannot be expected to beat out a full size mono-band beam in terms of gain, but it can be very effective, especially on the lower bands where yagis are prohibitively large for most installations. On 40 meters, the front-to-back ratio is sufficient to cut down broadcast interference on phone and stateside QRM while working CW into Europe. It varies from about 20 dB on 7.1 MHz to about 7 dB on 7.3 MHz. The gain appears to be 3 to 5 dB with respect to a dipole. By cutting the antenna for any particular frequency, performance will be optimized for that frequency.

From the performance of the 40 meter array, similar results can be expected on the other bands. All in all, this is a very effective antenna system, in both performance and cost.

THE DIAMOND ARRAY

This efficient little antenna features several popular designs all in one. It is one element of a 40 meter quad-one full wave length in a diamond or square, two upright Vee antennas fed in phase, an upright Vee derived from the design of the familiar coax dipole, and it has a gain of 3½ or 4 dB over a standard dipole. It is rather broad in frequency response, and non-directional.

The antenna can be oriented 90 degrees, that is, fed from one of the high angles, with no noticeable change in performance. However, by using one of the low angles for the feedpoint, it may be possible to keep the feedline away from the field of the antenna, and also even use a more direct feed to the rig.

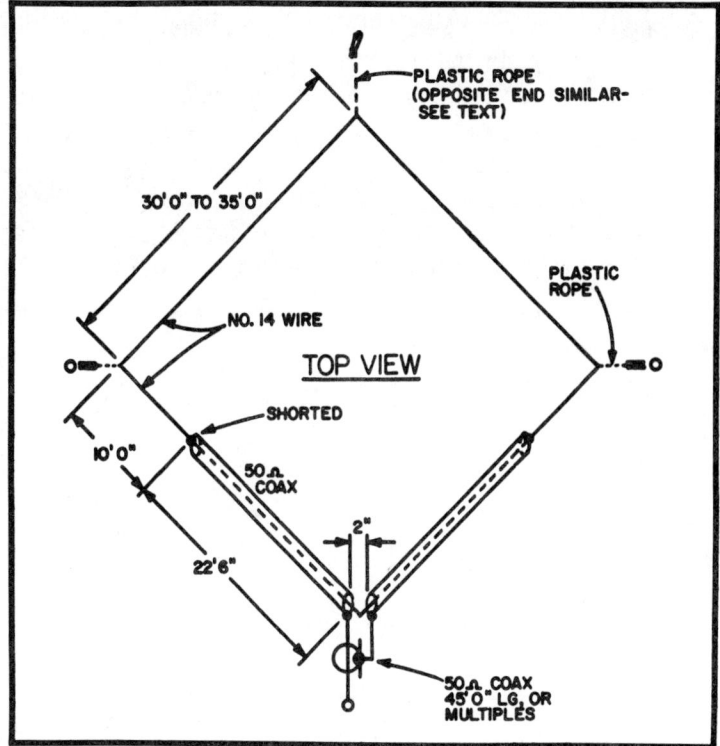

Fig. 4-19. The diamond array.

Plastic or glass lines are used for support and are run thru pulleys for ease of erection and tuning. The lengths of the single wire on the opposite Vee from the coax fed half of the array is not critical, but the actual tuning for best SWR and frequency of the antenna is done with the 10 foot ends of this first coax half.

The two low opposite angles are pulled down and out to supports at each side of the lot, resulting in approximately 20 feet above ground for these angles. The opposite high angles of the diamond are around 45 to 50 feet high.

The far end of each half of the dipole coax section is shorted, with the single wire extending around the diamond from these points. The 50 ohm coax shielding only is opened up for 2 to 3 inches and the feedline connected at each section of shielding. This folded dipole effect gives the design a good flat SWR throughout the band. To reduce the strain at the feedpoint, a short bridle or yoke of plastic or glass line is wrapped around each side of the coax and tightly taped with the tie-line brought out from this spot.

In tuning for the best SWR or for the best center frequency, always use an SWR meter with an exciter for low power—100 watts is ample.

If the low frequency end of the band shows the best SWR, or it results in an increasingly better reading although still far from 2 to 1 or better, the single wire section is too long, and a

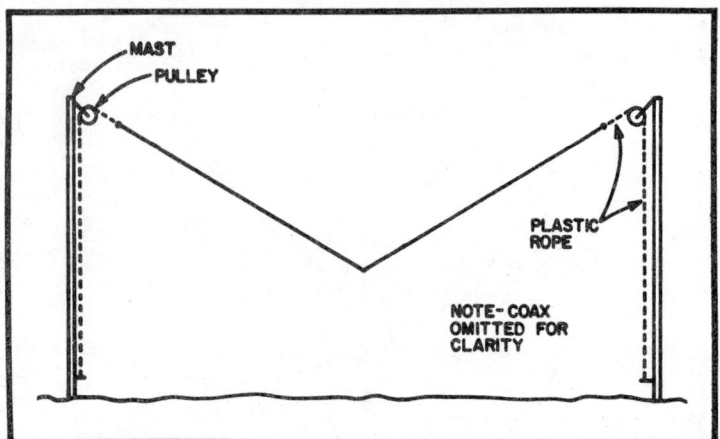

Fig. 4-20. Antenna supports.

foot should be cut off from each end of the 10 ft. extensions. The SWR reading then should be checked and if improvement is noted—continue cutting and testing until satisfaction is reached or the tuning is correct.

Now, should the SWR show improvement by tuning towards the high end of the frequency of the 40 meter band, and still be in excessive of, the antenna is too short. At least a foot or more should be added to each single wire end for the next check point on the SWR.

No balun is necessary, just keep the coax feed in the clear and away from grounds.

The SWR figures were as follows:

    7.3 mc—2 to 1
    7.2 mc—1.25 to 1
    7.1 mc—1.05 to 1
    7.1 mc—1.05 to 1
    7.0 mc—1.22 to 1

MULTIBAND LOG PERIODICS

Through the use of computer-aided "mathless" LPD design, three antennas for use in the amateur bands have been developed. The dimensions for the three are given in Table 4-3. All three antennas exhibit a forward gain of 13.5 dB with a front-to-back ratio of better than 15 dB over the specified frequency range. The swr is better than 1.8:1 over the specified frequencies.

The first antenna covers the range of 21 to 55 MHz; the second antenna covers 50 to 150 MHz; and the third covers 140 to 450 MHz. These antennas are designed with a 5% frequency overshoot at the low end and a 45% overshoot at the high-frequency end to maintain logarithmic response over the complete frequency range specified. In log periodic antenna operation, approximately four elements are active at any one specific frequency, thus the necessity for the low- and high-frequency extensions. All three antennas are designed for a feedline impedance of 50Ω for use with coax such as RG-8/U. All of the antennas are design-rated for 1 kW, 100% modulated. The alpha, or logarithmic element taper, is 28° for all three antennas.

Construction

Construction is straightforward, and various means may be used as far as fastening the elements to the boom, and in the choice of dielectric spacer configurations.

Heliarc welding can be used for securing the elements and fiberglass may be used for the dielectric.

Element lengths for the highest frequency antenna were calculated for the elements to be inserted completely through the boom, flush with the far wall. The two lower frequency antennas have element lengths calculated to butt flush against the element side of the boom. If the elements are to be inserted through the boom on these other two (21—55, 50—150 MHz), add the boom diameter to each element length shown before cutting the elements.

Table 4-3. Spacing and Dimensions for Log Periodic VHF Antennas.

Element	21–55 MHz Array			50–150 MHz Array			140–450 MHz Array		
	Length, ft	Dia, in.	Spacing, ft	Length, ft	Dia, in.	Spacing, ft	Length, ft	Dia, in.	Spacing, ft
1	12.240	1.50	3.444	5.256	1.00	2.066	1.755	0.25	0.738
2	11.190	1.25	3.099	4.739	1.00	1.860	1.570	0.25	0.664
3	10.083	1.25	2.789	4.274	1.00	1.674	1.304	0.25	0.598
4	9.087	1.25	2.510	3.856	0.75	1.506	1.255	0.25	0.538
5	8.190	1.25	2.259	3.479	0.75	1.356	1.120	0.25	0.484
6	7.383	1.00	2.033	3.140	0.75	1.220	.999	0.25	0.436
7	6.657	1.00	1.830	2.835	0.75	1.098	.890	0.25	0.392
8	6.003	0.75	1.647	2.561	0.50	0.988	.792	0.25	0.353
9	5.414	0.75	1.482	2.313	0.50	0.889	.704	0.25	0.318
10	4.885	0.75	1.334	2.091	0.50	0.800	.624	0.25	0.286
11	4.409	0.75	1.200	1.891	0.50	0.720	.553	0.25	0.257
12	3.980	0.50	1.080	1.711	0.375	0.648	.489	0.25	0.231
13	3.593	0.50	0.000	1.549	0.375	0.584	.431	0.25	0.208
14				1.403	0.375	0.525	.378	0.25	0.187
15				1.272	0.375	0.000	.332	0.25	0.169
16							.290	0.25	0.000
Boom	25.0	2.0	0.5	16.17	1.5	0.5	5.98	1.5	0.5

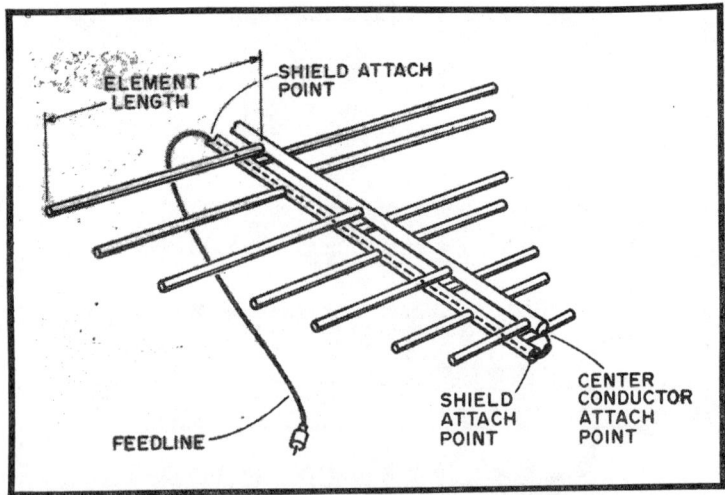

Fig. 4-21. Typical log periodic antenna. Note that the bottom is fed from the coax shield while the top boom is fed from the center conductor.

Two booms must be constructed for each antenna as shown in the isometric view of Fig. 4-21. Also, in supporting a log periodic antenna from a metal mast, the two booms must have a dielectric spacing from the mast of at least twice the

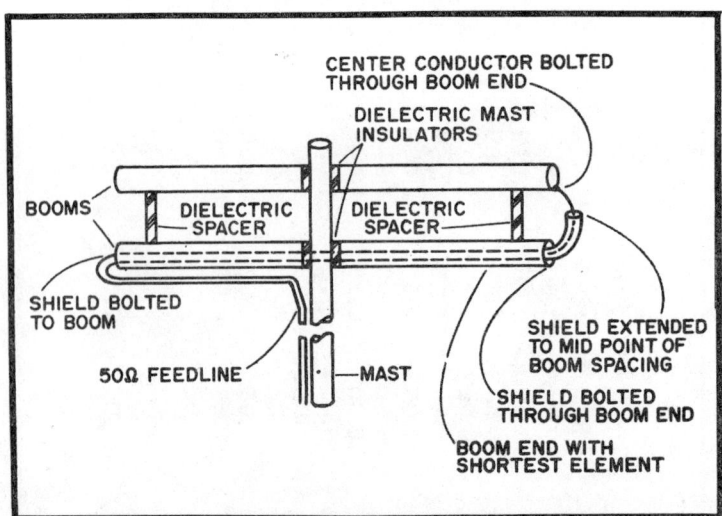

Fig. 4-22. Feeding the log periodic is relatively simple. Just remove the outer plastic jacket from feedline for the entire length of the boom, so that the coax shield is permitted to short itself inside the boom as well as the solid electrical connections at each end of the boom.

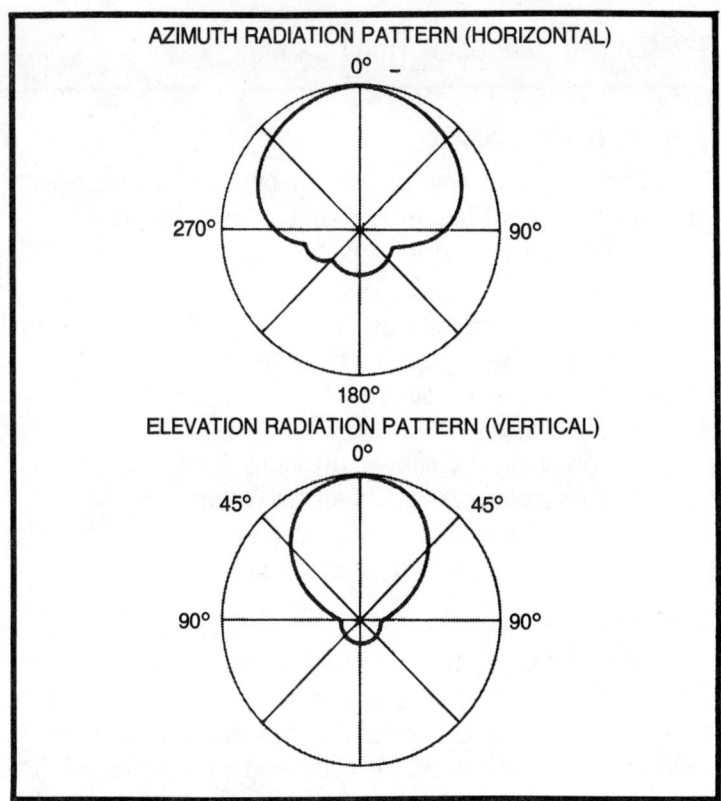

Fig. 4-23. Typical log periodic antenna patterns.

boom-to-boom spacing; otherwise, discontinuities will be introduced into the feed system.

Feedline insertion and connection are shown in Fig. 4-22.

Notes

Any change in the listed boom diameters will necessitate a change in the boom-to-boom spacing in order to maintain the feed impedance at 50Ω.

The location of the support mast is not critical; ideally, it should be at the array's center of gravity.

The antennas may be oriented either horizontally or vertically, depending on the type of polarization desired. The horizontal beamwidth of a typical log periodic antenna is approximately 60°, while the vertical beamwidth is on the order of 100°. These are the −3 dB points shown in Fig. 4-23.

TRIANGULAR LOOP BEAM FOR 7-28 MHz

Antenna Design Criteria

The following criteria for an amateur antenna were traded off in evolution of the triangular-loop-beam:

- Operation on 7 thru 28 MHz bands
- Rotary beam to maximize effectiveness
- 3 to 5 dB gain on 7 MHz and 8-9 dB gain on 14, 21 and 28 MHz as compared to a dipole
- KW power capability
- High radiation efficiency
- Withstand 85 mph winds coupled with ice loads
- Turning radius of 17 feet maximum
- Minimum weight and cost commensurate with design capable of amateur construction.

Loop Beam vs Helix

The multiple loop beam antenna is merely a special case of the axial mode helix antenna in which the helix pitch is zero. The other extreme of the helix is a straight wire, when the helix is stretched out until its diameter becomes zero. The multiple loop parasitic antenna and the axial mode helix when both of one wavelength in circumference exhibit equivalent gain when the boom length is less than about ½ wavelength. For longer boom lengths the helix outperforms the parasitically driven antenna. For the less than $\lambda/2$ boom length, the greatest differences apparent between the two antennas are that the helix has a bandwidth of almost 1.7 to 1 of the design wavelength (much broader than the loop) and the loop has a better front to back ratio than the helix.

The helix requires a ground plane of 0.8 wavelength diameter behind it to be really effective. If one considers use of the helix for 14 MHz a ground plane 56 feet in diameter becomes a real structural problem for the amatuer. The parasitic loop beam antenna uses a reflecting loop instead of the ground plane and is somewhat easier to build.

Research has led to the conclusion that a four-element parasitic loop antenna using loops of one-wavelength in circumference and a boom length of 0.4 wavelength should turn out an honest 10.5 dB gain as compared to an isotropic radiator or about 8 dB more than a dipole. A boom length of 0.4 wavelength at 14 MHz was therefore adopted as meeting the design criteria. This boom length is 0.2 wavelength at 7 MHz and is satisfactory for a two-element folded dipole beam having a gain of about 3 dB over a dipole.

Element Apertures, Gain and Radiation Resistance

Many amateurs are aware that the gain of an antenna is proportional to its "capture area," (also called aperture, intercept area, or cross section). Apertures and gain of several element configurations are tabulated in Table 4-4 together with radiation resistance. In comparing antennas or antenna elements it is well to bear in mind that as the radiation resistance of the antenna increases, the power radiated to a distant point as opposed to the power stored as a space charge around the antenna increases. If one were to select a beam antenna element from only the data of Table 4-4, the 1½ wavelength loop would be the logical choice; however, the 1½ wavelength loop for 14 MHz on a 28 foot boom requires a clear turning area of 23 feet which is more space than many of us have available. The turning radius criteria of 17 feet incidentally resulted from consideration of space available on an average metropolitan area lot to swing a beam without invading neighboring air space or encountering obstructions when working with it on the tower.

Square vs Triangular Loop

The question arises: How do the triangular and square loops compare in performance? Table 4-4 shows that the triangular loop of one wavelength periphery has 96.5 per cent the gain of the square loop. Comparison of the patterns and gain of the two loops on near the DX signal reception over a six month period of time revealed the following information. The triangular loop when oriented with one triangle apex down and fed at the lower apex (horizontal polarization) has two major lobes concentric with the loop axis in the horizontal plane and broadside to the loop plane. Since it is a single loop, it radiated in two directions like a dipole and it has two main lobes in each of these directions about 20° off the loop axis. When oriented with one apex of the loop-up and fed either at the to apex or the center of the lower horizontal leg of the triangle (horizontal polarization) one broad lobe can be detected at right angles to the loop plane in two directions. This pattern is similar to a dipole.

The square loop (horizontally polarized) exhibited two lobes in the horizontal pattern about like the triangular loop oriented with one apex down. Stations about 10 miles away could be completely nulled with either the single, square or triangular loop, although the triangular loop seemed to be slightly better than the square loop in front to side ratio and also slightly broader in pattern than the square loop. One other point of interest was noticed; the triangular loop is better than the square loop on QSB when oriented with one apex straight up.

The diamond configuration square loop and the triangular loop were also mounted on a tilting fixture and gain was measured at various angles of inclination of the plane of the loop to the vertical. The tilting operation originated from a desire to see if the one wavelength diamond loop really acted like a rhombic as has been hypothesized in some diamond quad articles. It does not from the measurements taken. The effect of tilting up or down from an initial position with the loop vertical is to decrease the low angle radiation of the antenna because the horizontally radiating aperture is decreased. Also, the one wavelength diamond loop is not a uniform traveling wave antenna like the rhombic or the helix; it is simply a resonant, standing wave radiator and its gain over a dipole results from the larger aperture of the one-wavelength of wire (see Table 4-4).

Wind Loads

The primary structural load on the multiloop beam is wind force. For practical design purposes the wind force on an antenna or tower is given by:

$$F = (v)^2 C_d A/391$$

Where: F = wind force on structure in lbs

Table 4-4. Antenna Parameters.

Antenna Element	Effective Aperture (Square Wavelengths)	Directivity *	Gain Over Isotropic Source (DB.)	Radiation Resistance (ohms)
Isotropic Source	0.08	1.0	0	
1/2 Wavelength Linear	0.13	1.64	2.15	73
1 Wavelength Linear	0.142	1.8	2.55	93
1 1/2 Wavelength Linear	0.158	2.0	3.0	106
1/2 Wavelength (Open End) Folded Triangularly	0.126	1.59	2.0	75
1 Wavelength Triangular Loop	0.145	1.83	2.63	140
1 1/2 Wavelength (Open End) Folded Triangularly	0.2	2.51	4.0	110
1 Wavelength Square Loop	0.147	1.86	2.7	140

Directivity = Maximum effective aperture divided by maximum effective aperture of isotropic source. An isotropic source is one which radiates power equally in all directions.

v = wind speed in mph
C_d = drag coefficient which should be taken as 1.7 for amateur antennas or towers
A = Area in square feet of antenna in a vertical plane (that is at right angles to a horizontal wind force)

Using an 85 mph wind criteria for the antenna yields a force per square foot of vertically disposed antenna area of:

$$F = (85)^2 (1.7) (1)/391 = 31.5 \text{ \#/sq. ft.}$$

Wind loads for various wind velocities are tabulated on Table 4-7.

Antenna Boom Design

Three commonly used designs for beam antenna booms are (in their order of increasing complexity of construction): (1) the self supporting type fabricated from 6061-T6 aluminum or mild steel tubing; (2) tubing strengthened with outrigger tension members consisting of solid rod, steel cable or nylon rope and (3) the truss. As an example of strength of the tubing boom and of the wind forces which are exerted on the loop antenna, a two-element quad on a 10 foot boom of 1.5 diameter × 0.058 wall steel tubing is stressed to the bending point of the material (elastic limit) in a steady state wind of 60 mph (calculated).

In section 3.0 it was stated that a 0.4 wavelength long boom is required for a four-element loop antenna to achieve the 8.0 dB gain stated in the design criteria of section 2.0. At 14 MHz the 0.4 wavelength is 28 feet. Applying the wind load of 31.5 lbs/sq ft for an 85 mph wind to design of a 28 foot long steel tubing boom to support four square loops reveals that a 4 inch O.D. × 0.134 wall is required and that the antenna will weigh 190 lbs. While the tubing boom is simple to construct, 190 lbs weight is excessive for many towers, including the author's home brew tilting tower. Truss construction is attractive as a means of reducing weight because the truss places direct axial tension and compression loads on the framework members (elimination of bending loads) and thereby achieves a maximum strength to weight ratio. If one uses the triangular loop to decrease loop weight by 25 per cent over a square loop and loop wind forces by 30 percent over a square loop, a structure such as shown in Fig. 4-24 is required

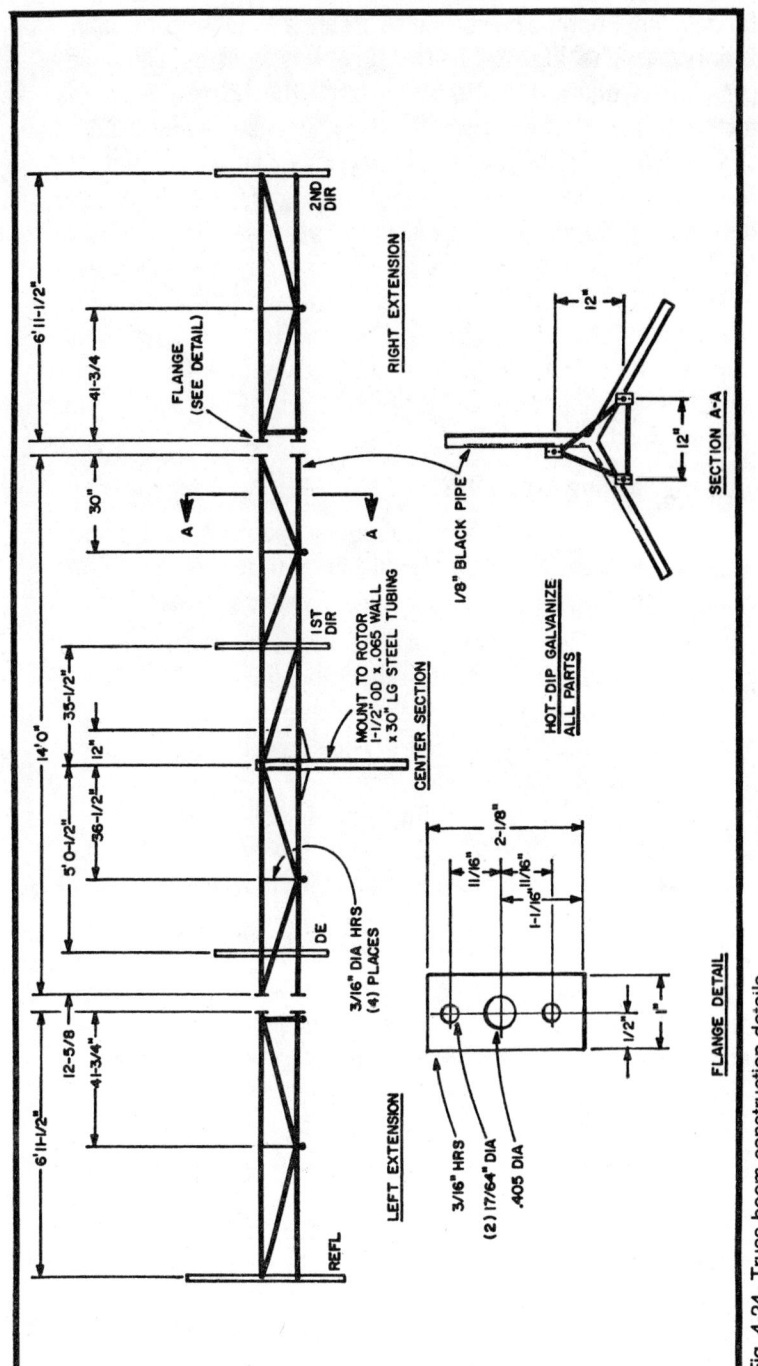

Fig. 4-24. Truss boom construction details.

for a 28 foot boom. The complete antenna weighs 65 lbs. using this truss. The Fig. 4-24 structure was built as a first approach. It required 70 hours to build the boom which was forthwith completely ruined by the galvanizer when handled with a bundle of heavy tower sections. Time was not immediately available to make a second truss, therefore, the outrigger construction was utilized at a sacrifice in boom weight. The design shown in Fig. 4-25 uses three outriggers attached to the boom three feet from the end so that the boom carries a combined bending and column (compression) load. This boom required 30 hours to build and the resultant antenna weighs 77 lbs. It is designed for 85 mph wind loads and an 80 lb total ice load.

Element Spacing Trade-Offs

Any multi-band beam represents a compromise between element spacings for the various bands in terms of antenna gain and bandwidth. The basic trade-off factors are as follows:

- An element spacing of 0.12 yields maximum gain for up to three elements on the beam.
- With a fourth element added a spacing of 0.3 wavelength between first and second directors seems to yield optimum gain.
- As the element spacing is increased gradually over a practical range from 0.12 to 0.2 wavelength the gain drops and the antenna bandwidth increases.
- Decreasing director spacing and increasing reflector spacing from the 0.12 wavelengths optimum will reduce gain and increase front to back ratio.

Table 4-5 shows the element spacing used for a 7, 14, 21 and 28 MHz band *compromise*; two elements on 7 and four elements on the other bands.

Table 4-5. Element Spacing.

	Spacing for Maximum Gain (Wavelengths)	Actual Spacing in Wavelengths			
		7.15 MHZ $\lambda = 137.5'$	14.17 MHZ $\lambda = 69.4'$	21.25 MHZ $\lambda = 46.2'$	28.7 MHZ $\lambda = 34.2'$
Reflector to Driven Element	0.118	0.204	0.118	0.177	0.238
Driven to First Director	0.12		0.12	0.18	0.243
First Director to Second Director	0.3		0.167	0.25	0.338

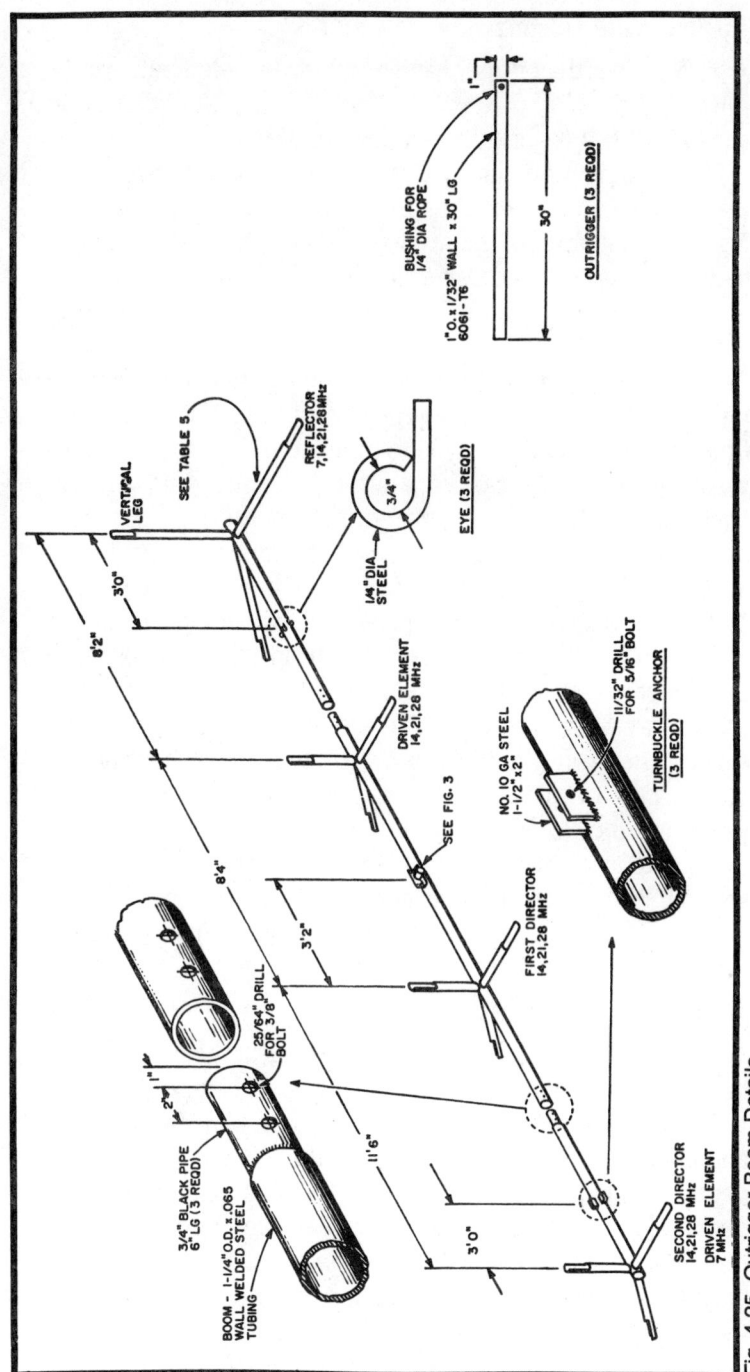

Fig. 4-25. Outrigger Boom Details.

Element Wire Lengths

Wire lengths of one wavelength driven elements can be calculated from $1 = 11800/f$ where $1 =$ length of wire in inches and $f =$ resonant frequency in MHz.

Sufficient bandwidth for the amateur bands covered by the antenna is obtained when the reflector wire length is made 5 percent longer than the driven element; the first director is made 2.5 percent shorter than the driven element and the second director is made 2.5 percent shorter than the first director.

Table 4-6 shows wire lengths and frequencies used by the author. Wire lengths were calculated and strung on the frames with no attempt made to tune them since the loop at its resonant frequency has very little inductance and any extraneous capacitance, introduced from measuring equipment or the human body, is sufficient to throw it off frequency. The accuracy of the wire length formula had been checked previously in building the single loops and the wire length variations for reflector and directors resulted from tabulation of much data by others. The two 7 MHz elements are not loops; they are ½ wavelength wires folded into an equilateral triangular shape. The 7 MHz antenna uses the two end spiders on the boom. The 14 and 21 MHz antennas use one-wavelength loops and the 28 MHz antenna uses 1½ wavelength wires folded into an equilateral triangular shape with the upper ends separated. A 6 inch spreader is needed between legs of the 7 and 14 MHz wires on the two end elements to keep them separated; ¼ inch diameter lucite works well.

Feed Point Impedances

All driven loops of the antenna are fed at the center of the bottom, horizontal wire. Dependent upon height of the antenna and proximity to surrounding objects, impedances of the antenna will be found to be close to the following: 7 MHz-45 ohms; 14 MHz-50 ohms; 21 MHz-80 ohms and 28 MHz-55 ohms. Many methods of feed have been published and will not be repeated here. One fact is very pertinent concerning feeding loop antennas; that is, in relation to nearby sources of RF interference the loop will respond only to the magnetic component of the interfering field *if it is balanced*. (That is, it will not respond to the electrostatic field and will therefore pick up less interference with balanced feed.) The feed system consists of a double shielded 125 ohm twin lead coax from the transmitter to an antenna switch at the top of the mast. The switch completely isolates those antennas not in use. The 75 ohm, twin lead ¼ wavelength lines (not shielded) run from the antenna selector switch to the 7, 14 and 28 MHz antennas. A ¼ wavelength line from the switch to the 21 MHz driven loop is formed from three pieces of 300 ohm TV lead in parallel which yields a 100 ohm section. Lengths of the ¼ wavelength matching sections from the antennas to the switch are: 7 MHz-24.75 ft; 14 MHz-12.32 ft; 21 MHz-8.21 ft; 28 MHz-6.08 ft.

Table 4-6. Wire Lengths in Inches.

Element	Frequency-Band			
	7.15 MHZ Dipole	14.17 MHZ Loop	21.25 MHZ Loop	28.7 MHZ $\frac{3\lambda}{2}$ Folded
Reflector	846	883	582	635
Driven	806	833	555	602
1st Dir		812	541	586
2nd Dir		792	528	572

The above method of antenna feed results in close matching across the bands, a low SWR and the feed to the antenna is balanced for low noise reception of DX signals. If an antenna switch is used, it is important that it switch both sides of the transmission line completely isolating the driven elements not in use. Switching of one wire, such as the center conductor of unbalanced coax with all of the shields of the coax antenna feed lines remaining connected, results in degraded performance over complete isolation.

Construction Notes

Both the truss boom of Fig. 4-24 and the outrigger boom of Fig. 4-25 are constructed in three pieces for ease of handling, galvanizing and assembly of the wire on the frames. The center post used can be any size suitable to match your rotor or extension mast. The 1½ O.D. × 0.065 wall low carbon tubing shown in Fig. 4-24 is only strong enough to extend six inches from the rotor and still meet the 85 mph wind load

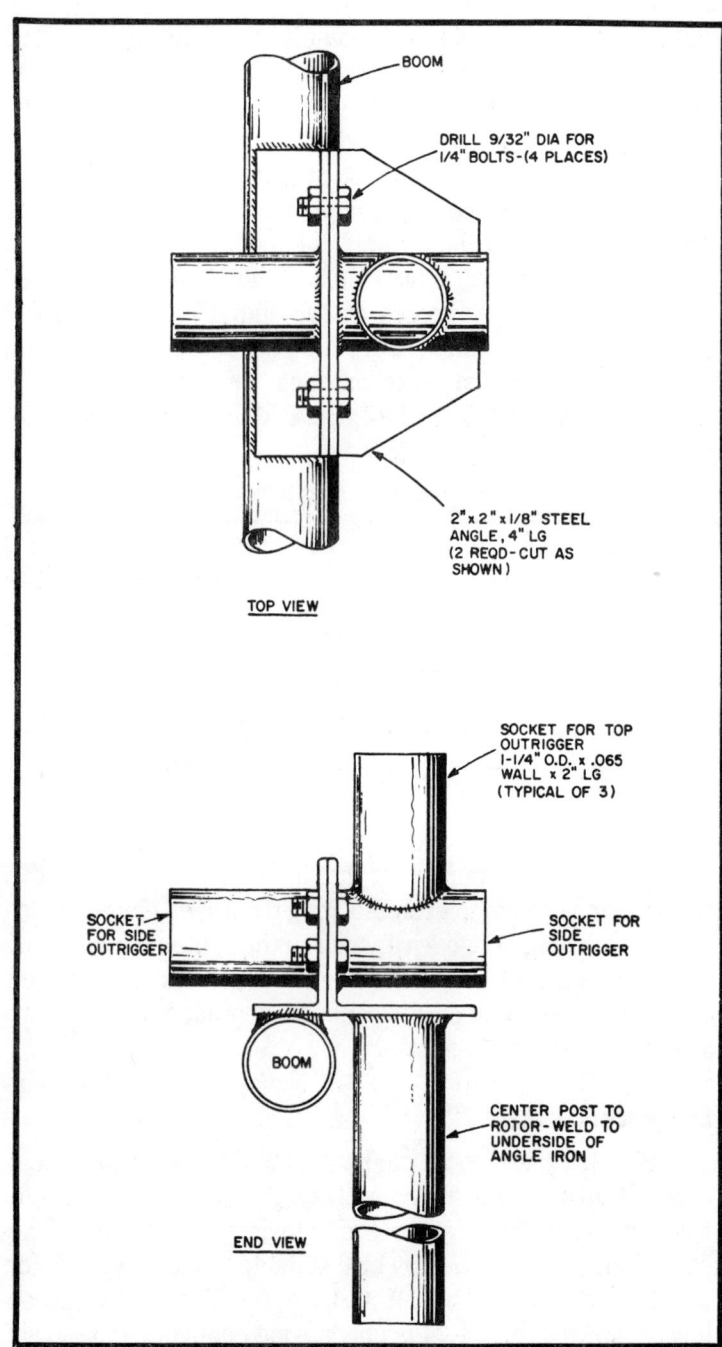

Fig. 4-26. Center post to boom construction.

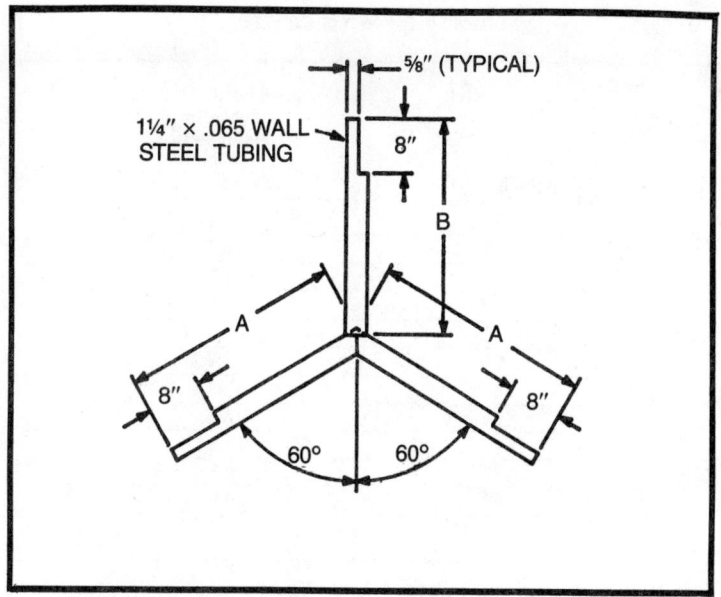

Fig. 4-27. Details of spider construction.

design criteria. The spider construction for both types of boom is shown in Figs. 4-27 & 4-28. The 1¼ O.D. × 0.065 wall tubing used for the spider is cut back for 8 inches along the tubing center line to receive the fiber glass arms which are fastened in place with two hose clamps per arm.

The detail of the center post to boom construction of the outrigger boom is shown in Fig. 4-26. The outriggers are 30 inches of 6061-T6 aluminum as shown in Fig. 4-25. The outriggers fit loosely in the sockets of Fig. 4-26 so that they will not be loaded eccentrically. A ¼ inch diameter nylon rope is used for tension members and it slides through the bushings in the outriggers. The bushings can be made of nylon, or they

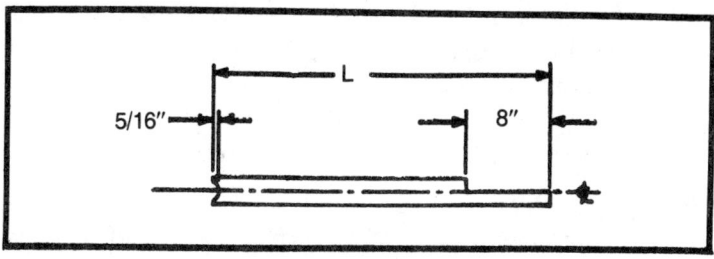

Fig. 4-28. Spider detail.

289

Table 4-7. Power of the Wind.

Wind Velocity (MPH)	Horizontal Force on Antenna or Tower (Pounds psf)*
30	3.9
35	5.3
40	7.0
45	8.7
50	10.9
55	13.3
60	15.7
65	18.4
70	21.3
75	24.3
80	27.8
85	31.5

*Take area as largest cross section of member. For example, tubing cross section equals diameter x length.

can be any non-rusting material. One 5/16 inch turnbuckle is used in each tension member, positioned at one end of the boom for a tilting tower or the center of the boom for non-tilting towers. All stainless steel hardware was used except for the aluminum turnbuckles. Advantages of this antenna can be summarized as follows:

- It provides a four-band rotary beam capability.
- Directivity and discrimination against rear and side signals is excellent on all bands. Front-to-side ratio better than the quad and gain is equivalent.
- It is less susceptible to QSB than the square loop.
- Cost of arms is reduced along with antenna weight and wind load area over the square loop.
- Appearance is good.
- It will stay up.
- If fed with a balanced line, it is very quiet on reception.

Disadvantages are:

- It requires 1.3 feet more turning space (radius) than the square loop (with 28 foot boom).

- It is more sensitive to interference between ground and sky waves than a plain wire. (Also true of the square loop.)
- Hams will knock on your door and tell you that part of your quad has fallen off.

LONG WIRE ANTENNA

The radiation pattern of this simple single wire antenna will generally be as described in all antenna books. So, if it is a halfwave long, the maximum radiation will be at right angles to the wire. If it is a full wave in length, and center fed, it becomes a double Zepp and the maximum radiation is still at right angles to the wire. The antenna problem is how, by using one wire, to radiate East and West to cover the United States, and Northeast and Southeast to cover Europe and South America, with good efficiency. A long wire antenna for twenty meter operation would be the thing to cover Europe and South America. If it was made a wave and a half long it could be fed at the center current loop, be run North and South and have a fine East-West pattern when used for the forty and eighty meter bands. When excited with a twenty meter signal, on the East Coast, one of its main lobes will cover Europe, and another South America. Don't be concerned about its impedance due to the use of a tuned transmission line (TV twin lead) and an antenna coupler. Feeding it in the center will make adjustment of the coupler simple and broad enough to cover a large segment of each band without retuning. When it is used on twenty meters, a gain of .8 dB over a dipole and 3.8 dB over a vertical is realized. Also, its cone shaped pattern off the ends makes it less sensitive to height for low vertical radiation angles. This antenna was cut and strung North and South at a height of about 22 feet. Tests proved that it operated just as planned and out-performed vertical antennas on every occasion. It also makes a surprisingly neat appearance.

As shown in Fig. 4-29 the antenna is 50 feet 6 inches long on each side of the feed line. The twin lead can be any length, and seems to be lossless for all practical purposes. For parallel tuning on all three bands, it should be 73 feet long. Figure 4-29 also shows the construction details of the center insulator and feed line connector. It is made of circuit board material, preferably fiber glass, because of its strength. The three fillers and two outer plates are cemented together with epoxy to make sure it is sealed against the weather. The one hole in each

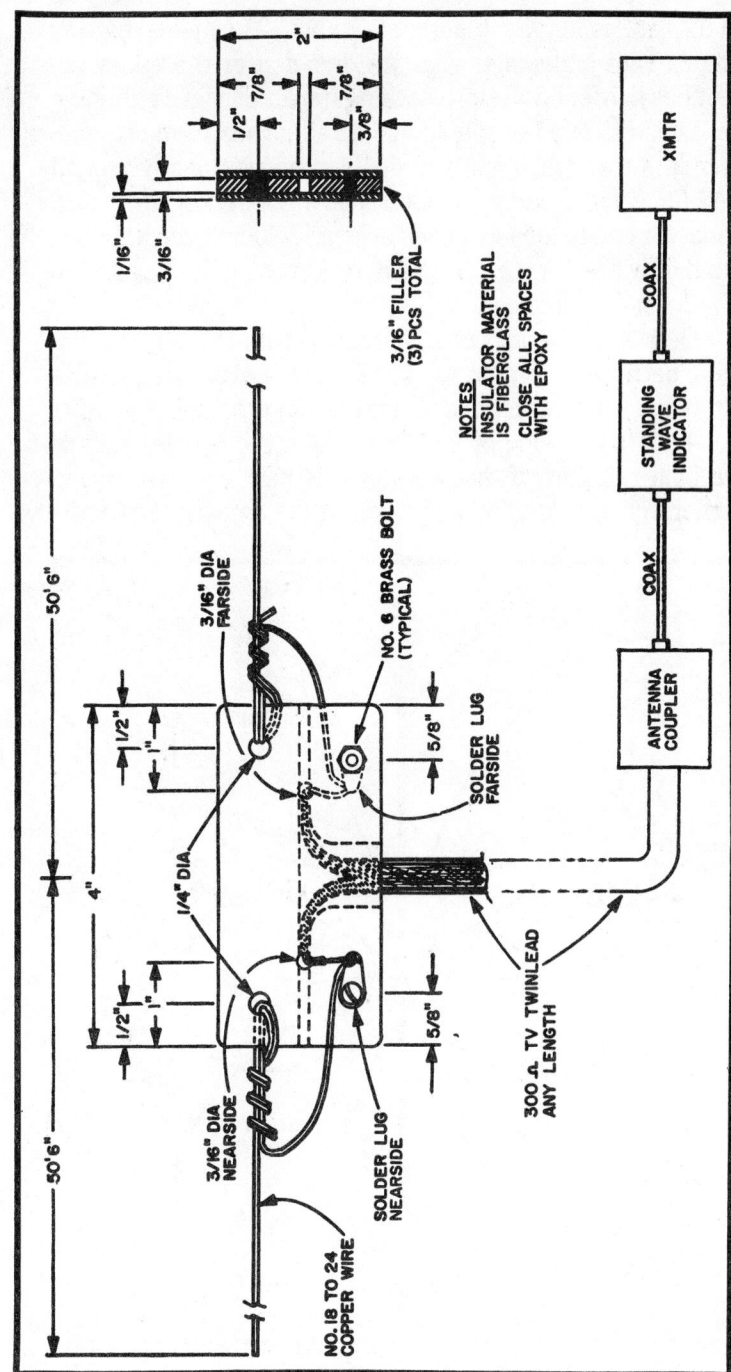

Fig. 4-29. Antenna dimensions and construction details.

outer plate is drilled before assembly. The holes for the antenna wire and solder lugs are drilled after the epoxy cement has hardened. Before passing the antenna wire through the insulator, bend it double about ten inches from the end. After it is through, wrap the doubled portion neatly around itself for about a inch. This will leave enough of the single conductor end to loosely bend back and solder to the lug along with the twin lead wire. The insulators at the extremes should be at least four inches long.

Details of the antenna coupler adjustments can not be given, because each type will have to be used according to its own operating instructions. Connect one terminal of a NE51 neon bulb to one side of the twin lead at the couplet output. The bulb will glow if the glass part is near the case of the coupler and give a good indication of the amount of rf at its

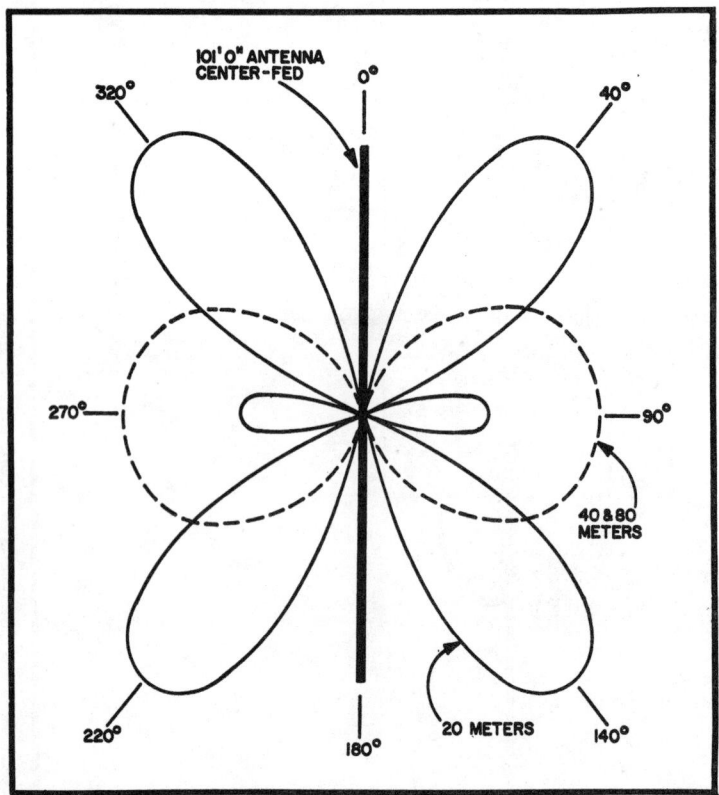

Fig. 4-30. Free space horizontal pattern.

output. Keeping one eye on the standing wave indicator and the other on the bulb will prevent adjustment to a false standing wave indicator null.

There is no reason why this antenna can not be used for the ten and fifteen meter bands, in fact, its gain will be improved as the frequency gets higher. If the twenty meter band is an only interest, it can be fed with 75 ohm coaxial cable instead of TV twin lead eliminating the need for the coupler. Carefully cut the antenna for the lowest VSWR in order to compensate for its surroundings, but, after completed, its pattern will be the same as with the twin lead. Figure 4-30 shows the free space antenna pattern when excited with a twenty meter signal. All kinds of interesting results can be obtained by tilting the wire. This will tend to move the top part of the main lobe parallel to the ground, giving a very low angle of radiation. The lower angle will bounce the signal a little further.

The materials used for constructing this antenna are very strong, but light in weight. This permits the assembly to be held in temporary positions with heavy fishing line for experimenting or permanently fastened to withstand the heaviest weather.

Chapter 5
Quad Antennas

AN INEXPENSIVE QUAD

This cubical quad antenna is quite directional, has high gain, and is inexpensive when compared to other beams.

The antenna is mounted with spreaders running horizontally and vertically rather then diagonally. This enables the metal spider brackets to be welded with greater ease and may also add some strength to the assembly. The spider brackets should be made of ⅛" × ½" × 2' aluminum angle (4 each required). Weld each pair on centers and at right angles. The spider to boom bracket should be made of ⅛" × 1" × 2' aluminum angle (2 each required). Weld in the center and at a right angle to the ½" wide legs. The metal may be obtained from a junk yard, some supply houses or any welding shop; take the materials to a welding school or high school metal shop to be welded.

The boom to mast support bracket should be made of ⅛" × 1½" × 2' aluminum angle (2 each required). These two pieces should also be welded to each other at right angles and on centers (see Fig. 5-3).

The boom is made of 2" × 2" lumber. One piece is 11 feet long and the other is 6 feet long. These two pieces should be nailed together with the shorter piece centered below the longer piece.

Center the aluminum boom to mast bracket on the boom, drill at least 8 nail holes through the horizontal leg and nail the assembly together.

Obtain the bamboo from a carpet store as carpets often come wrapped around bamboo poles. Try to get unsplit, straight poles 13' long, the thinner the better. You will need 9 poles.

Cut up a couple of coat hangers into 3" lengths and form into wire hooks as shown in Fig. 5-3 inset.

Lay out the bamboo to the dimensions shown in Fig. 5-1. Drill holes through one side of the bamboo and install the wire hooks into 3 legs of the spider. On the fourth leg drill the holes

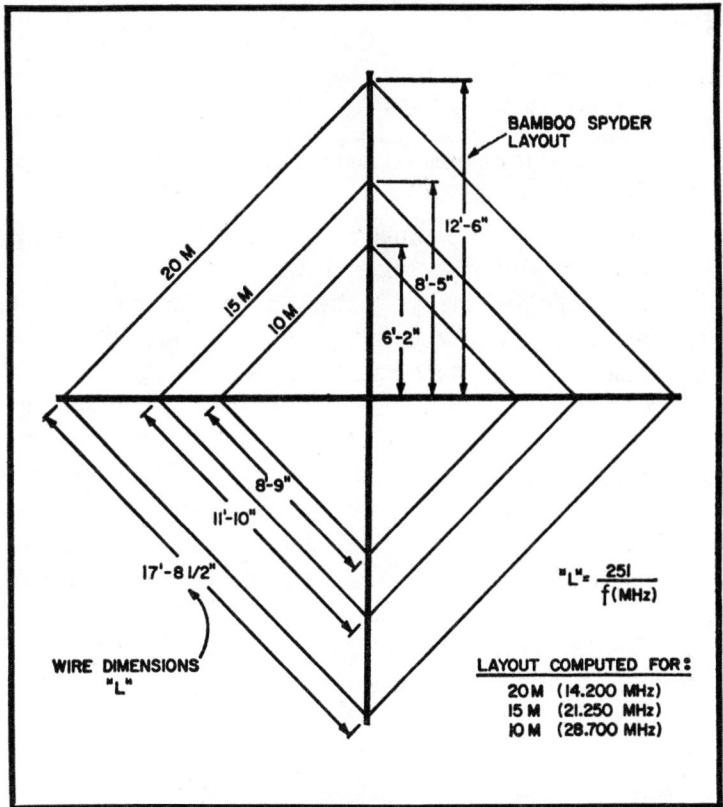

Fig. 5-1. Complete layout of the three band, two element quad; bamboo poles and a wooden mast provide very economical construction.

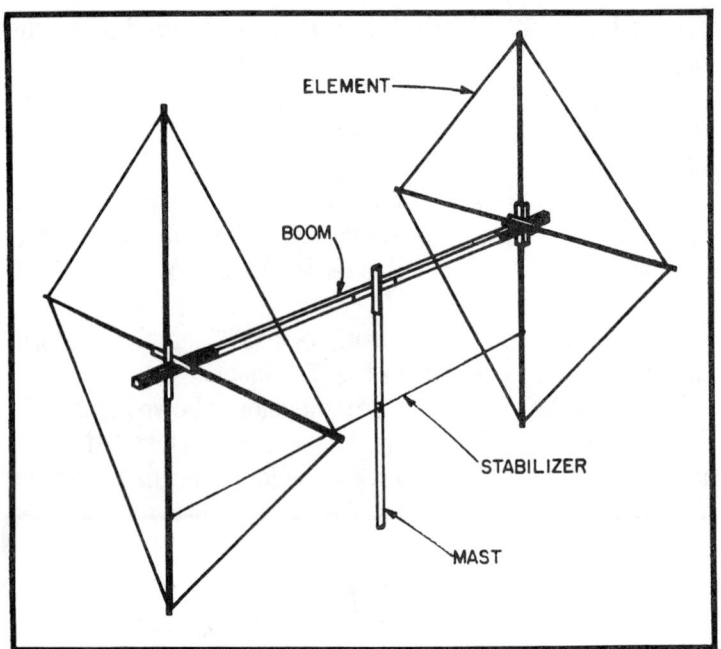

Fig. 5-2. Overall view of the two element quad showing the layout of the boom, mast and stabilizer.

all the way through the bamboo 1" above and 1" below the laid out dimensions for each spider assembly.

Assemble the bamboo to the spiders using 2 small hose clamps for each pole. Most auto stores have an ample supply of hose clamps in assorted sizes.

For each band, attach one end of the wire through the upper hole on the fourth leg. Wrap the wire around the spider and attach the end through the bottom hole. Attach the feed line to the wire ends on the driven element and solder. Short the wire ends together on the reflector element. Tape over the wire hooks to make sure the wire stays in place as it has a tendency to stretch with time.

Assemble the spiders to the boom with large hose clamps. (This is the toughest part.) Space the elements as shown in Fig. 5-3.

The last bamboo pole is the stabilizer. Cut it 9′ long, drill a small hole through two small hose clamps and screw them to each end of the pole. Attach the stabilizer about 8′ down on the fourth legs between the elements and parallel to the boom.

Number 17 bare stranded copper wire is adequate for a QRP station, but where higher power is used, #12 copper wire should be employed.

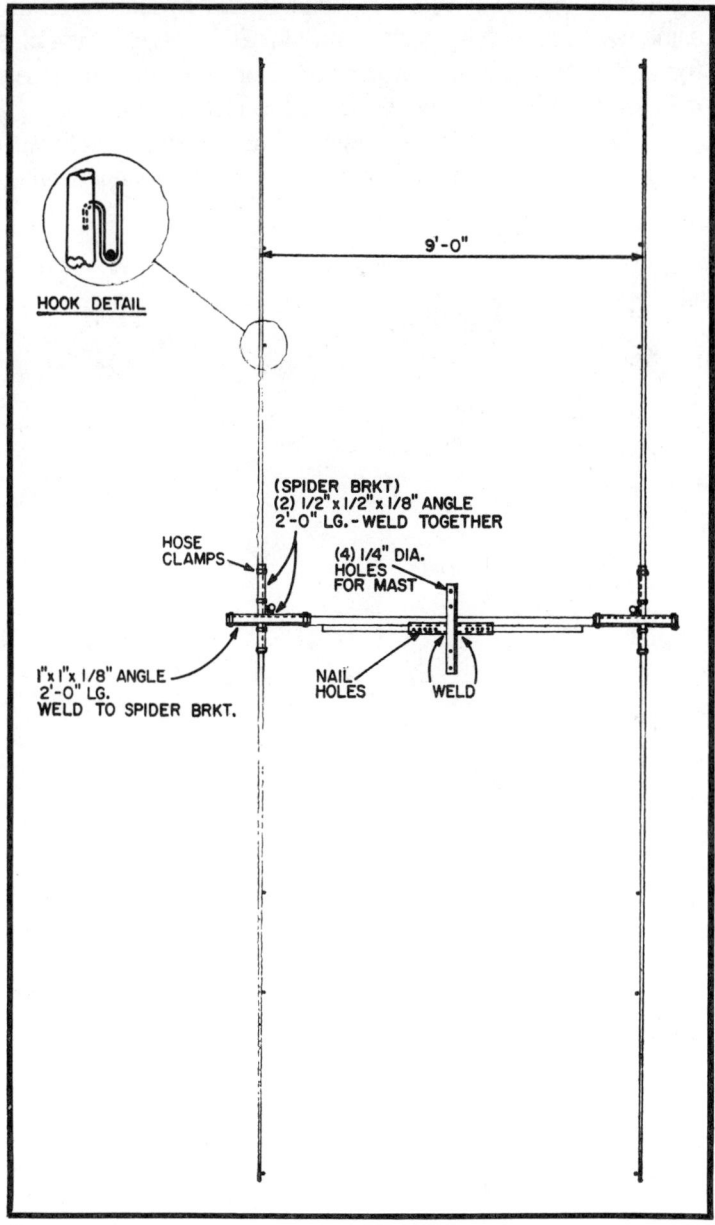

Fig. 5-3. Constructional details of the two element, three band quad.

Feed the array with 52 ohm coaxial cable. It was found that two of the bands could be fed with the same feed line without appreciable loss, but the other band had to be fed with a separate feed line; this array has 10 and 15 meters on the same feed line with a separate line for 20 meters. It was also found that tuning stubs on the reflector were not absolutely necessary and were omitted from the installation.

The antenna has been mounted on a 20 foot tower with good results, but much better results are obtained when mounted on a 40 foot tower.

THE MINIQUAD

The Miniquad has two unusual features. 1) It is of all-metal construction, thus eliminating the problems of treating bamboo and welding spiders, only to have the whole antenna come tumbling down in a year or two, and 2) It is miniaturized, taking up less than *half* the space of a normal two-element quad. Added features of the Miniquad are its low cost, extremely light weight, and general ease of construction. The Miniquad can be built from parts of an old beam, or it can be fabricated from scrap aluminum. It is light enough to be turned by a low-priced TV rotator.

Theory

The antenna illustrated in Fig. 5-4 is essentially a two-element quad with .12 wavelength spacing. Note that the two loops are insulated from the booms and thus from each other.

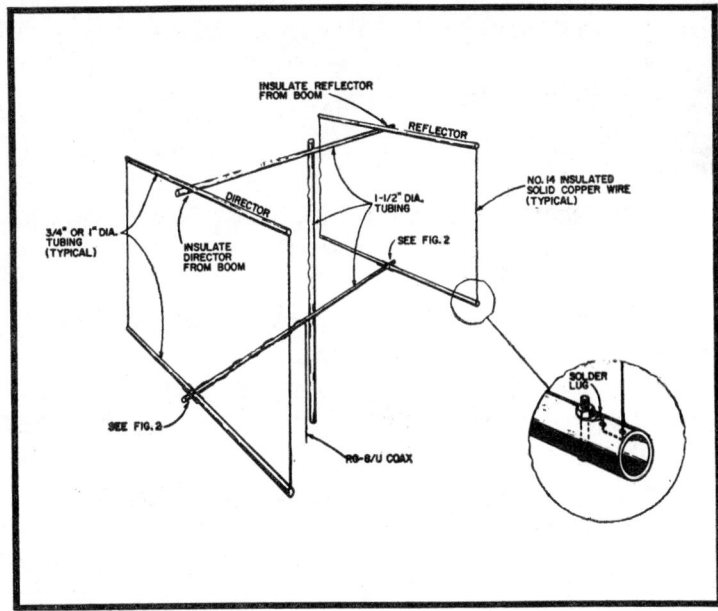

Fig. 5-4. Construction of the miniquad. For operation at 14250 kHz, element spacing is 100 inches, the horizontal supports are 208 inches long and the vertical distance between the horizontal supports is 104 inches. The upper supports are insulated from the boom with standoff insulators.

The horizontal dimension is .25λ, while the vertical dimension has been reduced from the usual .25λ to .125λ. The difference is made up with loading coils at the bottom of each of the two loops. The Miniquad is thus *rectangular*, rather than cubical, in configuration. The 52-ohm transmission line is inductively coupled to the loading coil on the driven element.

Construction

The Miniquad lends itself to much flexibility in construction. The original version was built at almost zero cost from the parts of an old Telrex beam. However, eight tenfoot sections of tubing of almost any material and any diameter provide elements. Center mounts can be constructed of aluminum angle irons with standoffs as insulators. Masts and booms are made of TV masting. Standard antenna hardware is used for mounting the booms to the mast.

The vertical portions of each loop are of insulated number 14 solid copper wire, of the type commonly used in electrical house wiring. The wires may be attached to the ends of the horizontal elements by any convenient means. The wires should be tightened so that the top and bottom elements "bow" slightly toward each other.

Coils

The exact number of turns for the two loading coils and the coupling link depends on many factors and therefore differs for each Miniquad. The original twenty meter Miniquad has coils each made of about three inches of two-inch-diamater B&W coil stock. The driven element coil should be grid-dipped for the center of the desired operating band. Be sure to make this measurement in the absence of stray inductances.

The reflector coil is adjusted, in the usual manner, for either maximum front-to-back ratio or best forward gain. Note that no tuning stub is required on the parasitic element of the Miniquad, as the element already has a loading coil. Thus the reflector coil will simply have somewhat more inductance than that of the driven element.

The 52 ohm coax is coupled to the driven element by winding about five turns of insulated #14 solid copper wire

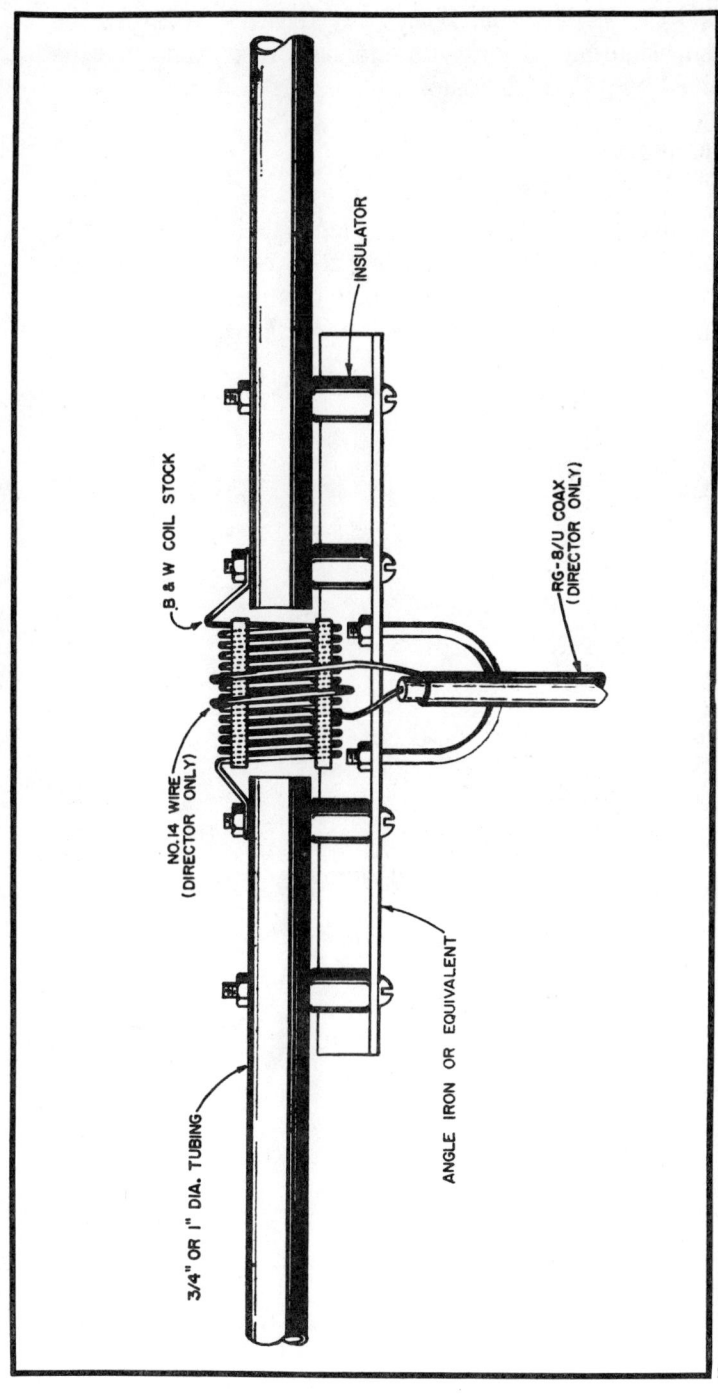

Fig. 5-5. Loading coil for the miniquad: a piece of coil stock two inches in diameter and three inches long is about right. The feedline is coupled into the antenna with a two or three turn loop around the loading coil.

around the loading coil. Since only this link is across the transmission line, a very low standing wave ratio may be obtained by proper choice of the number of turns.

Performance

The SWR of the original twenty meter Miniquad was less than 1.5:1 for the entire band, and close to 1:1 over much of the band. Transmitter output tuning is quite broad, with retuning required only for large frequency changes. Front-to-back and front-to-side ratios seem quite satisfactory.

THE THREE ELEMENT QUAD

The gain of a 2 element is comparable to that of a three element yagi. The front-to-back ratio ranges from 25 to 30 dB, with the front-to-side ratio reaching as much as 40 dB. For the size of the antenna it packs a mighty punch in the roughest of pile-ups, and amateurs around the world will attest to its performance. Another favorable aspect of the quad is the relatively low construction cost. Obviously, a 3 element quad will provide even greater performance.

Construction

Spreaders: Fiberglass poles make excellent, durable spreaders, but are quite expensive. Bamboo poles also suffice, but do not weather well unless they are protected. A few coats of Spar varnish will last several years, but if the poles are fiberglassed, they will last indefinitely. Fiberglass resin and fiberglass cloth are available at most boat centers and sport shops. The bamboo poles were cut to a length of thirteen feet, and wrapped from the small end to the butt with three inch wide fiberglass cloth. About half a quart of resin was mixed at a time, and applied with a paint brush. About two days are required for the poles to dry.

Supporting crossarms: Three foot sections of one inch angle aluminum were used to hold the poles to the boom. The muffler clamps used between the angle stock and the boom are three inch. The bamboo poles are held to the angle aluminum by one and a half inch hose clamps.

Boom: The boom consists of twenty-one feet of three inch irrigation tubing. The spacing from the reflector to the driven element is ten feet, and the distance from the driven element to the director is eleven feet.

Stringing the elements: Number fourteen wire was stretched out and marked at 69' 1" for the director, 70' 1½" for the driven element, and 72' 1½" for the reflector. After the crossarms had been assembled, the spreaders were staked out perpendicular to each other, and each element was strung.

Assembly and tuning: Each element was fastened to its re-

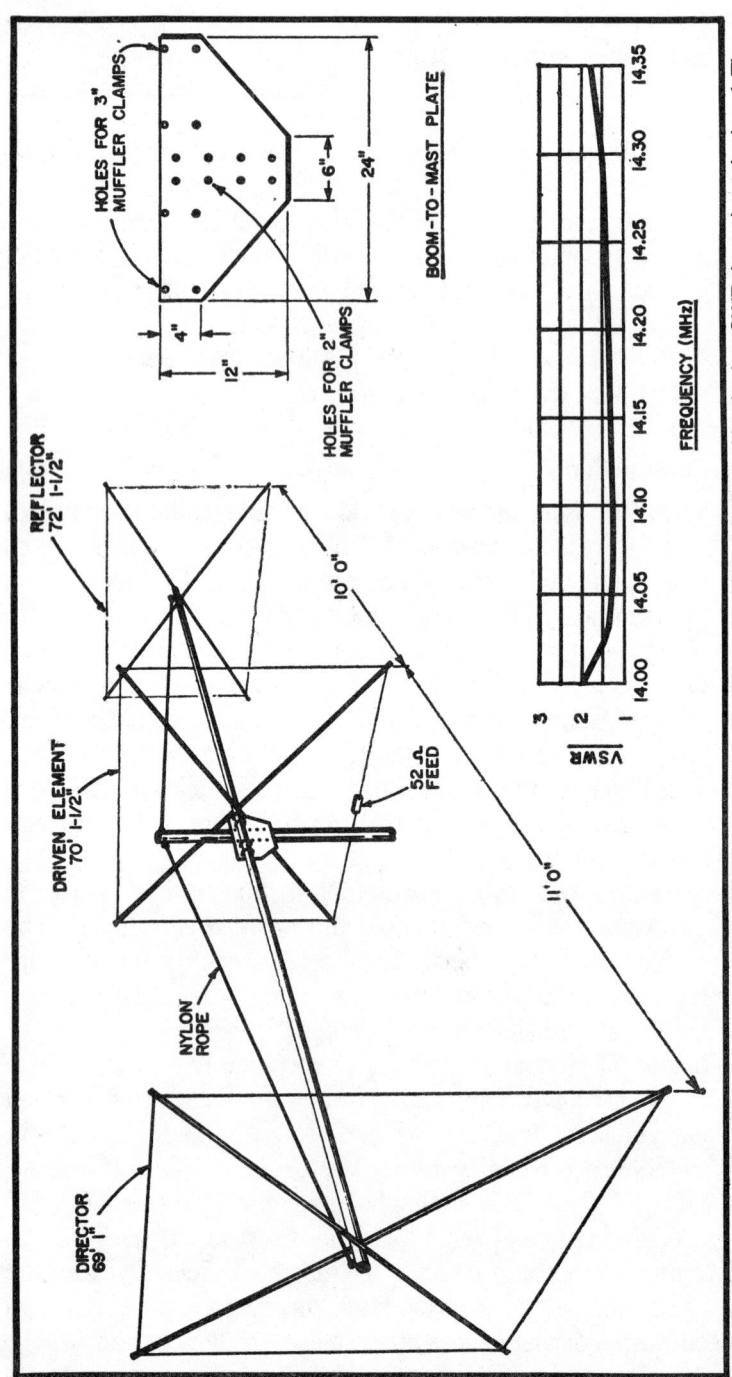

Fig. 5-6. This three element quad for twenty meters provides extremely wide bandwidth as indicated by the low SWR throughout the band. The construction of this quad is straight-forward and only requires a boom length of 21 feet.

spective position on the boom, and 52 ohm cable was attached directly to the driven element. Tuning stubs were fashioned out of #12 wire and fastened to the director and reflector. The antenna was raised to approximately twenty feet, and the stubs were adjusted to give maximum s-meter readings on a receiver beneath the antenna.

Repeated comparisons with a nearby station, on the long and short haul DX, seem to indicate that the three element quad is comparable to a four element yagi. The front-to-back and front-to-side ratios are as good if not better than the two element quad.

A LIGHT FOUR ELEMENT QUAD

Most people don't have the room or the money to put up a six element beam but a four element quad doesn't take too much space and is considerably cheaper than the beam.

In the past, four element quads have been bulky, very heavy and parts were hard to come by. Today a four element quad can weigh under 50 lbs. By using light-weight fiberglass arms (16 of them) each weighing only a pound and by using a light-weight aluminum boom and using aluminum braces to mount the arms to the boom all totals up to a lightweight four element quad that can be supported by most towers capable of supporting a tri-band beam.

A 2" × 30' × .065 wall tubing spec. 6061ST-6 boom will support the lightweight arms and has a lower wind resistance than the 4" boom used in the past.

The least expensive arms for a quad are made of wood or bamboo, but arms can crack, split, warp and are easily broken during construction or by the wind after construction.

Commercial fiberglass is now available within the price range of the average income. This fiberglass is made especially for quad antennas and is manufactured by several companies. Quad arm mounts are also available.

Boom to mast clamps are available but be sure that they are big enough to support a four element quad. The support should be 12" × 2" O.D. for the boom and 6" × 1½" O.D. for the mast.

The clamp must be capable of supporting both the 2" boom and the 1½" mast above and below the boom (Fig. 5-9).

About 50' of ½" galvanized guy wire, 2 heavy duty turnbuckles and 8 cable clamps are needed to give the boom additional support (Fig. 5-7).

There is no real difference between a quad shaped like a diamond and one shaped like a square. The important thing is the feedpoint (Fig. 5-7). Feeding the antenna at any point marked H results in horizontal polarization and feeding at any point marked V will result in vertical polarization.

The most widely used polarization is horizontal. It is easier to tune the antenna and the tower will have less effect

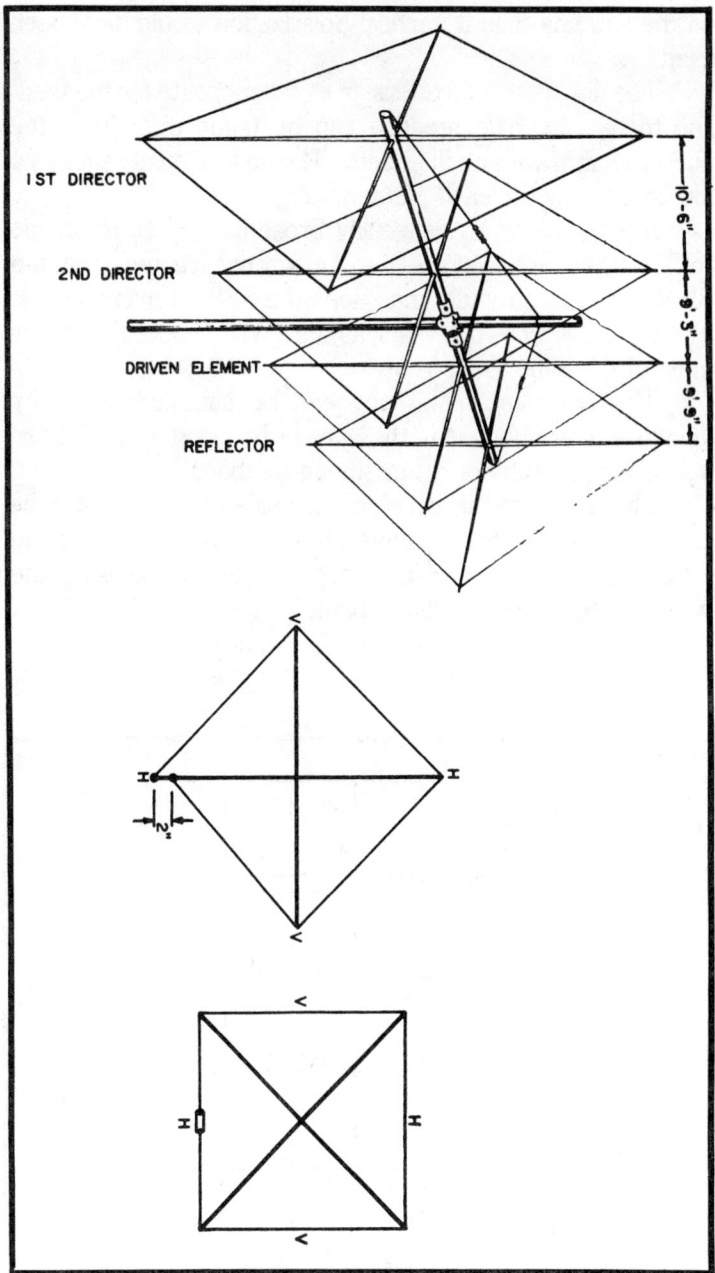

Fig. 5-7. Overall view of the four element quad for 20 meters; the total boom length is nearly 30 feet. Either diamond or square construction may be used as shown on the right, but the square model requires insulators—one in each element.

on the antenna than if vertical polarization would have been used.

The diamond quad offers an excellent route for the feedline to be run. The feedline can be taped directly to the fiberglass arm with no ill effects. The square quad requires 4 insulators, one for each element (Fig. 5-7).

A quad is not an especially broad-banded antenna; the SWR can be tuned to almost 1 to 1 at any one frequency but the SWR rises more rapidly than does the SWR in many beams. This is not a serious drawback as the SWR is under 2 to 1 over most of the band (Fig. 5-8)

The resonant frequency can be changed easily by lengthening or shortening the wire or by using coils, shorted stubs, tuning stubs or other tuning methods.

The length of the wire on each side of the quad can be calculated by using the formula 248 or the formula 984 for the total length of each loop. The approximate total length of the wire for the 20 meter phone band is:

    1st director   66' 6"
    2nd director   68' 3"

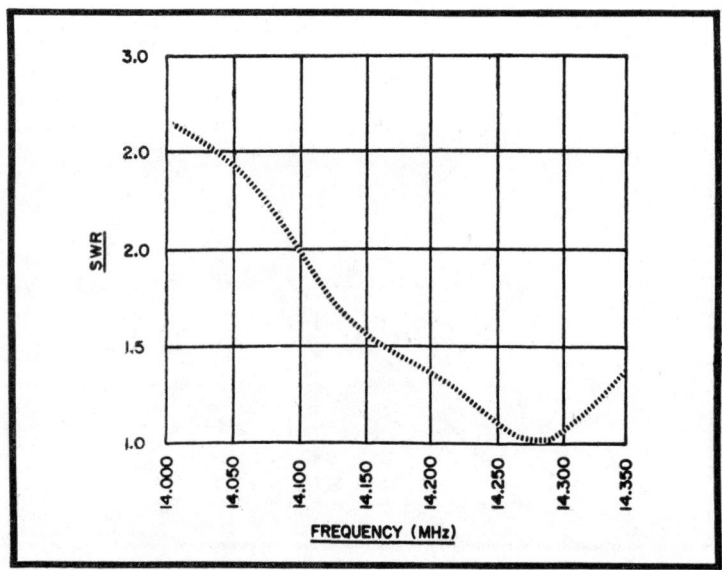

Fig. 5-8. Standing wave ratio of the four element quad. Although this unit was tuned for minimum SWR in the phone section of the band, even at the low end the SWR barely exceeds 2.5:1.

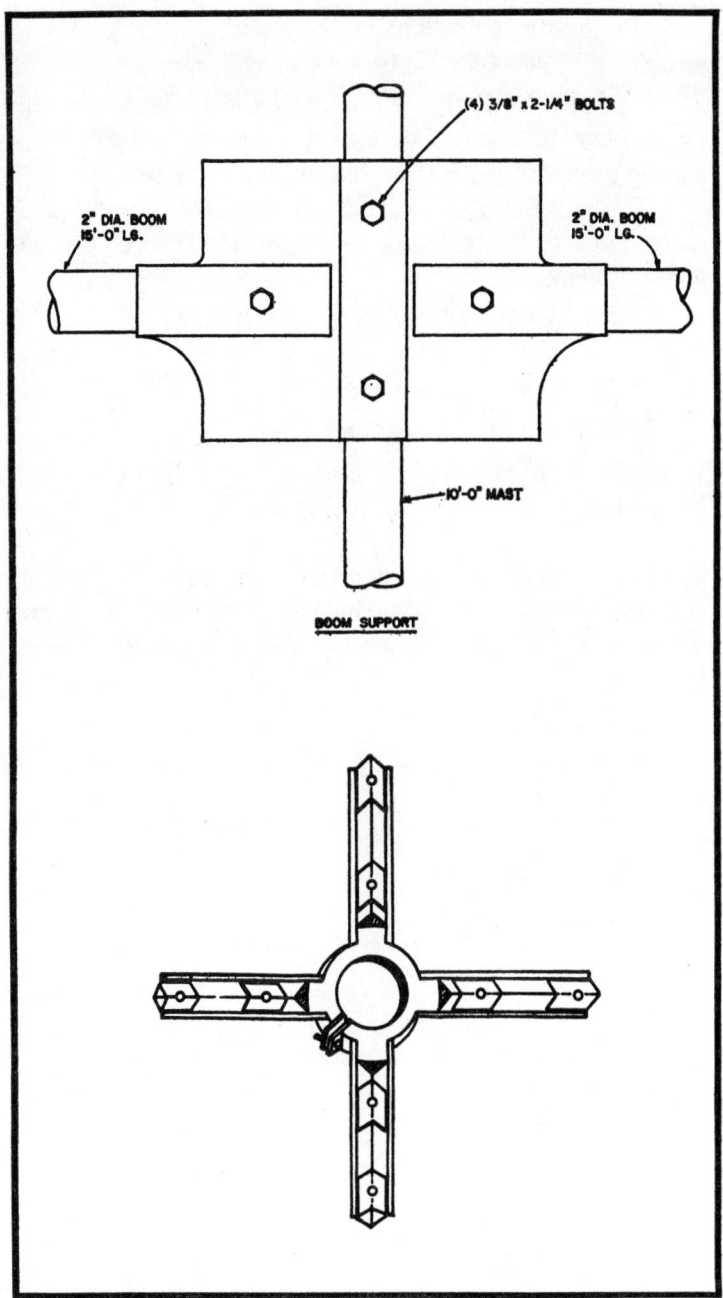

Fig. 5-9. The commercial arm mount and boom support used with the four element 20 meter quad. This particular design has withstood winds up to 80 miles per hour.

driven element	70' 0"
reflector	71' 9"

The above lengths were calculated by increasing the driven element length 2½% for the reflector and by decreasing the directors 2½% from the driven element size.

The radiation resistance of the quad will be close to 50 ohms and the greatest gain will be achieved when the elements are spaced:

1st to 2nd director	10' 6"
director to driven	9' 3"
driven to reflector	9' 9"

This uses all but 6" of the 30' boom.

1. Select a flat place in the yard to lay out the quad arms. Mount the arms to the arm mount. Do this to each of the 4 mounts.

2. Drill 1/16 holes in the arms approximately 12' 1" from the center of the boom; drill a 2nd hole at 12' 3" in the bottom arm of the driven element. Be sure to drill all holes straight and in a level plane.

LONG, CIRCULAR QUADS

This discussion is intended to give some insight into new types of VHF and UHF antennas which are usable on 2 through ¾ meters and into the UHF television spectrum. Frequency considerations of bandwidth, patterns plotted, and element tapering are discussed with curves from Project OSCAR tests and observations and UHF television on channel 32. Propagation on channel 32 was limited to a faint forward scatter signal. The other antennas use parasitic elements cut to the same dimensions as the driven element (DE). The round element antennas were tested in the shop as well as in the field; with precise matching on 432 MHz, with the "business end" of the array pointed up. It was found to be interesting to note that the quarterwave circular stub presented about 60° (approx. 1 radian) arc-length when unity SWR was obtained on 432 MHz. Use of RG-58/U was preferred for this band, while 300 ohm line was chosen for use on Channel 32 reception. Reception was so good that about double the signal strength was observed using the 11 element Long Circular Quad as compared with 23 element Long Yagi.

The following paragraphs show, without mathematics, the properties described. RG-9/U is a minimum size coax to be used, with preference for Foam Heliam as in commercial installations. Silverplated RG-9/U has about the same rating as RG-8/U but has a non-contaminating jacket: and the shield-braids (2) remain interwovenly conductive within. Both RG-8/U *and* RG-9/U are adversely affected by dew, and as much as 20% variation in reflected power can be expected, even with a perfect match! For experimental use and testing, the RG-58/U and an Amphenol coaxial termination are a necessity, mostly because of element length-to-surroundings scale, etc. Heliax is the only answer on 100-ft. runs.

On the 18 element Long Circular Quad, RG-58/U was made exactly ¼ wavelength *on the outside* for proper connection from the screen (*rf* ground). Preliminary comments on the circular quads show a nearly perfect major lobe whose half-power-beamwidth gives 15 db gain over a matched half-wave dipole. This means 18 db over a point source. These calculations were taken from the pattern itself, plotted through a research laboratory antenna range. These patterns were traced by an automatic pen recorder which was apparently servo-controlled. Some breakdown into vertical and horizontal modes were observed, but this model was not very well balanced or completely matched when tested. Calculations are accurate, since the minimum amplitude for each graph is shown to be above "zero" or presents a plotted minimum value on the same scale factor as the peak pen travel.

Also in this article, construction is detailed thoroughly for the home craftsman. Air-dried seasoned oak is preferred for the 18 element antenna 12-ft., 6-in. boom; however, a sturdy fiberglass quad arm may be better in the long run. With a heavy round quad arm, a spring brass U-bracket could be fashioned, similar to an auto radiator hose clamp with tightening pressure similar to a metal TV stand-off insulator (this would be from the underside). The other solution is to use epoxy as described later on.

Long Quad Development

First in the discussion of these new antennas is the 8 element long quad of more conventional design. The approach was this: Make a regular quad with a screen reflector on the APX-6 transponder and add elements to make it "long," meaning in Yagi terms to increase the boom length with extra elements in excess of a total of 5. Using approximately 984 mha *rf* from the transponder with a Heath Tunnel Dipper as an FS meter (diode position), it was found that maximum gain and directivity was obtained with the eight elements staggered, aperiodic and quad-shaped; the directors tapered to 95% of the driven element circumference, combined to product optimum results!! All of the directors were the same 95% circumference of perimeter of the DE. The elements were cut to 984/FMHz or, at this frequency, 1 foot, for convenience. The reflector screen was moved back-and-forth to match the array the first time; the second try was to build a ¼ wave stub (shown in the quad development photo) and vary its spacing from the soldered feedpoint connections. Of course, 984/

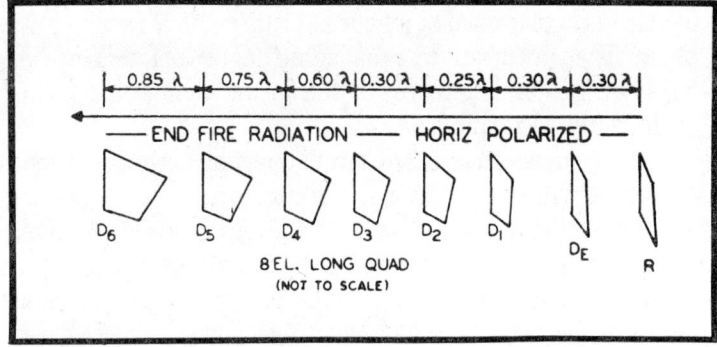

Fig. 5-10. 8 element long quad.

FMHz is free-space wavelength, and a 4% shorter length was used for the DE. Unity SWR was not consistent with maximum forward gain. The six directors were staggered as shown in Fig. 5-10 with the closest director spacing being between D2 and D1 (not between D1 and DE as would be expected). Two db gain was measured over an equivalent 8 element Yagi of the same boom length, or 13 db.

Circular Quad Development

The next step was to make the elements circular. The APX-6 was cranked up to about 1160 mhz, where one wavelength was close to 24 cm. and one-quarter wavelength was approximately 6 cm. None of the expected characteristics of the long quad or long Yagi were to be found. Instead, it turned out that one-quarter wave-length spacing intervals, and staggering of the dimensions were of prime importance. The parasitic directors were cut to 95% of the DE length and odd-numbered groups of director elements, with quarter-wave inter-group spacing, gave the highest gain for the array. Also, the staggering of these groups of elements allowed a "gap" or space between for the DE and D1 reflected image to appear in phase in the right "place" on the wooden boom for more gain. Keep in mind that the reflector was untuned and presenting an electrical "mirror image" of the elements discussed. See Fig. 5-11.

Polarization was semi-circular, in an axial mode (end-fire directivity) and not sideways from the boom. There was considerable horizontal radiation similar to that obtained by feeding a regular quad at its apex. Little vertical radiation was noted. The conclusion was that some degree of imbalance was imparted by standing waves on the line or a non-perfect match at the antenna terminals.

No balun was used. The largest portion of the radiation is horizontal with an additional "circularizing effect" possible because of the circular shape of the directors themselves. Also, the vertical and horizontal components are similar which makes the antenna fine for OSCAR and vhf Moonbounce work. Incidentally, "Circular Quad" is a misnomer when used to show how the antenna was derived. With the circular quad there is no 30 db winding sense loss, a characteristic of the Helical Beam Antenna. Faraday rotation, which is simply defined as rotation of an incident wave upon passage through (the Earth's) magnetic fields, should be no problem.

Mechancial Details and Performance

To prove that this antenna has promise for the amateur and experimenter, as stated in the introduction, a fringe-area uhf model of the circular quad for Channel 32 was constructed, using 11 elements; and the results were excellent. Working at 579 mhz with the driven element cut to the picture carrier, it was found that the gain/aperture was at least equal to two commercial corner reflectors, yielding a snow-free picture of this elusive signal.

Performance

Two lobes were reported, with gain figures based on beamwidth plotted from the *VHF Handbook*, by Orr & Johnson. Note that gains of 14 db horizontal and 8 db vertical were calculated. This plot is at 5% below the normalized frequency of 1,000 MHz. At the normalized frequency, only a single plot was obtained, composed of both lobes equally yielding 16 db power gain over a dipole. The screen reflector is about 12 by 15 inches square (not critical) and an N connector with hood is at the base of the decoupling sleeve, with RG-9B/U inside it. The SWR was high. L-shaped steel corner braces hold the N connector and base of the decoupling sleeve securely, as well as the screen.

At 5% above the normalized frequency for design-center, there is a 16 db power gain over a dipole. The major lobe is clean and very little energy went into the others. The effective aperature for the array is approximately 6.5 wavelengths squared. Aperature is an interesting measure. Physical aperture of the Long Quad is twice that for any other array. Or, it is like expanding a 2-wire folded dipole into a circle. The physical aperture is at least 1.4 times that of any Yagi; and perhaps as much as 2.5 times, taking into account "wavelength factor." The equipment used consisted of a 25-watt, 2-meter tranceiver, a diode varactor and a Cesco Reflectometer.

Performance and Theory

Assuming a standing wave of current to exist on the structure looking "down" on the vertical plane of each loop, a reflection from the imaginary "screen image plane" gives the expected 3 db gain over a single dipole. See Fig. 5-12. Based on the assumed stationary wave phase, one of the several current maxima allows the amateur (or professional) to select his mounting point, depending on mast material. As the traveling wave progresses, with each peak displacing ¼ wavelength more, gain increases. If it's difficult to visualize, simply start at the 6 db "bar" and follow the standing wave as a "point" moving from the first current negative maxima, up through zero, then positive maximum, and back to zero. The zero point on the sixth element plane is in the same plane of the bar marked 9 db, below. Observe that it is only displaced 90° *relative* to the repeating (transcendental) maxima of the assumed standing wave, which does not move. And notice that the first negative maxima referred to is positive-going, while at the sixth element the amplitude is negative-going. What's

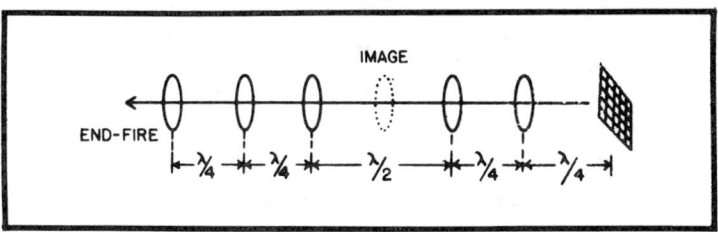

Fig. 5-11. Untuned reflector gives mirror image of the elements.

wrong? Nothing. *Both* wave motions are still from left-to-right, and add in time, giving the 9 db circular gain figure which results. In the first place, it can be asked why the 6 db current bar was picked as reference. Well, the dashed-in 0 db bar could not be used because it exists as a non-reinforced reference, without physical dimensions; however, this image *and* real element above do add to produce 3 db. This leaves the 6 db bar the choice, to eliminate confusion.

Several models of the Long Circular Quad were built. After much testing and refinement of use of instruments, starting with an APX-6 on 984 MHz, enough data was accumulated from building a uhf reflectometer and the 1000 MHz scale model to conclude the information presented here. Further study has led to the conclusion that the parasitic directors must be shortened a "mystical 5," as some have said, for maximum gain.

The first group of directors from the end are all spaced one-quarter wavelength gap; then a three-quarter wavelength gap; then the next group of three loops, each at one-quarter wavelength spacing, with a one-half wavelength gap; and on the D1 and DE, each at one-quarter wavelength. The reflector (R) is made of "expanded" aluminum and is made approximately 5/2 wavelengths, minimum, square. In this model, the feed-point is at 12 o-clock, with heavy 200 ohm twinlead feedline. The boom length is 3⅓ wavelengths at 579 MHz. Estimated gain is 12 db, including a 3 db polarization loss from transmitted horizontal to receiving circular. There appears to be no sense to the antenna, as there are two helices. Please note that we are looking into groups of 5, then 3, and then 2 physical one-wavelength loops from the opposite end of the boom on which rests the sheet screen reflector, R. The staggering is ¾ wavelength gap *between* the groups of 5 and 3, with ½ wavelengh gap between the next group of 3 and 2; and ¼ wavelength from the DE to R. More gain can be obtained by placing 7 elements with a gap of one wavelength ahead of the 5 shown; and 9 can be added with a gap of 1 and ¼ wavelengths ahead. The limit is what beamwidth is practical for accurate steering. Matching is not a difficult problem with a DE loop configuration.

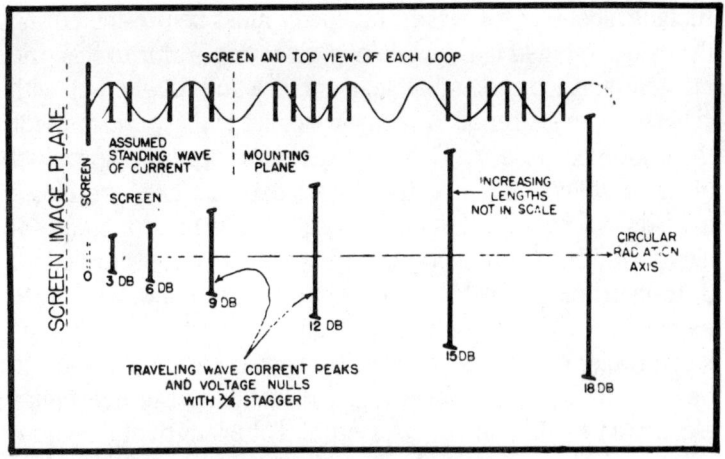

Fig. 5-12. Relative phase and gain of 18 element quad.

18 Element Long Circular Quad Construction

The 18 element Long Circular Quad is mounted on a piece of 1¼ inch steel TV mast with U-bolts and nuts. A front "nose" brace of wood keeps the array from drooping, and it is lashed to the seasoned oak boom with filament tape. The feedline is RG-8/U, with N connectors. The boom length is 12 feet (a little short for all 18 elements); however, allowance was made for 6 inches "over-length" by crowding the elements ⅓ inch each, with no loss in performance or SWR hike. The reflector screen is connected to the boom in paddle-wheel style, using ¾ by ¾ inch air-dried seasoned oak as the cross members. The reflector itself is made of "expanded aluminum," 2-feet square, which is about 5/4 wavelength on a side. One very important thing about this type of mounting is that ¼ wavelength ahead of the second group of elements is a voltage null which permits the use of metallic support. At other distances on the boom there are nulls that can be used for U-bolting a vertical conductive support; but this position is nearest the balance-point of the array. It wouldn't be a bad idea to use a V fiberglass support arm. Don't forget to balance the antenna with the coax weight on it. Foam Heliax is heavy.

The elements were glued to the boom with epoxy, but it was first necessary to coat the rounded No. 6 hard copper with a thin layer of epoxy diluted with a very small amount of rubbing alcohol. Of course, the epoxy must be pre-mixed and the alcohol should not contain any glycerine. Prior to this, the element surface should be clean and dry and roughed-up with cabinet paper. After all 17 loops are coated, allow two days for the undercoat to dry. This method is the only simple way of gluing that works. Once the undercoat was dry, the No. 6 hard-drawn copper elements were inserted and a final coat along with the wooden boom, which was previously drilled for 3/16-inch holes. The No. 6 cannot be bent by hand; take these elements to a sheet metal shop where they have a roller press just for this. An Amphenol or Andrew coax termination should be used for properly balancing and matching the one-radian electrical length of the circular stub. Total length of the circular stub and balun section is 90°, or one-quarter wavelength of No. 12 tinned copper, with 30° of the 90°, approximately, being used for balance. Reiterating: Just ahead of the first group of three parasitic elements is *the* first complete null for mast mounting with metal U-bolts and clamps. With a sturdy fiberglass mast, the mount can be made anywhere on the boom not contained within a group of elements.

Conclusion

Looking back, there are several points which should be summarized: (1) The Long Quad is semi-circularly polarized when correctly matched and thoroughly *balanced*, with some breakdown into vertical and horizontal lobes which can occur when the antenna is held less than 17 feet above ground; approached with less than 50 feet from the measuring site, and used too far below design-center. With 1.00 to 1.05 on the reflectometer and no standing waves *on the outside* of the feedline, it is semi-circular, at least relative to the reflecting screen. (2) The gain of the array, when considering a boom length sufficient to hold 27, 38, or 51 elements (which are the next groupings of elements) places it in the category of a small dish, but with less physical aperture. (3) Frequency limits of +5% and −10% make it useful over a band of frequencies, and the loop wavelength can be calculated by 24,500/FMHz, with lengths in centimeters. This formula is for a one-wavelength loop, physically. (4) There is no winding sense rejection such as characterizes a Helix, and a single support arm through the reflector screen material may be used with a counter-weight to aim properly. For an in-the-air mount, use an aluminum Y hanger from a separate U-bolted clamp with old garden hose binding the clamp to the upper portion of the mast.

A FULL SIZE 7 MHz QUAD

If one takes a simple half wave folded dipole antenna (0 dB gain), stretches it into a square or rectangle in a horizontal plane, and feeds it at an appropriate place, 3 to 6 dB gain can be obtained (Fig. 5-13A) and 5-13B). This is similar to taking a 2 element parasitic beam and bending the elements 90° at a point ¼ of their own length back from each and joining them.

If two identical antennas are properly spaced, 3 dB additional gain can be obtained; theoretically, this array has 6 to 9 dB gain. Suppose, however, for convenience in feeding the elements of Fig. 5-14A **a** and **c** and **b** and **d** are not connected mechanically as in Fig. 5-13 but are connected inductively and capacitively. Also, instead of connecting the two feedlines to **a** and **b** together, omit one feedline and connect **a** and **b** mechanically by moving the bent tips into a vertical position. Assume that the same degree of coupling can be obtained regardless of the type, mechanical or inductive. *Do* the same with **c** and **d**; the result is shown in Fig. 5-14C, the standard cubical quad. It can be seen that the radiating portions of the antenna have not been moved nor their length changed; only the method of feed has been changed. Thus, the gain should remain the same, about 6 to 9 dB.

A full size 7 MHz quad requires supports for two squares of wire that are about 40 feet on a side and spaced 30 feet apart. A structure of several tons could conceivably be used to support these two loops of wire but a more practical way is to build a rotatable support system of minimum weight, bulk, and cost that will withstand wind, bird and ice loading.

The mast is a six foot length of 1¼" solid steel shaft; a surplus motor system with a gearbox and selsyn are mounted at the base of the shaft. The shaft is coupled to the motor through a slotted section of pipe mounted on the rotor. A greased sleeve mounted in the top of the Tri-Ex tower acts as a thrust bearing. Just above the top of the tower a 24" × 12" × ¼" steel plate is mounted on the 1¼" shaft by means of four U bolt muffler clamps. On the back of this plate four home made 4" U-bolts made from ⅜" Redirod were

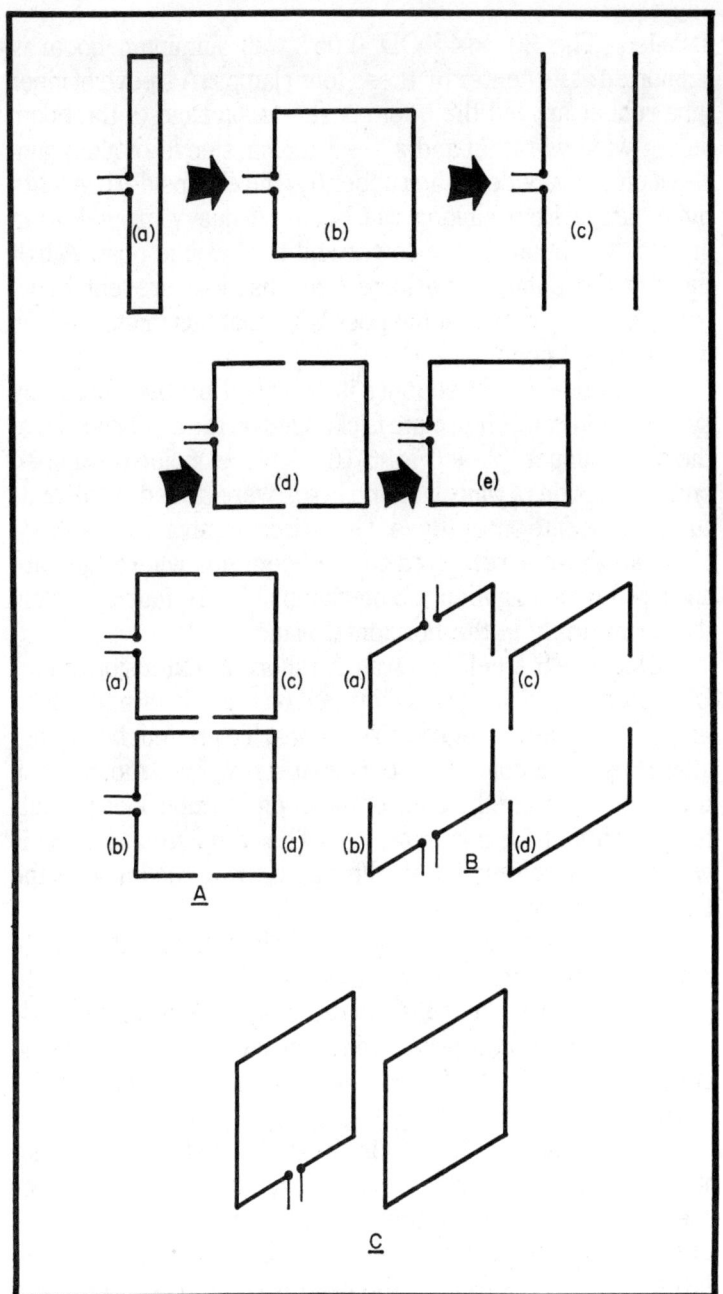

Figs. 5-13 & 14. Development of the cubical quad from the folded dipole into the two element quad.

installed. The 30′ × 4″ OD 0.06″ wall aluminum boom is supported at its center by these four clamps. A sleeve of inner tube rubber around the boom restricts slippage of the boom during wind vibration and a semicircular sleeve of aluminum sheet on the outside of the rubber (on the clampside) prevents tight clamps from kinking the boom. A heavy ground strap grounds the boom to the tower and to a ground post. A bolt through the boom could have been used to prevent boom rotation but there was some possibility that this might weaken it at a critical point.

Additional boom support is provided by the boom guy system. A second steel plate is clamped on the mast just above the boom support plate (Fig. 5-16). A piece of pipe is clamped on the backside of this plate and holes were drilled at the ends to accommodate screw eyes. Guy wires of galvanized #9 wire run from these screw eyes to the boom tips where they are anchored with irrigation tubing clamps. These four guys hold the boom firmly in the horizontal plane.

Since the 6′ steel shaft was too short, an extension of 1¾″ OD water pipe, 6′ long was mounted on top of it with two bolts going all the way through both sections. To prevent boom sag, guy wires were connected between screw eyes mounted on top of the mast and the ends of the boom; turnbuckles permitted adjustment for minimum sag. The complete assembly is well balanced and spins on the ball bearing mount with the twitch of a finger.

Many spider systems were considered, but the one finally adopted consisted of a piece of 24″ × 12″ × ⅛″ (10 gauge) steel plate bent 90° at the center. A piece of 4″ ID water pipe was sliced into 1″ wide rings; each of these was slit along one side. Two of these were welded to each plate as shown in Fig. 5-16.

On each side of the slits in these clamps an over-size nut was welded; a bolt could be run through these nuts to tighten the clamp snugly around the boom. Diagonal reinforcing bars of 1″ × ¼″ steel strap were added as shown in Fig. 5-16. Two holes drilled at the base of the plate accommodate U-bolts; a 3″ length of 1″ wide, ¼″ thick angle iron is welded into the top of each plate flush with the 90° bend. Each section of angle iron

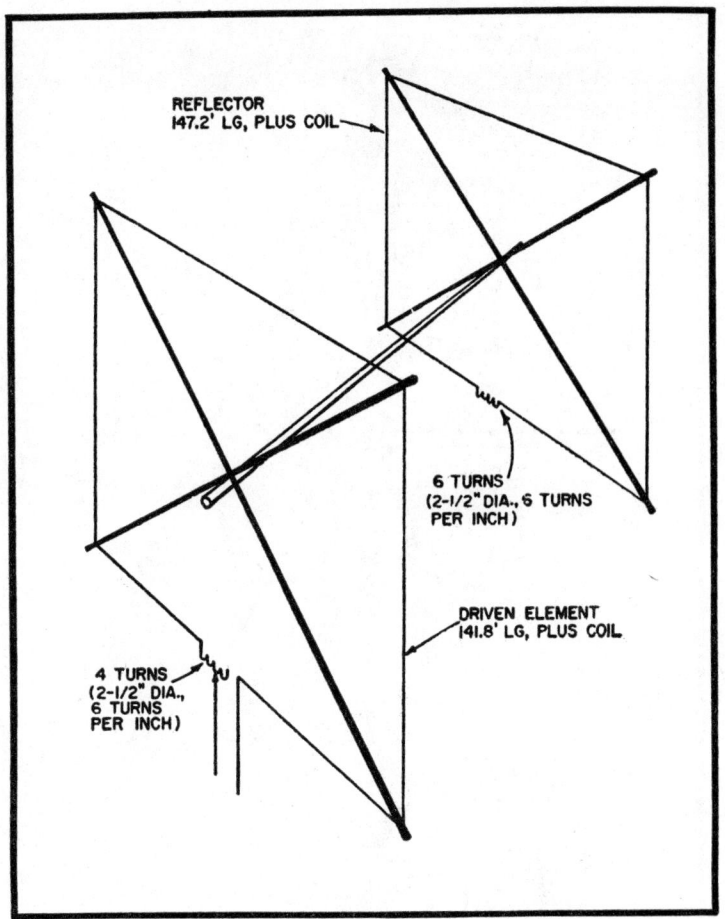

Fig. 5-15. The full size forty meter quad.

has holes drilled in it for a U-bolt and was mounted so as to have one side flush with the other side of the angle plate as shown in Fig. 5-16. A thirty foot piece of 2" OD aluminum irrigation tubing was used for the central section of the support arms. Five-foot lengths of 1⅞" OD tubes are telescoped inside the ends of this piece. Twelve foot lengths of high quality bamboo are telescoped inside the ends of the aluminum to give an overall length of about 56 feet. For maximum strength these bamboo poles should be wrapped with a layer of surgical gauze and coated with fiberglass.

The tubing to tubing and tubing to bamboo joints are made as shown in Fig. 5-16. Sleeves of sheet aluminum are

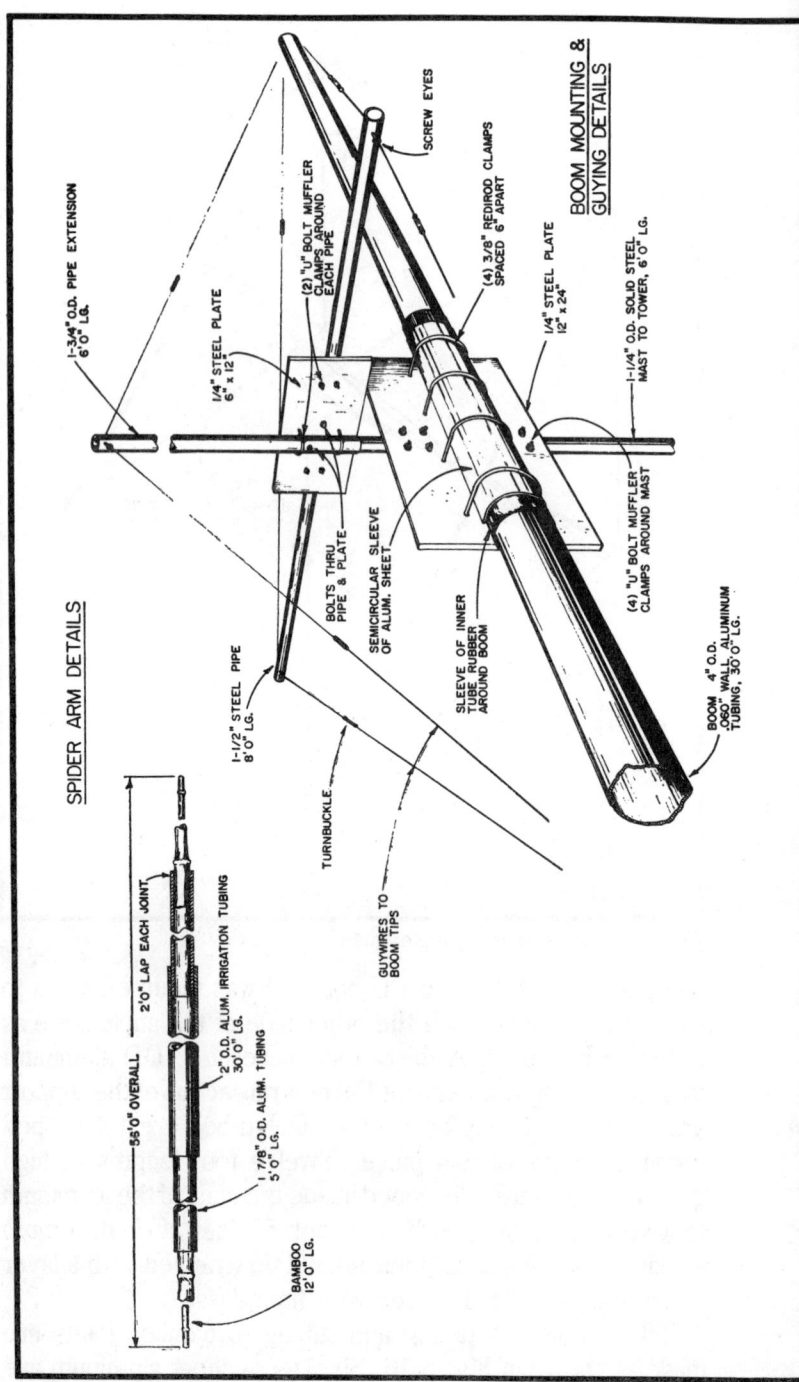

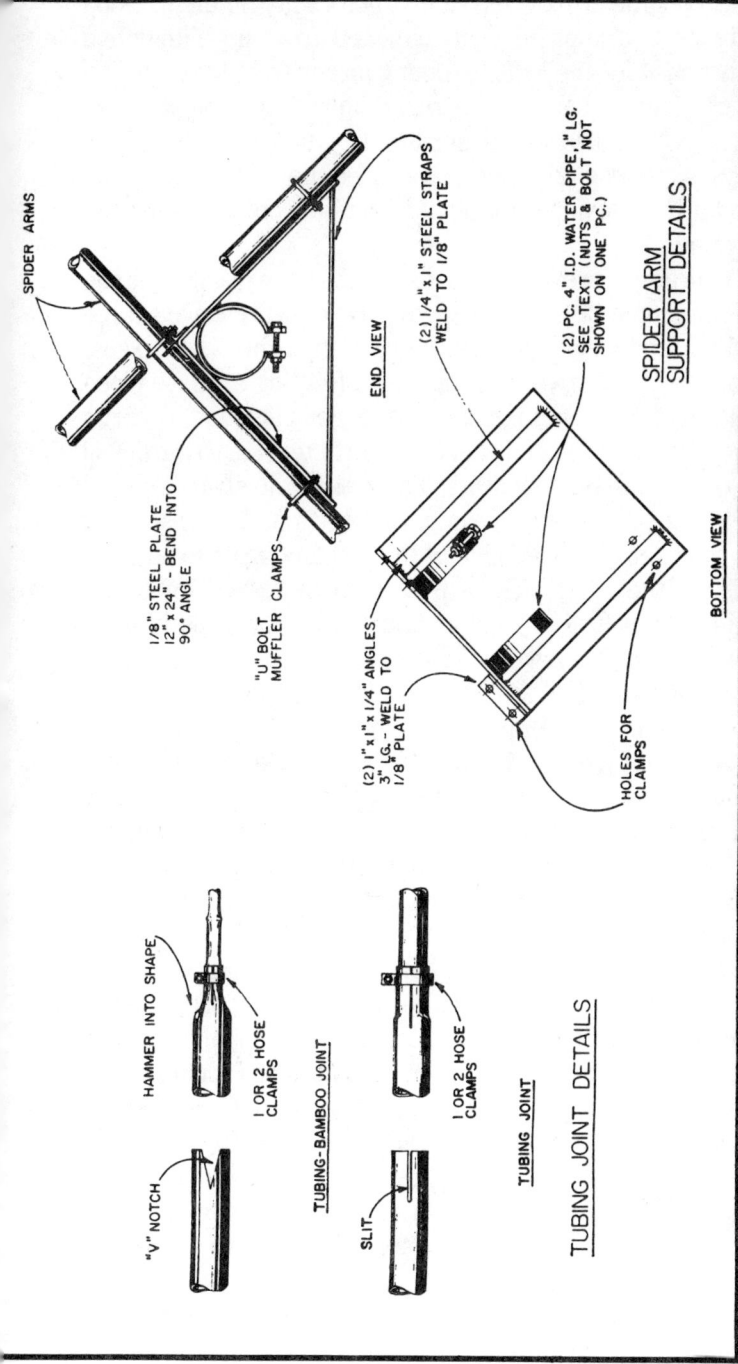

Fig. 5-16. Construction details of the forty meter cubical quad. This type of construction results in a light weight assembly that will withstand wind loading up to 60 mph.

placed over the arms where the U-bolts will grip them. Good quality hose clamps may be obtained from an automotive supply house for the joints. About 6 inches from the upper tip of each piece of bamboo the wood should be wrapped with tape, a hole drilled and a small pulley tied on. These two pulleys support the upper corners of the wire square and permit it to be lowered for adjustments; nylon cord is used in the pulleys.

In order to prevent the spider arms from breaking in the wind they must be properly guyed. Holes were drilled in the two 1" straps welded across the bottom of the spider plates and at the tips of two 5 foot pieces of ¾" pipe; these pipes were bolted on as boom extensions. Nylon parachute cord run from the holes in the end of each boom extension to the tips of each of the four element arms. Three separate strands should be used in parallel for maximum protection. Likewise, four guys run from the other side of the element arms to an extra spider on the inner side of the boom (reserved for future additions). For further support of the arms, particularly when the wire square is lowered, nylon rope was run in a square around the bamboo arms 6 feet from their tips; the total arm guy system is shown in Fig. 5-17.

Assembly requires strategy but this leaves room for ingenuity. First of all, get something to stand on 15 feet away from the tower and at least 20 feet high. A large pulley is placed on top of the mast to haul up parts. Most of the antenna can be assembled by one person but two people make it a lot simpler.

The items are mounted in the following order: rotor, ball bearing, gears, thrust bearing, mast, boom support plate, and boom guy plate. Then the spiders, guy extension arms, and boom guys with extra lengths of cord attached for manipulation from the ground. The boom is raised into place by pulley and then tied down to hang beside the boom support plate. The U-bolts, sleeves and rubber are added and the bolts tightened. The extra lengths of cord on the boom guys are used to pull the guys into place for anchoring; then they are temporarily tightened.

The four element arms are completely fabricated on the ground and their guys attached. One man can carry one up the

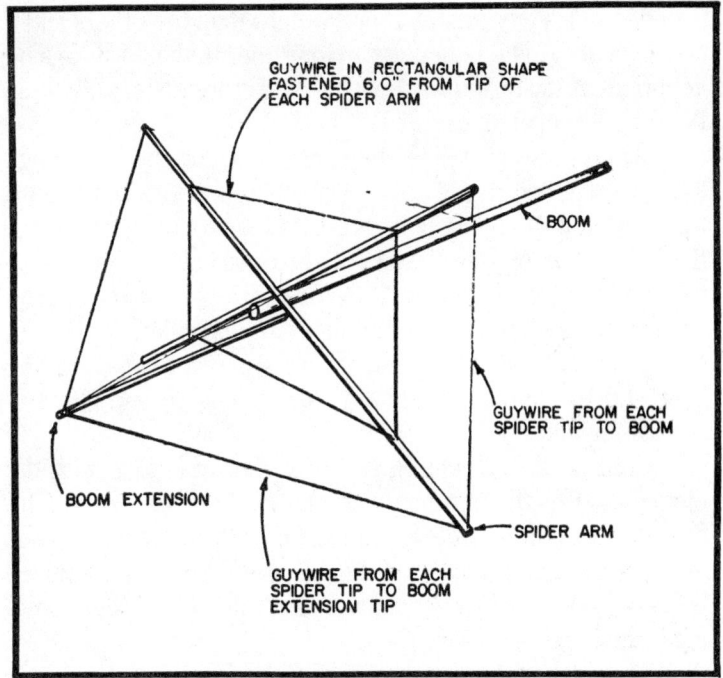

Fig. 5-17. Guying system used with the spider arms. Without this system the arms will break in winds of 30 to 40 mph.

ladder and tie it temporarily to the boom while adjusting the U-bolts and sleeves. However, it may be more convenient to have an assistant on the ground to hand up one side of the element arm while another manipulates the other side by means of a rope (don't use guy ropes tied to the tips of the bamboo for manipulation). Make sure the pulley tips of the arms are on the ground side.

The other ends of the guy ropes are now fastened. At this point only the square of guy ropes between the individual bamboo arms will have one or two ropes unfastened 35 feet in the air. The extra spiders are 7.5 feet out from the center of the tower, but by wearing a lineman's belt and leaning way out from the tower, they may be reached to tie the guys on.

The second set of spider arms is now added. If convenient, the boom may be rotated around to put the other boom end above your ladder; the second set of arms and guys is then added in the same manner. The pulley tips of the bamboo are close to the ground and the guy rope between them is securely fastened. This guy carries more weight than the others and if possible, should be of heavier nylon rope. It should always be kept tight as it distributes the wind load more equally between the four diagonal arms.

When these guys have been adjusted and the other guys have been added to the pulley arms, the 4 boom U-bolts are loosened and the boom is rotated 180° on its own axis to put the pulley tips high in the air. The remaining guys are then fastened. The wire elements are prefabricated on the ground and may be pulled up with the pulley ropes and tied down. In one case it was found convenient to hang a nylon rope from the boom tip to support the heavy gamma match and RG-8/U coaxial feedline.

Considerable adjusting and tinkering was done with the elements after the antenna was mounted on the tower. The final dimensions include a coil at the bottom center of each element for length adjustments. Little efficiency is sacrificed by using coils and they are much more convenient than changing large lengths of wire.

The resonant frequency was 7.0 MHz when the boom was 25 feet off the ground; at 56 feet the resonant frequency was 7.2 MHz with the following SWR across the band: 7000 kHz, 1.8; 7100 kHz, 1.4; 7200 kHz, 1.25; and 7300 kHz, 1.4. The resonant frequency shifted about 50 kHz during rotation due to proximity of nearby objects, but this was of little consequence because of the broadbanded nature of this quad. The gamma match uses two #12 wires 4.6 feet long and spaced 6 inches apart with plastic spacers; a series 200 pF capacitor was used to tune out the inductive reactance of the gamma match.

The front to back ratio runs from 5 to 40 db depending on the direction, angle of radiation and skip characteristics. Numerous observations of 7 MHz broadcast stations and hams showed average gains of two S-units or better over a 1000 foot long-wire aimed at Europe.

The antenna has stood up well in 50 to 60 mph winds, snow, and ice as long as all guys were kept tight; if the guys are not tight the arms will break in winds of 30 to 40 mph. This antenna is a joy to operate on 40 meters.

GIVING THE QUAD A NEW LOOK

The cubical quad antenna has long been popular among hams because of its inexpensiveness, excellent performance and light weight. However it suffers from two major drawbacks—it is cumbersome and not very rugged. The design described reduces these deficiencies and even further reduces the quad's weight and perhaps expense.

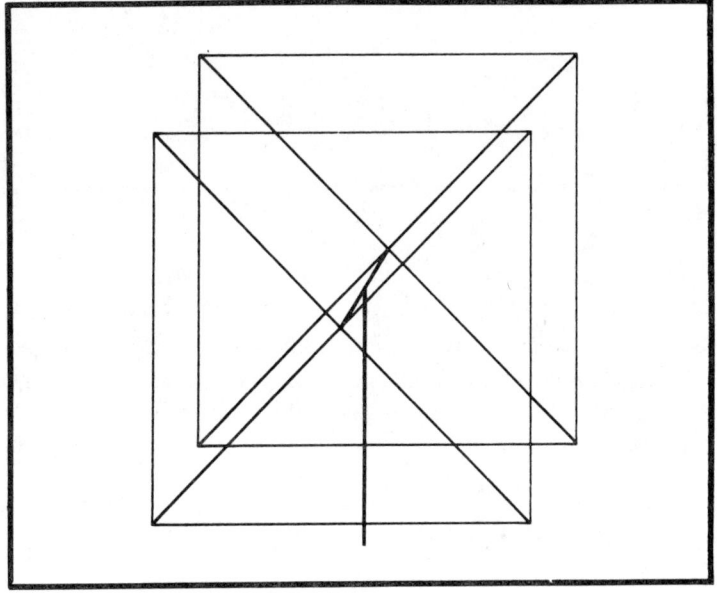

Fig. 5-18. The cubical quad.

The mechanical structure of the cubical quad serves to support two square forms of wire a short distance apart. The traditional method of construction is not particulary efficient, particularly as large torques are impressed on the boom and mast, leading to lack of rigidity, mechanical failure, etc. The design presented here (Fig. 5-19) overcomes these deleterious forces to a large degree. A quick examination shows the merits of the latter. First, it uses less support material, thereby reducing weight, wind loading, and stress on the mast. Second, the center of mass of each side of the antenna is much closer to the mast, resulting in reduced torque in the

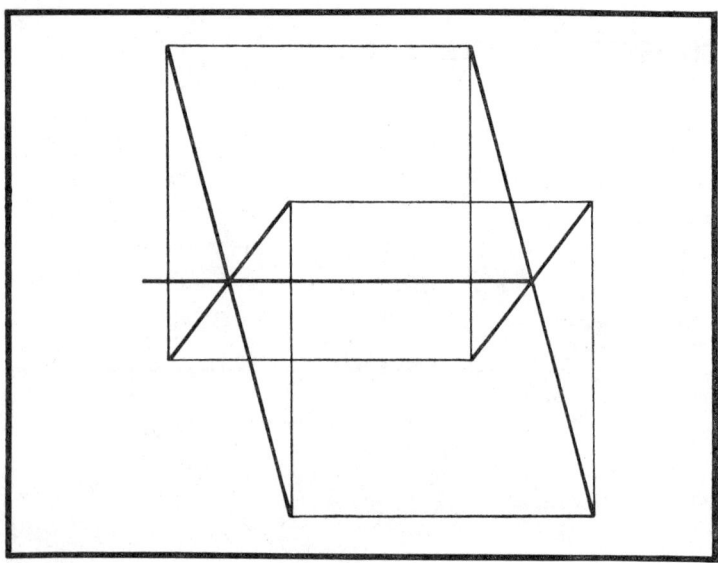

Fig. 5-19. A better mechanical mount for a quad.

center of the antenna, which might otherwise cause it to break. The electrical specifications are in every way identical with those of the traditional quad, and any of one's preferred techniques for constructing a regular quad can be used for this one, except of course for the fact that metal cannot be used for the arms.

Chapter 6
VHF-UHF Antennas

A 146 MHz MOBILE ANTENNA

The most used 2m antenna is the quarter-wave whip. This antenna leaves a great deal to be desired especially if it is mounted where shielding reduces its effectiveness. An antenna which has some appeal is the coaxial dipole, an efficient radiator, which could be elevated above the car roof to minimize shielding. However, this antenna usually has problems with feed lines in its standard form.

The normal coaxial dipole consists of a quarter-wave whip on top of a metallic supporting pole which is metallically and electrically joined to a quarter-wave sleeve. The coaxial cable center conductor is connected to the bottom of the whip and the braid to the pole and the sleeve. This sytem produces a strong ground wave but also produces standing waves on the supporting pole. By placing radials a quarter-wave below the bottom of the sleeve they act as an rf choke to reduce the standing waves on the pole. A secondary effect of these radials is to utilize the standing waves to reinforce the original radiated signal. Thus the radials add to the gain of the antenna.

If such an antenna could be used with the car body acting as the ground plane, it would be a very efficient mobile radiator. The feed impedance of a coaxial dipole antenna is a nominal 75Ω and normally it would be necessary to feed the coaxial cable up the center of the supporting pole to the feed point. This necessity would make the antenna a rather messy one to attach to a car. On studying the suggested antenna it was realized that the distance from the ground plane to the feed point is approximately a half wavelength.

One fact emerges from this discovery. Because impedances are repeated each half wavelength on a transmission line it is possible that a feed point impedance at the ground plane could be repeated in impedance at the junction of the whip and sleeve. However, because of velocity factor effects on transmission line, it would be impossible to use ordinary coaxial cable for this purpose. The transmission line must have a velocity factor close to unity.

A transmission line with a velocity factor of unity is air spaced coaxial cable. By now the reasoning may have become clear to the more astute. The support pole can become the sheath of an air spaced coaxial cable so that a wire fed centrally through its half wavelength will produce the required unity velocity factor half wavelength transmission line. The impedance of this line is not critical as it will repeat the impedance seen at one end to the other. This means that the materials used can be governed by the fittings and facilities of a home workshop.

In practice the inner conductor will need the support of two or three beads along its length. These could be pieces of poly from coaxial cable. This will tend to reduce the velocity factor very slightly. The bottom section of the antenna, due to end effect, is slightly less than a half wavelength. These two factors just mentioned tend to cancel each other out.

There are many ways of fabricating the antenna and one suggested method is shown in the accompanying sketches. For economy the PL239 plug assembly was chosen for a base connector. The half wavelength supporting tube is brazed or soft soldered to the tailpiece of the connector. Incidentally, pick a connector with an insulation material that is not susceptible to heat. Also note that the bottom section length should make due allowance for the length of the connector used. A brass spacer ring is brazed or soft soldered to the top of the support pole. This brass ring is drilled and tapped at three or four points to allow the brass sleeve to be screwed into position.

The inner conductor and whip is made from one piece of material. One end of this material is reduced to fit into the

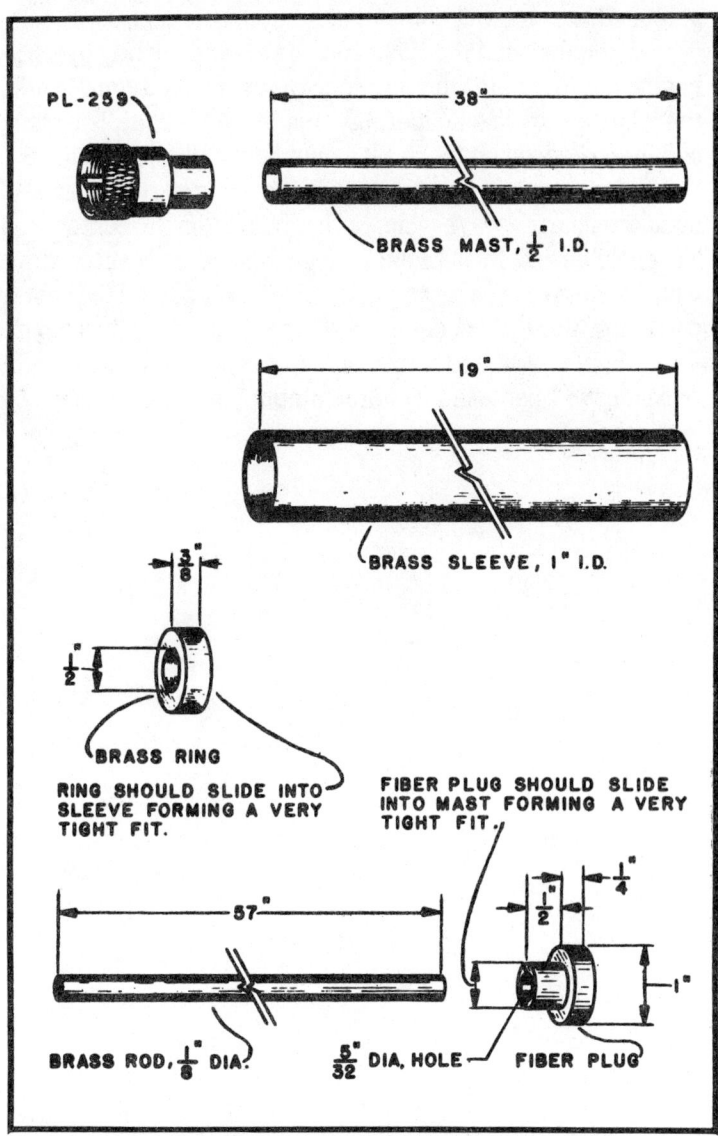

Fig. 6-1. Sub-assemblies for 146 MHz coaxial dipole items shown are not drawn to scale and most critical dimensions have been left out.

inner of the connector. Slip the support beads on the inner conductor, insert it into the support pole and solder the end to the connector. Next a small fiber, or similar material plug is fed over the whip end of the inner conductor and pushed to the top of the support pole. A generous application of an epoxy based glue at this point will complete the construction.

When installed the SWR may be shifted slightly by varying the length of the whip section. On the few antennas made, the whip length was deliberately made long, about 22 inches, and then reduced bit by bit till a minimum SWR was achieved.

In-operation tests were made by comparing against a standard quarter-wave whip, both mounted on the center of an automobile roof. In all tests, changing from the quarter-wave whip to the coaxial type antenna more than doubled the limiter current of the FM receiver used for signal strength comparisons. Some of these antennas with normal quarter wave radials have been used as home station antennas with excellent results.

TWO METER MOBILE HALO

A 2 Meter mobile installation that is capable of optimum performance and can be installed or removed in about 2 minutes was the objective, allowing the same transceiver to be used for both home and mobile operation. The principal feature of this installation is a new halo antenna held in place by a magnet. This antenna is a Halo designed especially for this installation which is light in weight and very simple to construct. The aluminum tubing used was 20 inches long (approx. ¼ wavelength) to allow placing the halo in the center of the car's metal roof. A surplus magnetron magnet holds the antenna to the car roof at all legal speeds and then some. Thin mylar type was used on the pole faces of the magnet and a plastic plug cap was placed on the tubing's lower end to protect the car.

The halo antenna's element is made from flat stock of 24 ST aluminum which is the springy variety. This allows bowing a flat piece into the circular shape. A simple gamma match is made from a piece of ⅜ aluminum tubing having a short piece of wire (inner part of a piece of RG-8/U) inside of it for capacitive coupling to the tubing. No adjustments are included for the SWR should be under 2:1 over the 2 meter band if it is constructed as shown. The nominal impedance of this halo is 50 ohms. The spacing between the ends of the main element will increase or decrease the frequency band at which the optimum SWR is obtained.

The performance exceeds that of the other 2 meter mobiles tested and has been completely satisfactory in all respects.

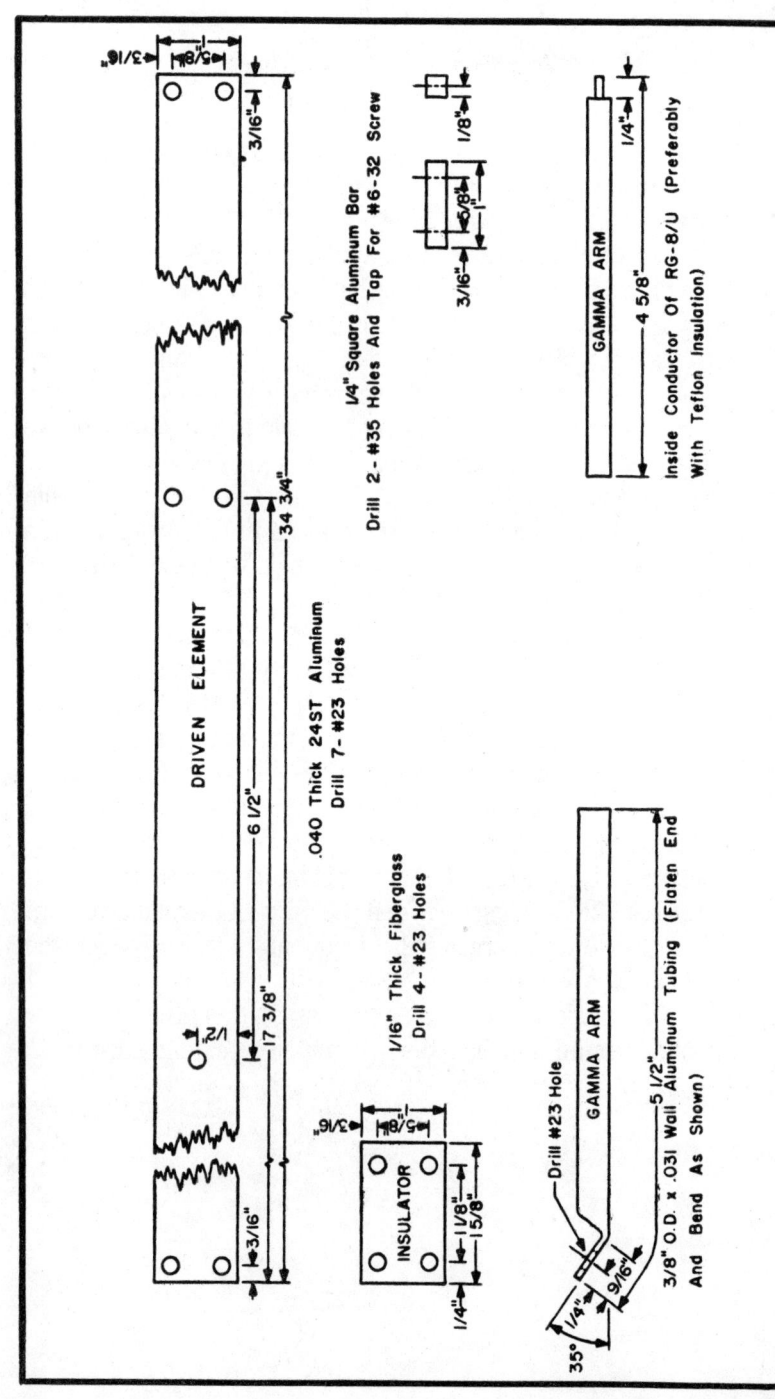

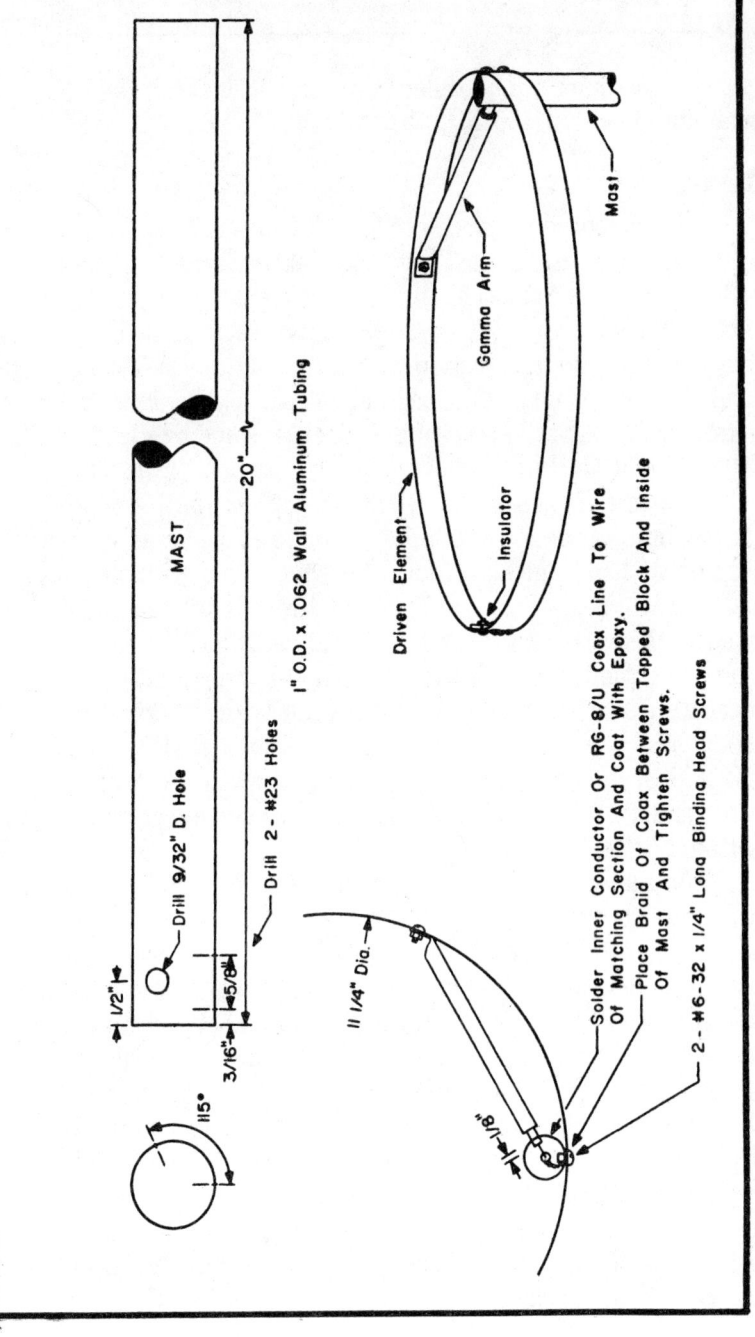

Fig. 6-2. Construction of a two meter halo.

6 db on 450 MHz

This antenna can be called a stacked collinear; it does a beautiful job in the 450 MHz band.

Construction

Start by cutting 5 pieces of RG-8 10 inches long. These will be the ½ wave sections. Strip the braid back until the braid is 8-11/16 inches long. The braid should be centered in each section leaving about ⅜" gap between the sections as in the drawing. Cut another section of RG-8 about one foot long; strip the rubber insulation back about 6". This will be the bottom section. Strip some of the braid back and overlap the center-conductor to the braid as before, placing this section below the already assembled 5 sections. Solder this piece on. On this bottom section, from the top edge of the braid, measure down 4-5/16 inches. Mark this spot. Tin it with solder and mark it again. This is where the 4 radials will leave the braid for their drooping 30 degree 5 inch run. Needless to say, start your radials with about 6½" lengths then bend them, measure 5" and cut them off. Then solder securely, binding them with copper wire first. For the radial material, brazing rod is ideal. It bends with no difficulty but won't accidentally bend, and it solders very well. Put the coax connector on the bottom of this section.

For the top section, cut a 12 inch length of RG-8. Prepare one end as before. Remove the rubber insulation. From the upper end, pull the braid back to the point 4-5/16" from its bottom edge. Mark this point . About ¼" further towards the top, cut the insulator and center conductor off. At the marked point, 4-5/16" from the bottom edge of the braid, carefully cut the polyethylene off leaving the exposed center conductor and the retracted braid. Cut the center conductor off leaving about 1/16" exposed. Fan each strand out to the edge as shown in Fig. 6-3. Pull the loose braid up over the cut center conductor and pull down tight as shown. Solder the braid to the fanned out edges of the center conductor, then solder the braid making it solid for about another 5½". Measure up from the

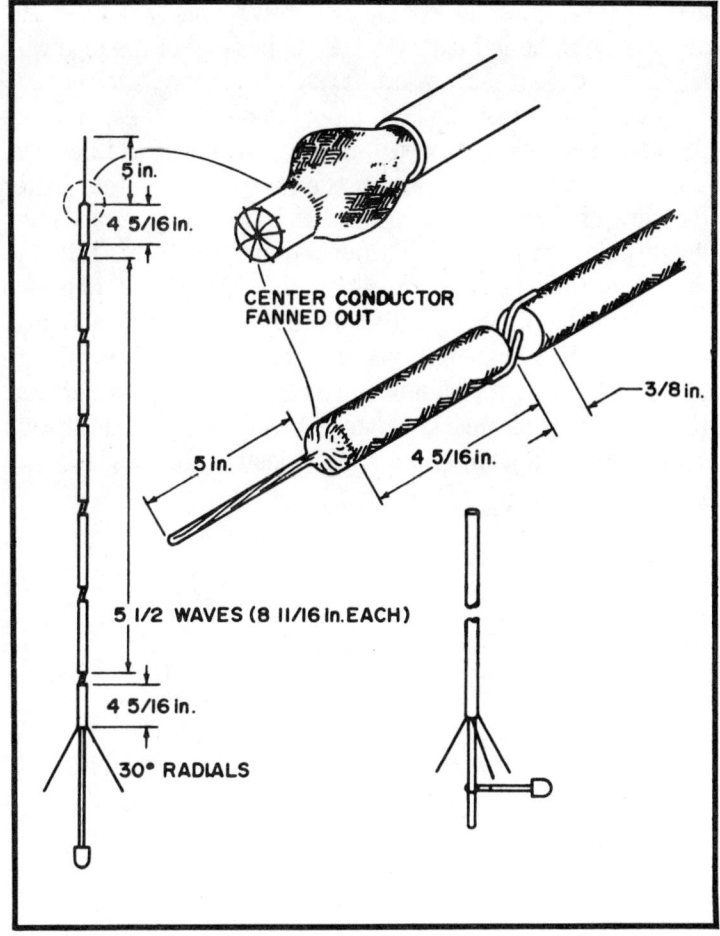

Fig. 6-3. Fan each strand of the center conductor out to the edge.

fanned out center conductor and cut the braid off at 5 inches. This 5 inches will not compute with the formulas but, for some reason, it works best.

This completes the construction of the antenna itself. Buy a 10 foot length of PVC rigid plastic pipe ¾" diameter. Lay the antenna down the floor. Place the PVC pipe beside it with the bottom edge at the top of the radials. Cut the pipe off about 4 inches longer than the antenna from the radials up. At the top section of the antenna, drill a small hole through the pipe. Place the antenna inside the pipe and thread a piece of nylon fishing line through the pipe, through the gap in the antenna and out the other side. Tie securely. This will keep the antenna straight and taut. Seal the upper end of the pipe with either the cap, plastic wood, fiberglass, epoxy, etc.

Take the remaining PVC pipe, cut off 12 inches, and take the rest and place below the antenna, overlapping 12 inches. Take PVC cement and cement the two pipes together. After they are dry, take the remaining 12 inches of pipe, and rip it lengthwise into 4 pieces. Throw two of them away. Take the other two and place them along the 12 inch overlap in a concave fashion. Cement them onto the assembled antenna.

More ½ wavelength sections can be added to this antenna as long as the number of sections is an odd number. However, going from 5 sections to 11 will only provide about 3 dB of additional gain, and a lot of construction headaches.

A TUNEABLE ANTENNA FOR 432 MHz

A novel and interesting tunable type of UHF antenna has been developed for indoor test use, antenna range work and amateur phased array use, on 432 MHz.

Figures 6-4, 6-5, and 6-6 show views of the single unit. It is particularly suitable for use with large screen reflector where many would be used, spaced a certain distance apart in proper phase; with each fed by a separate 50 ohm matched cable. Note that this is a tunable antenna with a knob and plenty of gain and front-to-back ratio, also. A one-to-one SWR can be obtained by juggling the spacing of the radiator from the reflector using the threaded rod in the center; adjusting the trimmer in the matching section; the position along the radiator of the trimmer; and the tuning or length of the radiator versus the "penny" tuning of the end capacitor C2.

The tunable feature can be very useful for working different portions of the band, such as on 440 for ATV. The average antenna cannot cover such a range properly. This one can be cut and tuned as needed and then retuned with a knob.

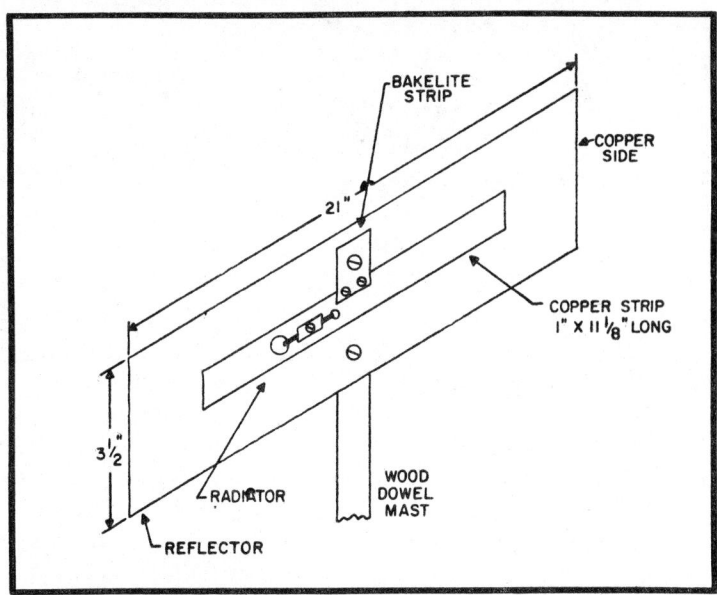

Fig. 6-4. Front view; 432 MHz test antenna.

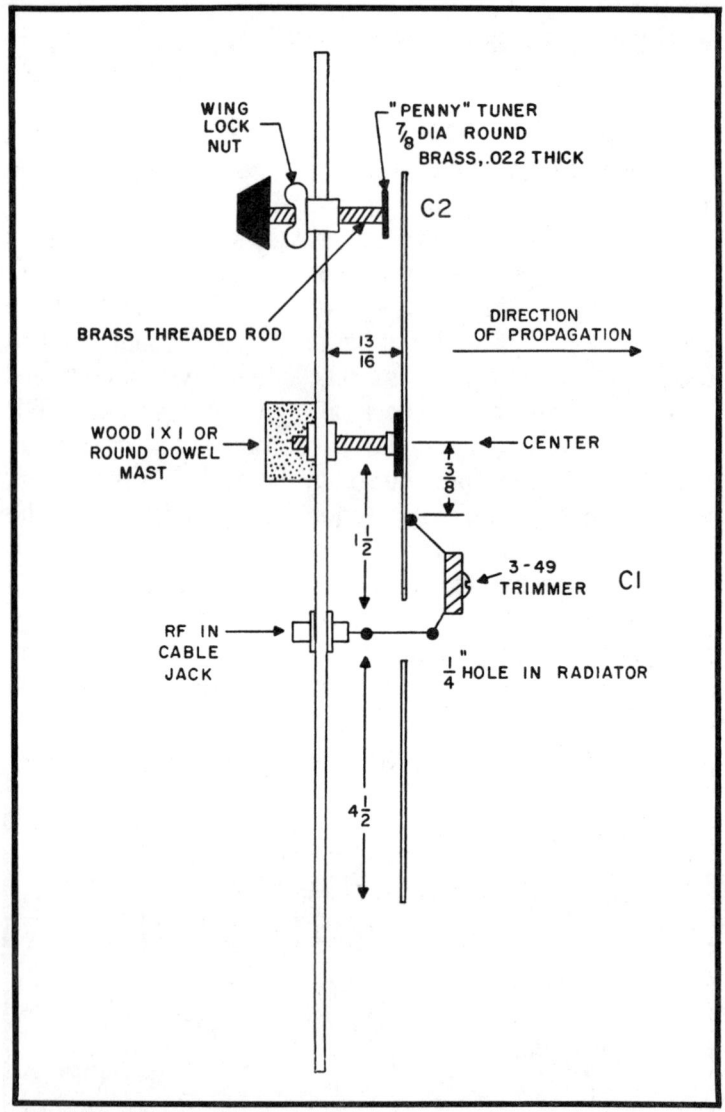

Fig. 6-5. 432 MHz test antenna.

Note that the tuning knob and input cable can be tuned from the back of the antenna. The matching capacitor C1 can also be turned around and tuned from the back through a hole in the reflector with a full-insulator screwdriver. These are useful features and it is quite surprising to see how dead it is in the back. Of course, a larger reflector will have an even greater front-to-back ratio. Note also the resemblance to the famous low frequency "Windham" antenna, with its single-wire feed.

For on the air indoor tests, and small antenna range work, a lamp bulb receiving antenna is shown in Fig. 6-7. The reflector and radiator assembly are somewhat similar to that of Fig. 6-4, except that the no. 49, pink bead, 120 milliwatts, is shunted across a section of the radiator which is a piece of solid aluminium wire. Both of these antennas can be used to table-top or on floor stands.

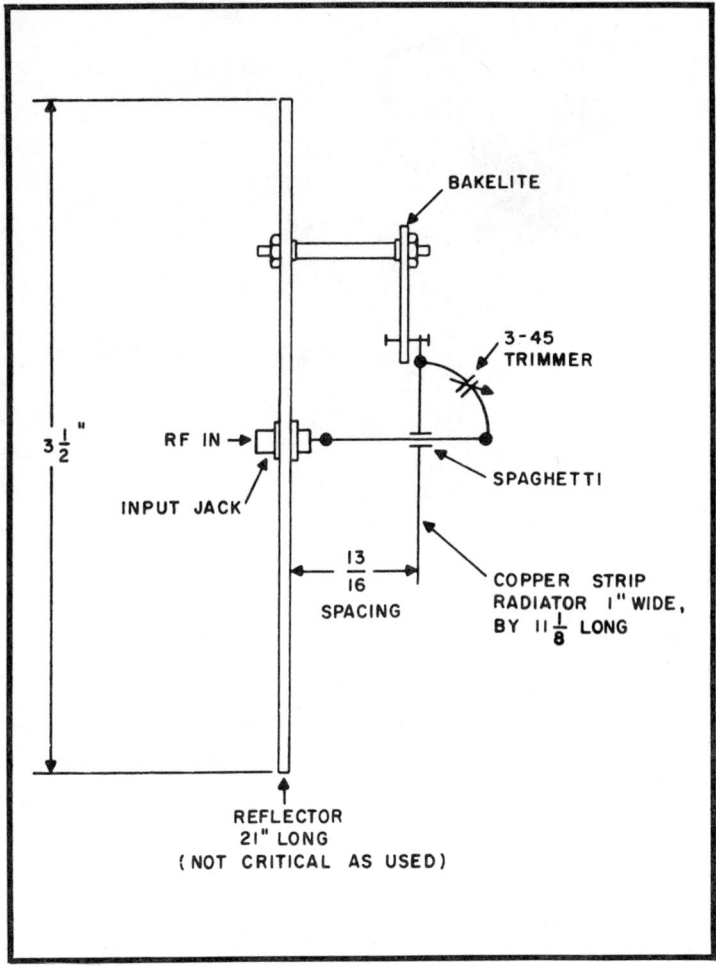

Fig. 6-6. Side view, 432 MHz test antenna.

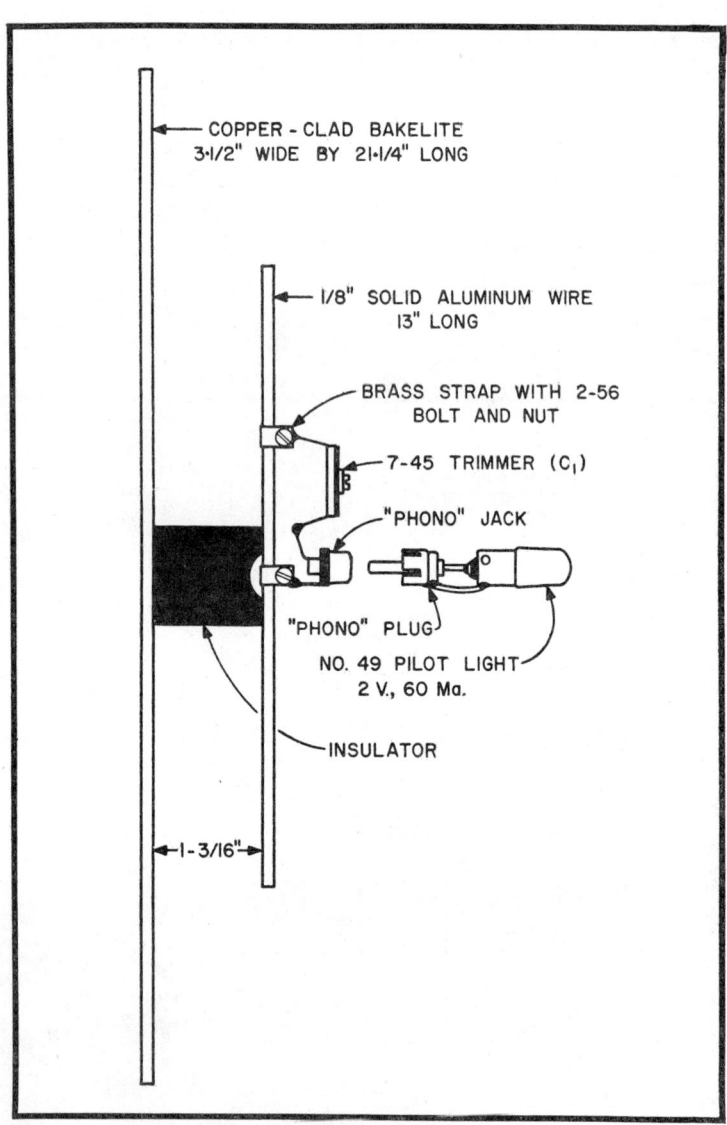

Fig. 6-7. Top view, 432 MHz test antenna.

COLLINEAR-GAIN ANTENNA FOR VHF-UHF REPEATERS

One of the most popular antennas in the amateur repeater world is the omnidirectional collinear coaxial stack, although it is seldom called by that name. Versions of this antenna are manufactured by several firms that build antennas specifically for the commercial bands.

Two of the reasons the collinear antenna is so popular are that it can be made to exhibit a great deal of omnidirectional gain at a very low angle of radiation and it takes up very little space. In its manufactured form, it resembles a long fishing pole with a pair of crossed fins at the base.

In spite of the fact that a great deal of painstaking effort is required to make the antenna and get it just right, the operation is surprisingly simple. And what makes it even more attractive to the amateur is its low price. About all that is needed is a good-sized length of 50Ω foamdielectric coaxial cable and some polyvinyl-chloride (PVC) pipe. For 2 meters, the pipe should be between 20 and 21 ft in length; for 450 MHz, an 8 ft length will suffice. The total omnidirectional gain (as compared with a reference dipole) will be about 6 db.

Building the Antenna

Ignoring the structural aspects, the antenna itself is nothing more than a series of precise lengths of coaxial cable soldered in an alternate phase-reversal configuration as shown in Fig. 6-8. A quarter-wave whip at the antenna's tip shorts the inner and outer conductors of the coax and becomes the terminal radiating element. At the lower end of the antenna, the last coax section actually becomes the feedline itself, whose length, incidentally, is not critical as long as the dimensions are followed with religious fanaticism.

A number of amateurs have managed to build antennas of this type, and diagrams have never been scarce. But few have handled the project successfully. Getting the antenna together is not difficult. The problems start to happen when it's time to turn the soldered-together pieces of coax into a structurally sound antenna. Applying wet epoxy, as in a fiber-glassing

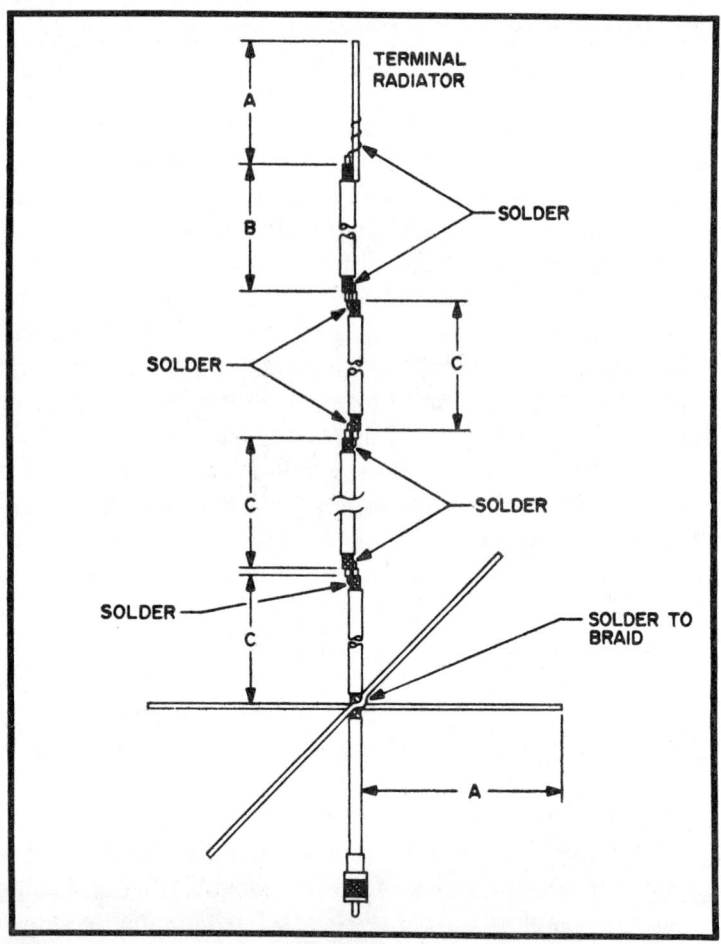

Fig. 6-8. The collinear gain antenna is made up of coaxial sections connected in a phase-reversal configuration. The bottom section (from the radials to the first joint) and the upper section (which joins the antenna to the shorted radiator) are half the size of all other sections.

scheme, doesn't work out. The problems may be attributable to some chemical interaction between the wet epoxy and the coax dielectric (changing the dielectric constant of the line) or because the hardened epoxy doesn't allow any flexing of the coax braid. Sealing the antenna with epoxy will not work. When the antenna is rigid and looks great, it will measure a very disappointingly high standing wave ratio.

The commercial antenna use fiber glass, but they do not use it to seal the antenna. Instead, they use an inert and flexible sealer, then encase the whole assembly within a preformed fiber-glass tubular envelope. At least one of the commercial suppliers uses beeswax as the inert sealer. Actually, there is no real need to immobilize the antenna once it has been placed inside the PVC pipe. The most important point in the construction process is to make it water-tight. Water drops inside a length of coax do bad things to antennas and feedlines.

The dimensional details of the antenna are shown in Fig. 6-9. Lengths have been calculated in the decimal system to the nearest hundredth of an inch. The 2 meter figures are based on an operating frequency of 147 MHz. The antenna is broadbanded enough to give an swr of close to unity regardless of the FM channel of operation. The 450 MHz frequency of operation is 441 MHz, exactly three times the frequency of the 2 meter version. The 450 MHz dimensions are just one-third those shown for the 2 meter version.

The 220 MHz dimensions are calculated for 220.5 MHz (half the 450 frequency).

To begin construction, cut eight lengths of coax from the reel. Each piece should be cut about an inch oversize, then trimmed down later so that all pieces are of exactly the same length. The dimensions given are end-of-braid to end-of-braid for any given length. (See closeup detail in Fig. 6-10.) The braid-to-braid distance should be approximately the same as the distance between the inner and the outer conductor of the

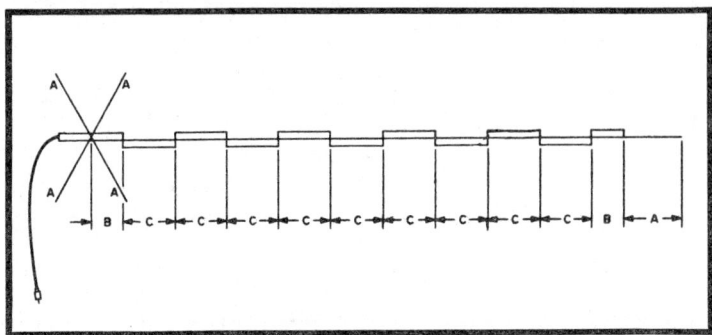

Fig. 6-9. Layout and dimensions of collinear gain antenna. The 2 meter dimensions are for a frequency of 147 MHz; the 450 MHz dimensions are for 442 MHz exactly. The 220 dimensions are for 220.5, just half the 450 frequency. The antenna is broadbanded enough to yield a low vswr on any frequency within a megahertz of that shown.

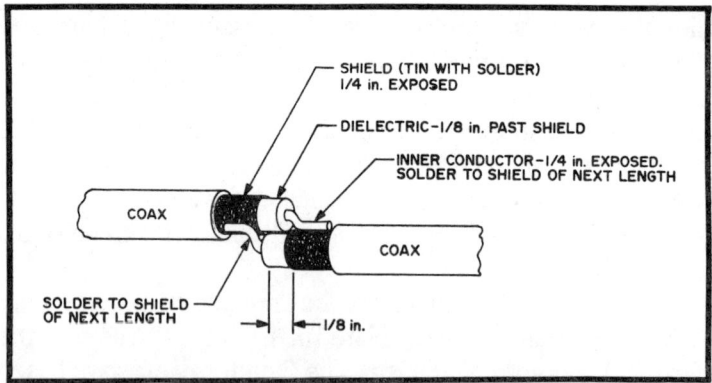

Fig. 6-10. The coaxial lengths should be soldered as shown. Keep the braid trimmed evenly all the way around and make sure the brad from one section doesn't make contact with the brid of the next section. If conductors are well tinned, problems will be minimized. Coax lengths are measured individually from braid to braid end.

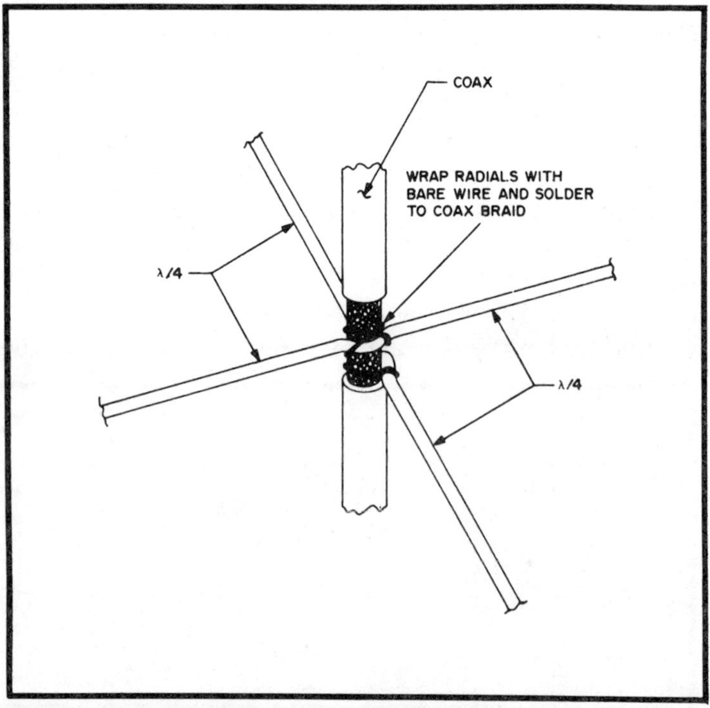

Fig. 6-11. Radials, of narrow-diameter copper tubing, should be cut to slightly longer than a half wavelength. The center should be bent to conform to the rounded shape of the coax braid so that on each radial a quarter-wave length extends outward from the coaxial braid. Tin the braid first. After wire-wrapping and soldering, wrap the joint well with electrical tape.

coax used, or approximately ⅛ in. This dimension is the only one that does not change with operating frequency or band.

When all the lengths have been cut and trimmed to the precise lengths, study Fig. 6-10 carefully, then tin all exposed braid and conductors. This tinning process is an important step and should be done as completely as possible.

Solder the lengths together, and use care to avoid handling the soldered pieces any more than is absolutely necessary. The braid can pull loose without much encouragement. When everthing looks correct, wrap the joints with tape. Be very careful in the handling until the antenna is safely inserted into its plastic pipe.

The quarter-wave radiator that goes at the top can be any good conductor, but copper is best. And the easiest way to get

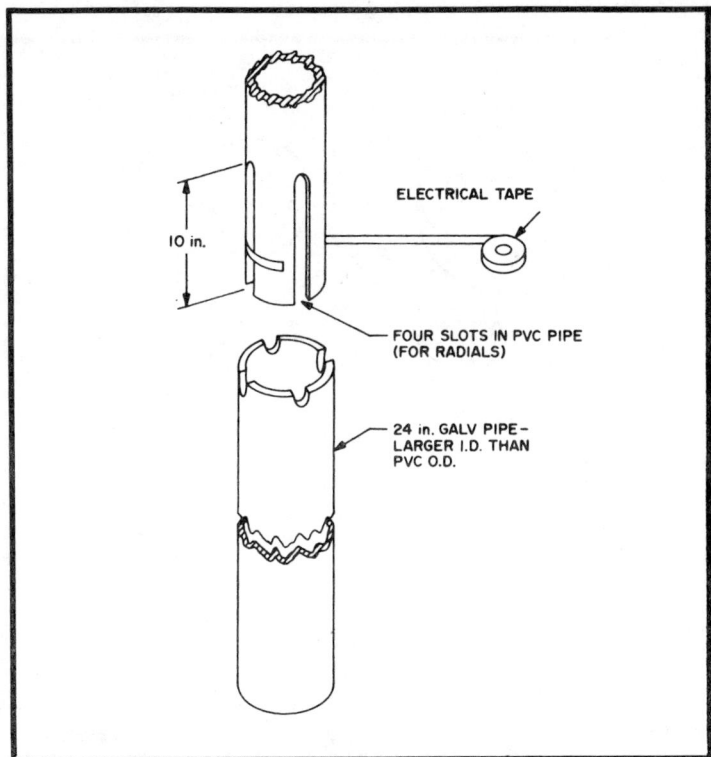

Fig. 6-12. Long slots in the PVC pipe will hold the radials in place with the antenna inserted. Wrap the bottom well with electrical tape after the antenna is installed in the fiber tube. Notch four matching places on a 2 ft length of galvanized pipe to seat, and try for a snug fit.

a good, stiff copper conductor is to buy some narrow-diameter (⅛ in. is ideal) copper tubing. The same material can be used for the radials at the base of the antenna.

Ground Radials

There is nothing critical about the manner in which the radials are attached to the antenna. Figure 6-11 shows the system used.

The slots (Fig. 6-12) are cut lengthwise into the bottom of the PVC pipe so that the radials can be held in place when the PVC is inserted into the mounting pipe (made of heavy metal). The metal pipe is notched gently to seat the radials. Before inserting the PVC into the larger pipe, the slots on the PVC should be taped up (after the antenna is installed in the PVC sheath).

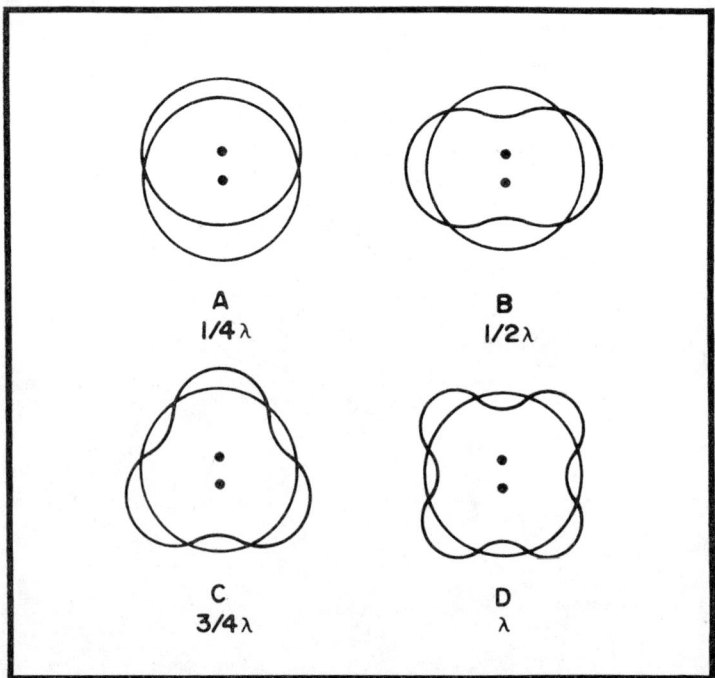

Fig. 6-13. By spacing the antenna the proper number of quarter wavelengths, some interesting radiation patterns can be obtained. In the patterns shown, the circles represent the 5.8 dB omindirectional gain achieved by top-mounting. The asymmetrical overlays represent the patterns obtained by side-mounting. Note that even though signal loss occurs in some directions, significant gain improvement is realized in other areas.

If the repeater doesn't give omnidirectional coverage, more gain than the 5.8 dB discussed may be obtained by merely spacing the antenna a prescribed number of quarter wavelengths from the tower. Of course this means that the antenna will have to be side-mounted rather than top-mounted. Quarter-wave spacing will result in a major lobe in the same direction as the antenna is from the tower mass, as shown in Fig. 6-13A. Each additional quarter-wave essentially adds a lobe that exceeds the 5.8 dB omnidirectional reference point.

ROOF MOUNTED VHF WHIP ANTENNAS

There are many books and papers published that go into the theory of antenna performance, but none have been found that show the actual pattern and gain in addition to valid comparisons between the different types.

Here are actual test results in three basic mobile VHF whip antennas. The three antennas are the one-quarter wave whip, a base loaded one-half wave "gain" antenna that most two meter FM'ers graduate to, and a balanced end-fed one-half wave "J" antenna. Antenna talk among professionals as well as amateurs is a highly emotional topic, due to all the different misconceptions that have been proliferated.

By measuring an antenna pattern in a realistic situation, one gets a realistic pattern and not an ideal pattern. Great care must be exercised in making antenna measurements to avoid mistaken results. To make meaningful antenna measurements, certain parameters have to be kept constant, like path distance, the receiver and transmitter frequency and gain settings and the losses in the test antenna feedline system. All of these details and more were taken care of by months of hard work, anguish, and finally success.

Test Setup

To simulate the roof of a car or station wagon, a 140 × 250 cm sheet of sixteen gauge brass was attached to the three-axis antenna positioner. The equipment hook-up is shown in the block diagram and the photos.

The Coordinate System

The coordinate system is shown in Fig. 6-15. The antennas under test are oriented along the Z axis when the ground plane is parallel to the earth. The X Y plane is in the ground plane when it is parallel to the earth; the X axis is in the direction of the transmitting antenna.

The azimuth or horizontal angles of rotation, ϕ, is measured from the X axis in degrees. The elevation on vertical angles, θ is measured from the Z axis downward toward the X

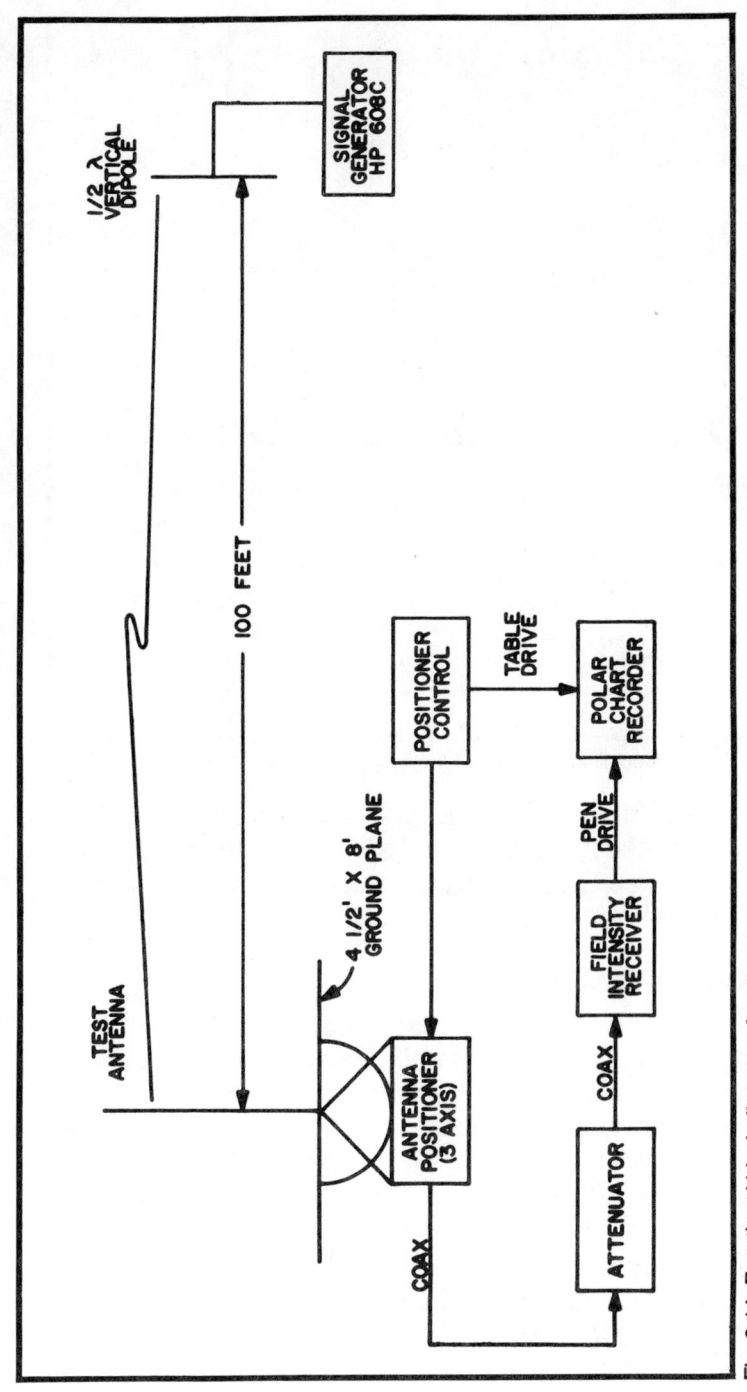

Fig. 6-14. Functional block diagram of test range.

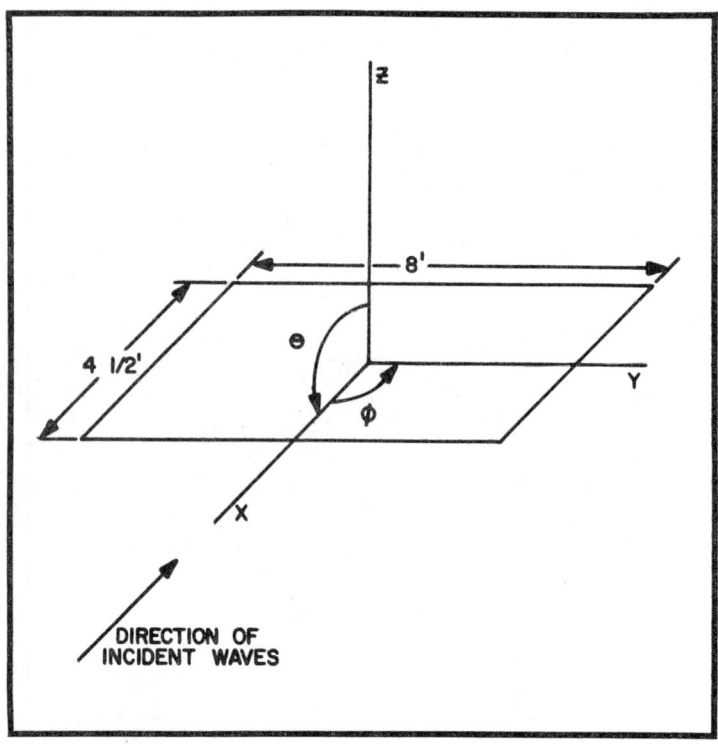

Fig. 6-15. Coordinate system.

Y plane. An example of the coordinate system: $\theta = 90°$ and $\phi = 0°$ references the ground plane parallel to the earth with the received signal arriving along the X axis or perpendicular to the long side of the gound plane.

Antenna Adjustment

The "J" antenna is basically an end fed one-half wave radiator. Only the upper one-half wave length of the antenna radiates, since the one-quarter wave matching section has balanced transmission line currents on it. A 4:1 one-half wave balun made from 50Ω coax is used to feed the 200Ω feed point impedance. The one-quarter wave whip was cut to resonance and connected directly into a bulkhead coax fitting on the ground plane. The one-half wave base tuned antenna (a Gam model TU-2) was also trimmed for minimum swr. The swr of all three whips was checked over a 2 MHz band centered on the 146 MHz and found to be less than 1.5:1. The reference dipole used was a commercially made, tunable, standard antenna used for field strength measurements. Preliminary measurements indicated that the "J" antenna was the most sensitive of the three, therefore it was used as the zero dB reference in the pattern measurements. The receiver and the polar chart recorder were checked using a calibrated attenuator to ascertain their ability to accurately track the signal strength.

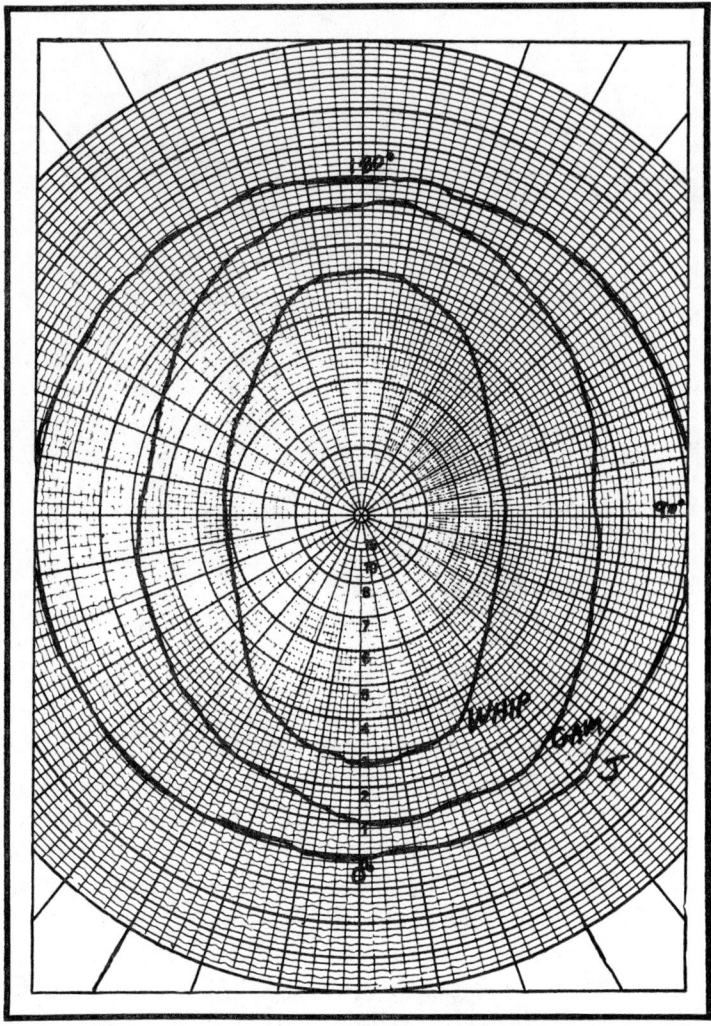

Fig. 6-16. Horizontal patterns. θ is held constant at 90°, while ϕ varies from 0° to 360°.

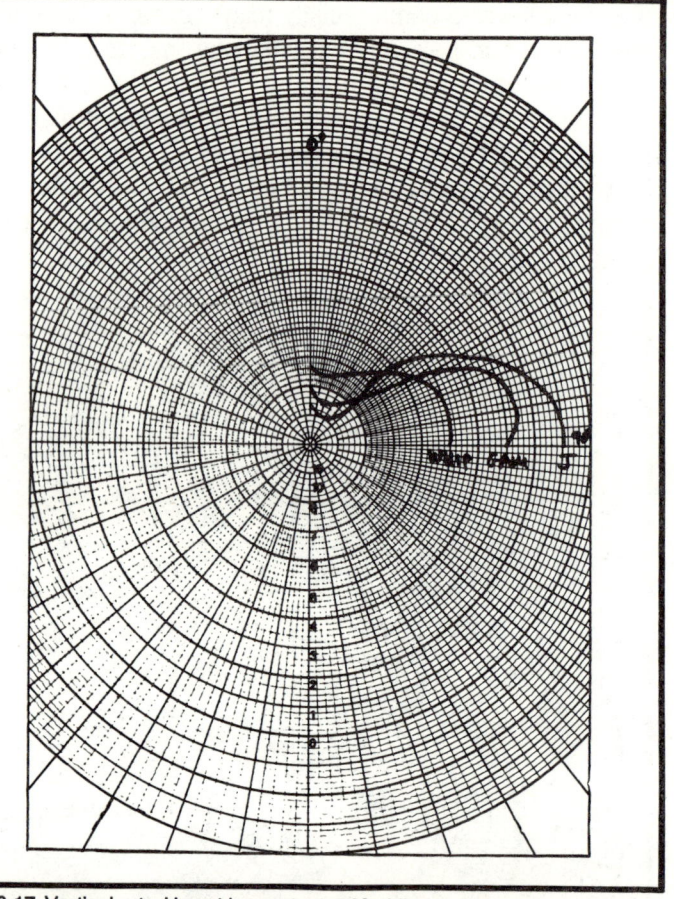

Fig. 6-17. Vertical cuts. Here φ is constant at 0° while θ is changed from 0° to 90°.

Test Results

First, all the antennas were rotated about the Z axis giving an azimuth on horizontal plane pattern. Referring to Fig. 6-16, it can be seen that the horizontal pattern for the "J" is within a ½ dB of being circular. The one-half wave vertical had an elliptical pattern. At φ = 0° and 180°, (along the short dimension of the ground plane) the one-half wave vertical was 1 dB down, but at 90° and 270°, (the long dimension) it was almost 3 dB down! For the quarter-wave whip, the elliptical pattern starts to approach a rectangle. At φ = 0° and 180°, the pattern of the whip was down 3 dB from the "J" and the 90° and 270° it was down about 7 dB. Notice how each succeeding antenna becomes more and more dependent on the ground plane.

Next ϕ was kept constant at 0°, 45°, and 90° and a portion of an elevation plane pattern was obtained by tilting the ground plane from horizontal. The antennas and the ground plane were rotated from 0° to 90° in the θ direction with ϕ held at 0°. Looking at Fig. 6-17, the "J" has its maximum lobe directly on the horizon in this direction. The half-wave vertical has a maximum lobe which is 13° up from the horizon in this direction. The quarter-wave whip has an almost circular pattern from $\theta \times 90°$ to 65°. The response of the quarter-wave whip varies about 3 dB from maximum to minimum in the ϕ

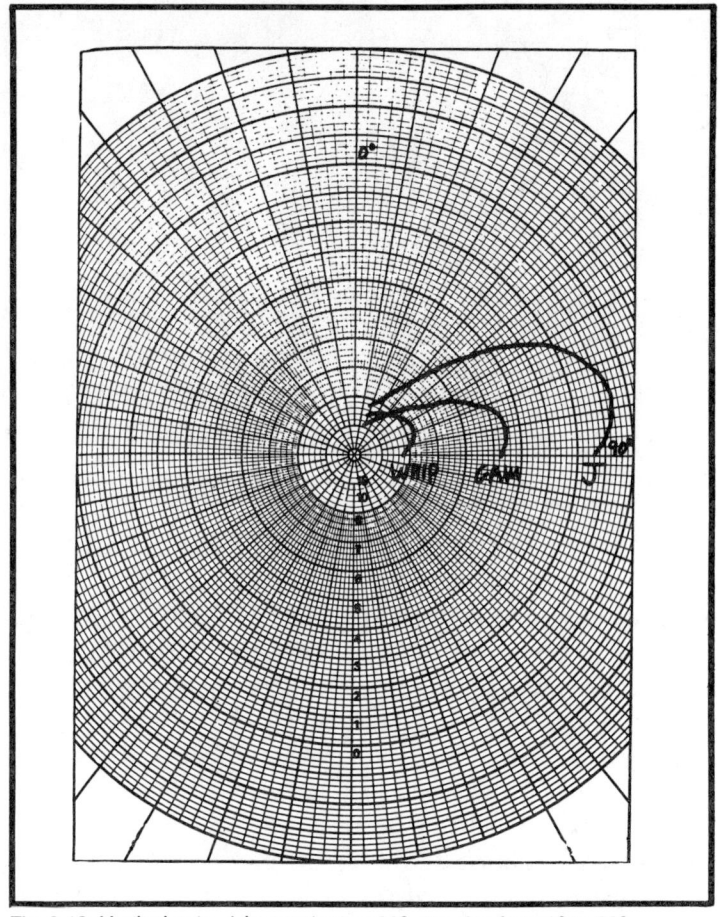

Fig. 6-18. Vertical cuts. ϕ is constant at 90°, θ varies from 0° to 90°.

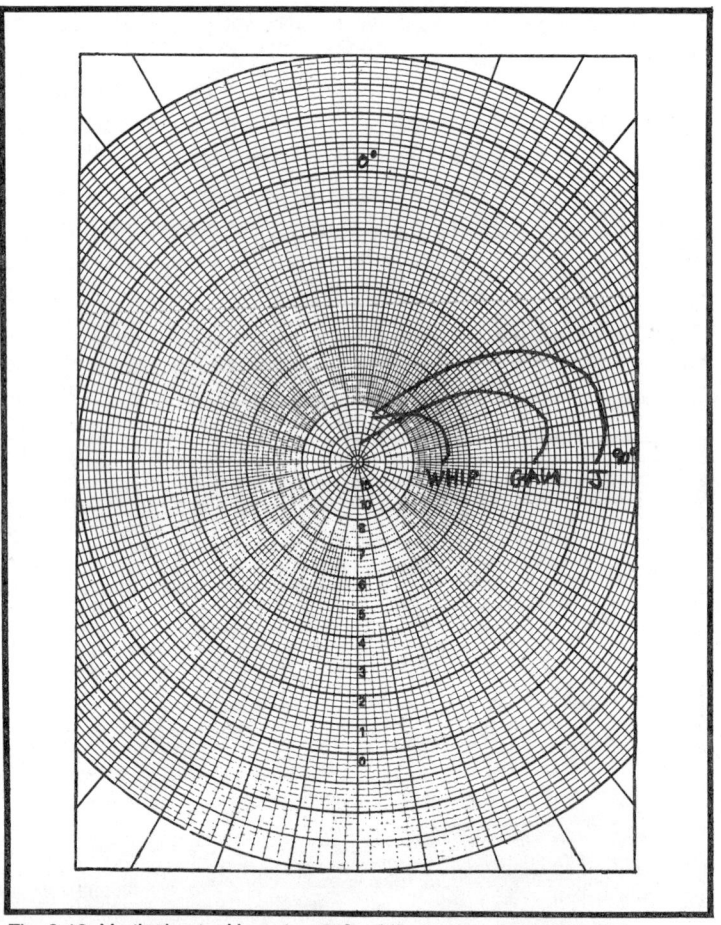

Fig. 6-19. Vertical cuts. Here $\phi = 45°$, while θ varies from 0° to 90°.

direction, whereas the response of the "J" varies about 15 dB from maximum to minimum in the $\phi \times 0$ direction.

As would be expected by this time, elevation cuts in the $\phi = 90°$ direction show an even more pronounced difference. Looking at Fig. 6-18, where $\phi = 90°$, it can be seen that the presence of the ground plane actually degrades the "J" performance by lifting the major lobe about 17° above the horizon! The direction of the major lobe of the half-wave vertical remains the same (about 10° − 15°), but the overall response in that direction is down 2 dB from the maximum. The quarter-wave whip begins to show a bit more directivity with a lobe up about 25° from the horizontal.

The θ or vertical patterns in the $\phi = 45°$ plane (Fig. 6-19) appear half-way between the extremes found in Figs. 4 and 5, and by symmetry can be reasonably reflected through the Z axis to give an accurate three-dimensional field strength pattern of these roof mounted antennas.

Since the "J" had the most symmetrical pattern and had the least dependency on the ground plane, it was compared with the reference dipole to arrive at a gain figure. The reference dipole major lobe field strength was compared to the "J's," using the same cable and connectors. This orientation is with $\theta = 90°$ and $\phi = 0°$. Negligible differences were found between the "J" and the dipole. By probing the incident fields

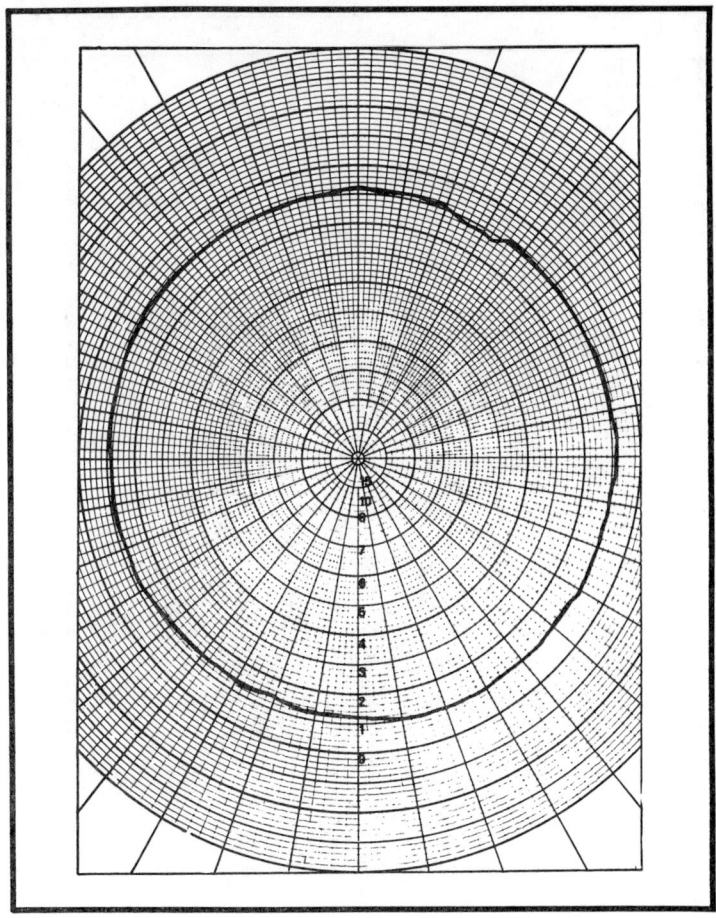

Fig. 6-20. Horizontal pattern for ¼ λ whip on a stick whip with radials.

over the area in front of the ground plane with the dipole and the "J," we found less than a ½ dB fluctuation in signal strength, indicating that reflections from surrounding objects were not sufficient to cause differences in the patterns from one antenna to the next. This means that the "J" can be considered to be a reference vertical dipole.

These results aroused a considerable amount of discussion in our local repeater group! In an attempt to keep from losing friends who owned half-wave Gams and quarter-wave Ground Planes, an attempt was made to find a set of conditions where the Gam and quarter-wave Ground Plane approached "specs." A classic quarter-wave ground plane was

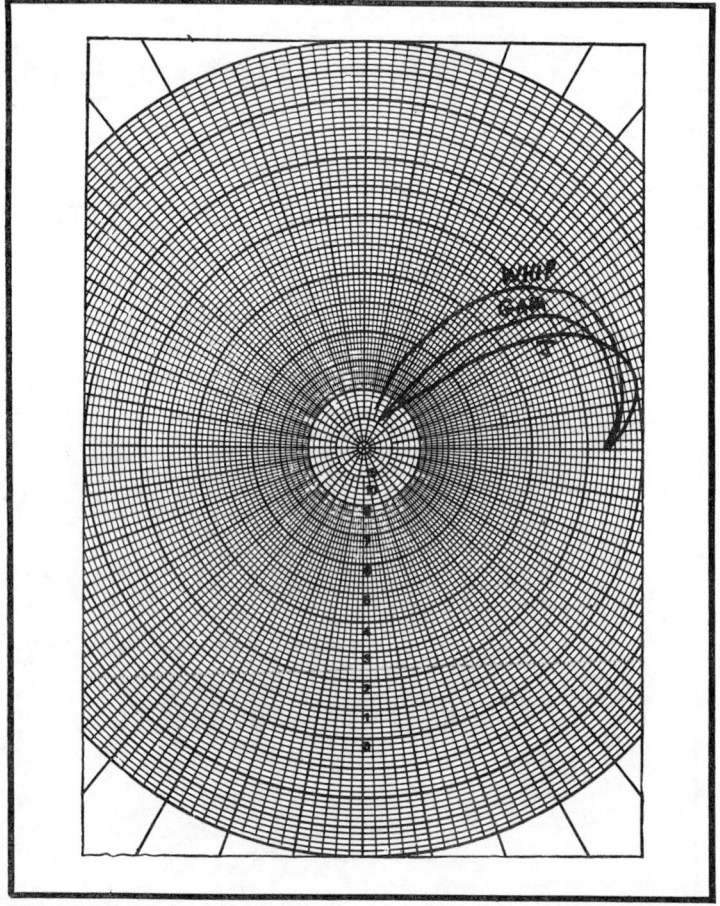

Fig. 6-21. Free space vertical pattern.

constructed using four drooping wires for the ground plane and the quarter-wave vertical radiator. This new ground plane antenna was mounted on a wooden pole well away from other metal objects (about 10 ft above the antenna positioner). Curiosity over the effect of radials on the horizontal pattern prompted a quick test. Figure 6-20 shows the results of the horizontal pattern for the ground plane. Delight followed the completion of such a smooth pattern! Next, the vertical patterns were compared in Fig. 6-21. The gain on the horizon for all three was essentially the same when they were in a "free space" condition. At best, all three antennas show the same gain as a vertical dipole.

Conclusion

We have shown that the effects of a nonresonant ground plane, such as a car roof, can be disastrous for antennas that "need" a ground plane to operate and can spoil the patterns of antennas that don't need ground planes for impedance matching. Further speculation is left to the reader as to just exactly how high above the nonresonant ground plane one would need to place a quarter-wave whip and the half-wave vertical.

TWO METER MOBILE WHIP

The need for a better two-meter antenna for mobile use has led many to try both five-eighth and three-quarter wave base-loaded antennas. From the schematic, it will be noted that two antennas are tried, one with a series coil, and the other with a grounded coil and a tap for impedance matching.

The parts required are:

1. Connector, of a type to be determined by individual need. However, in the grounded coil configuration, use a connector which makes a positive ground connection.
2. Plastic pill vial, size 1⅛" O.D. and 2¾" long.
3. #14 enamel wire
4. ⅜" dia. brass rod stock, 2" in length
5. 2 ounces of casting resin, catalyst (4 drops per ounce) and coloring if desired. Dye will give a transparent effect, pigment opaque. These items can be obtained from most hobby shops.
6. ⅛" diameter metal rod, 53" long, for series coil antenna.
7. ⅛" diameter metal rod, 48" long, for ground coil antenna (these rods are much longer than necessary).

Construction

Take the ⅜" diameter brass rod. From one end, drill a ⅛" diameter hole 1¼" deep. Measure ⅜" from this end, and at right angles to the previous hole, drill and tap for 6/32" bolt. Measuring from the same end 1", drill and tap for a second 6/32" bolt. On the opposite end of the brass rod, drill a hole ¼" deep, just large enough for #14 diameter wire.

The series coil is wound on a ¼" diameter form, using #14 wire, 11 turns. The coil length is 1½". At one end of the coil, bend the #14 wire at a right angle, then clip off a ⅜". At the opposite end, bend #14 wire to a right angle, and clip at a distance of 1½". Scrape enamel off both ends. Solder the short end to the brass rod. Take the pill vial, and, starting with a small drill, make a hole in the center of the bottom. Slowly increase this diameter of ⅜", being careful not to crack the plastic.

Fig. 6-22. Coil encapsulated in plastic pill vial.

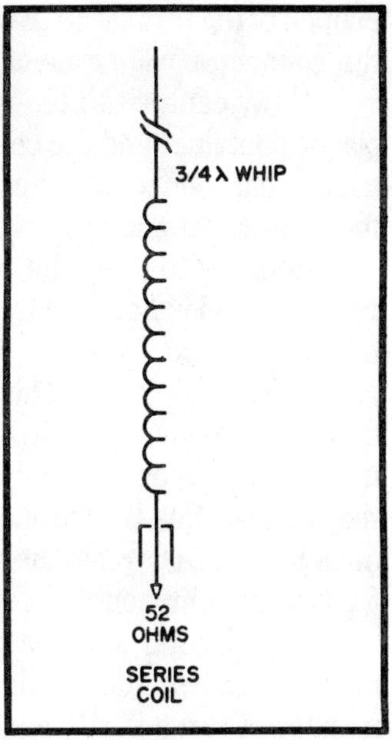

Insert the brass rod with the coil attached into the open end of the pill vial and slip it through the ⅜" hole so the rod will protrude from the bottom of the vial for a distance of 1⅜". Slip the coaxial fitting over the long end of the wire, and insert

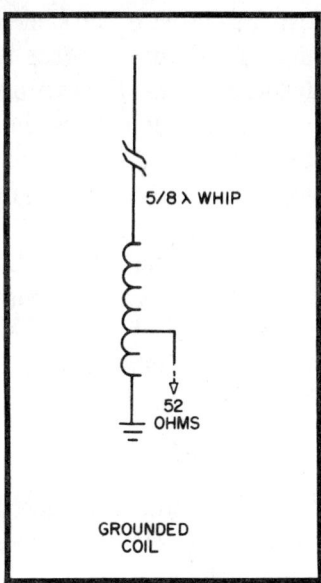

Fig. 6-23. Three quarter wave whip with series coil.

enough of the fittings so that it will be into the vial, making sure the connector will be clear to operate.

Now, center the brass rod and keep it parallel with the plastic container and the coaxial fitting. Solder the long wire to the coaxial connector. Check once again to be sure the fitting, the plastic vial and brass rod are all vertically in line.

In order to keep the plastic from escaping, and to keep the brass rod in line, it may be helpful to use either putty or a fast-drying glue. With the vial upright (coaxial connector up), pour the casting resin. This container will require two ounces of resin and eight to ten drops of catalyst. If coloring is desired, mix either pigmented color or dye with the resin before adding the catalyst. See the manufacturers instructions for details on mixing. Carefully pour the resin into the vial, and let it stand for twenty-four hours.

The second antenna is made in very much the same manner, except the coil is wound on a ⅝" form. The total number of turns is six, and the spacing between turns is the width of the wire. The end opposite the brass rod is soldered to the outside material of the coaxial fitting. A piece of flexible #16 wire is brought up through the connector and is tapped to the coil two turns from the grounded end.

To prune these antennas to frequency, use a piece of expendable wire, the same diameter as the whip, inserted into the opening of the brass rod and secured with 6/32" bolts. Insert a standing wave bridge in the coaxial line, tune the transmitter, and check the swr, which should be fairly high. Shut off the transmitter and clip approximately ¼" from the top end of the antenna. Turn on the transmitter, retune, and check swr. Keep using this procedure until swr is at a minimum. After finding the proper length, remove the whip and use this measurement for the permanent antenna, which will be made of spring metal. For best results, this pruning procedure should be done at the permanent position of the antenna on the vehicle.

Test Results

Tests were conducted over as flat a terrain as possible, trying to preclude the possibility of reflections. Three different test sites were used. The reference antenna was a

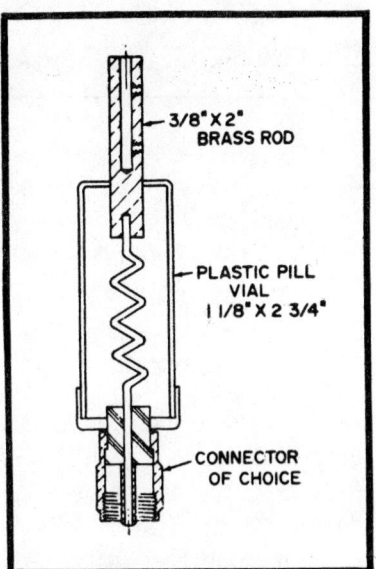

Fig. 6-24. Five eighths wave whip with grounded coil.

quarter-wave 19″ whip. All tests conducted showed the long antennas to give better signal strength and less mobile flutter both on transmit and receive.

It would appear that the three-quarter wave antenna has a higher angle of radiation than the five-eighths wave antenna. It would also appear that there was no advantage in one method of coupling over another. Although S-meter readings in most receivers are rather meaningless, they do give a relative indication of strength. The thirty tests made showed an average increase of over one S-unit. The standing wave ratio of various antennas was as follows:

Quarter-wave, 1.3 to 1
Five-eighths wave, 1.1 to 1
Three-quarter wave length, 1.2 to 1

Although it is not always a good idea to encapsulate the antenna coil, in this case the inductance was so small that no apparent differences were noted between this type of construction and air-wound coils. It was apparent the mechanical advantages were well worth any slight degrading.

THE TWO-METER GROUNDPLANE AS A GAIN ANTENNA

A groundplane is not always a unity-gain, omnidirectional antenna. Virtually *any* vertically polarized omnidirectional antenna can be used to provide gain and directivity *selectively*—without modification of the antenna itself.

To many, a true omnidirectional antenna represents the optimum approach. For the amateur who operates in the center of a metropolitan area, or the hilltop ham, or the centrally located net control—what could be better? But—what about the ham who lives between two cities and wants good, broad coverage in only two directions? Or the fellow at the foot of the hill who wastes all the rf by dumping half his output into it?

An omnidirectional antenna can still be the answer, but employed to provide gain.

The secret is not in the antenna itself, but rather in the mounting of the antenna. *Don't mount it atop a mast*. Place it near the top of a mast or tower, and adjacent to it so that the tower or mast itself becomes a part of the antenna system. Learn two simple rules to design omnidirectional antennas to give gain in practically any direction or directions chosen. The first rule is that for each quarter wave-length the vertical radiator of the antenna is spaced from the tower or mast, one major lobe is obtained. And the second rule: The bigger the mass of the supporting structure, the wider the frontal and side lobes. Consider the radiation pattern of Fig. 6-25. The solid round dot at the center represents an antenna supporting structure as seen from the sky. If an omnidirectional antenna were mounted at the top of the structure, the pattern would be roughly circular. The broken line represents this pattern at a relative field strength of 1.0. If the same antenna were to be moved from the top to the front of the tower and spaced a quarter wavelength from it, the pattern becomes more or less like that of the heavy asymmetrical line. (This is assuming the tower is between 6 and 8 inches in diameter and adjacent to where the antenna is mounted.) In the sketch, the antenna is represented by the small circle above the center dot.

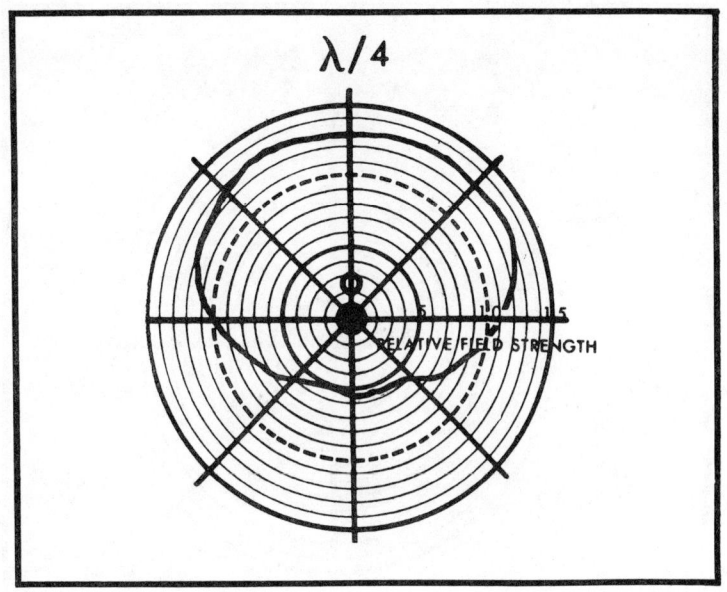

Fig. 6-25. Radiation pattern.

As shown, the result is an excellent 180-degree signal with no wasted *rf* off the back. And the bonus is a 30° increase in signal strength over 150° of that half-circle. Naturally, the amateur who uses this arrangement isn't getting something for nothing; whatever he gains in one place, he loses in another. This can be demonstrated by thinking of the broken line in the sketch as a closed loop of string. Manipulate the string and change the configuration of it, but for all practical purposes, the size remains the same.

For the amateur who wants good coverage in two general areas spaced roughly 180° apart, the best appraoch would be to mount the antenna a full half-wavelength from the support structure. A typical radiation pattern from this mounting method is shown in Fig. 6-26. It should be borne in mind that the mass of the tower affects the pattern substantially. A mast would yield a pattern with sharper, thinner lobes—more gain at the expense of horizontal angle of radiation. The half-wave pattern shows that the signal is reduced by 20° (from a top-mounted vertical) in a 90° area off the front of the antenna, and by about 35° in a 60° area off the back. But it is increased by as much as 150° laterally.

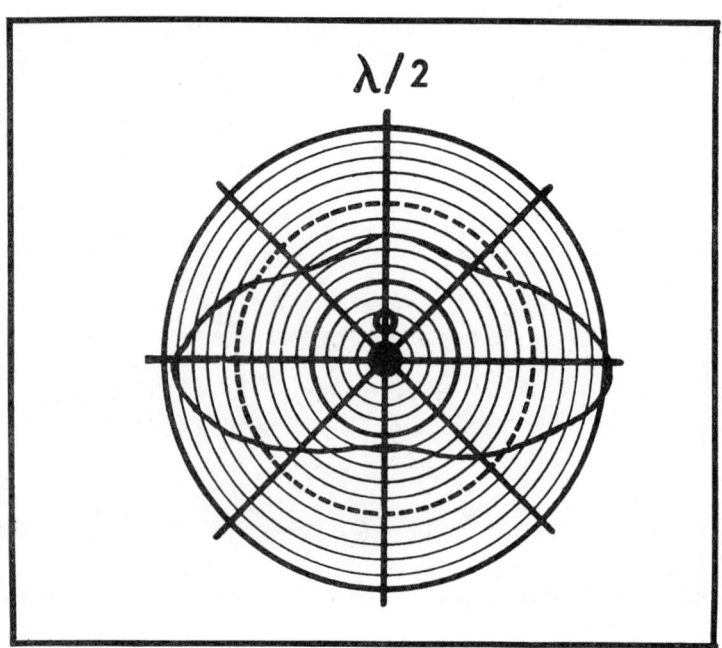

Fig. 6-26. Radiation pattern with antenna mounted one-half wavelength from the support structure.

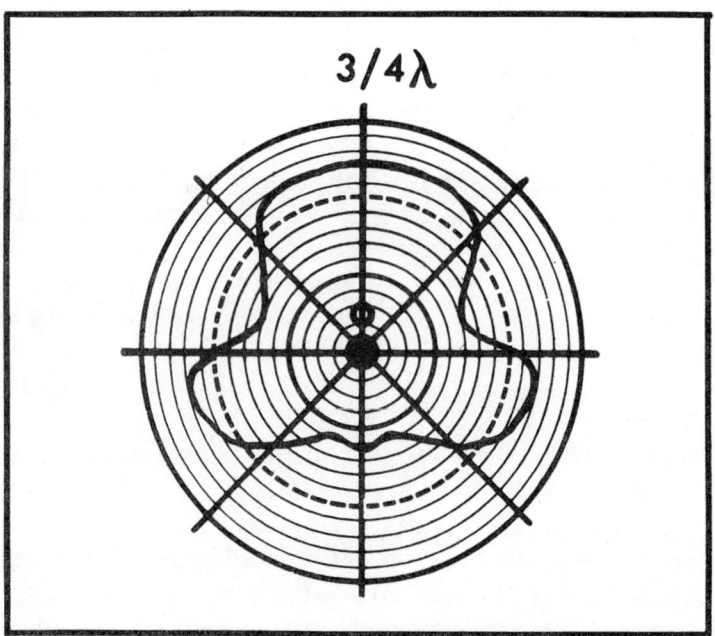

Fig. 6-27. Pattern with antenna three-quarters of a wavelength from its support.

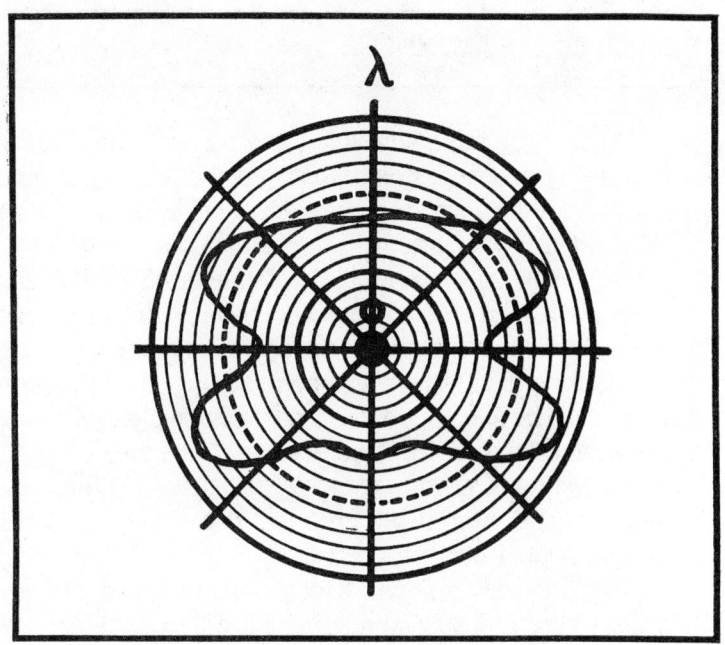

Fig. 6-28. Pattern with antenna a full wavelength from support.

A sort of cloverleaf effect can be obtained by spacing the antenna three quarter-wavelengths from the tower. As shown in Fig. 6-27, it results in a very broad fontal lobe with uniform gain over about 80 degrees. The two nulls slightly forward of both sides is compensated for by the gain just rearward of both sides.

It is probably impractical to consider mounting the antenna more than three quarter-waves from the tower. On two meters, a full wavelength would be in the neighborhood of six feet. But the sketch of Fig. 6-28 gives a pretty good idea of what the pattern would look like.

The important thing is that the theory is not restricted to any frequency. The patterns remain the same regardless of whether the operation is on six meters or 420 MHz. And the radiation patterns gradually shift from one to the other, so by experimenting with varying spacings, practically any desired effect can be achieved.

THE TWO DOLLAR AMPLIFIER

Whether you are a newcomer to 2 meters, an old-timer rediscovering the band, or a regular repeater user, you must admit that its wide open spaces are great fun. With virtually no QRM and small-size antennas, the band is a joy to operate.

Many hams, however, have made the mistake of sinking all their money into equipment and then, with little or no money left, find that they are seriously limiting their operating range by using dipoles or whips because they can't afford to buy a beam. The solution to this problem is simple. Build one. Construction of a beam for 2 meters is easy and cheap. Take a look at the antenna described below. It can be built for just a couple of bucks. You might even find that you have all the "makings" lying around the house.

Now that you know that a 2 meter beam is cheap and easy to build, you're probably wondering if it will work. Well, you should get about 6 to 8 dB of forward gain out of this antenna. This is roughly equivalent to raising your power about 4 times! Let's quit talking and begin to build. Your beam will be finished in a couple of hours.

You can begin with the boom. Try a piece of 1″ × 2″ if you like, but you will find 2″ × 2″ to be rigid, light, and not at all unwieldy. If you like, put a couple coats of shellac on the

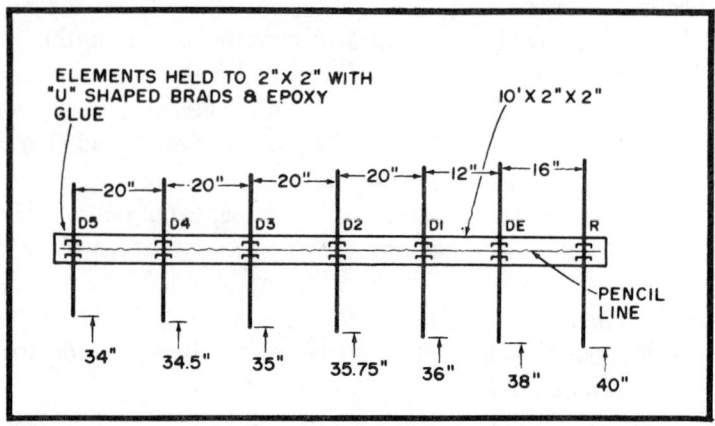

Fig. 6-29. Draw a pencil line across the boom where each element will be attached.

boom. Draw a pencil line longways through the center of the entire boom. Now measure the boom for the placement of elements by drawing a line across the boom at each point an element will be attached as shown in Fig. 6-29.

Next, prepare the driven element. Three methods are shown in Fig. 6-30. Figure 6-30A is a folded dipole made of one continuous piece of copper or aluminum tubing (auto gas line is suitable, but a bit heavy). In Fig. 6-30B, the folded dipole is constructed by attaching heavy wire to a piece of aluminum tubing, while Fig. 6-30C shows still another feed method which may require a little experimentation with the taps.

Now cut the reflector and 4 directors from a piece of aluminum clothes line. If you want to be real fussy about straightening them, roll them between two boards. After the elements are cut, and straightened, put a pencil mark at the

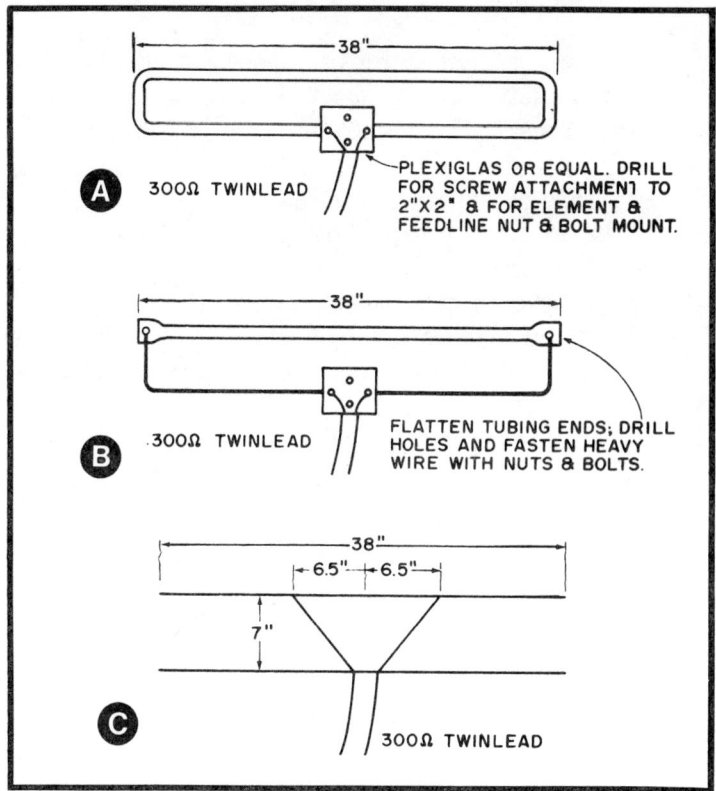

Fig. 6-30. Three methods for preparing the driven element.

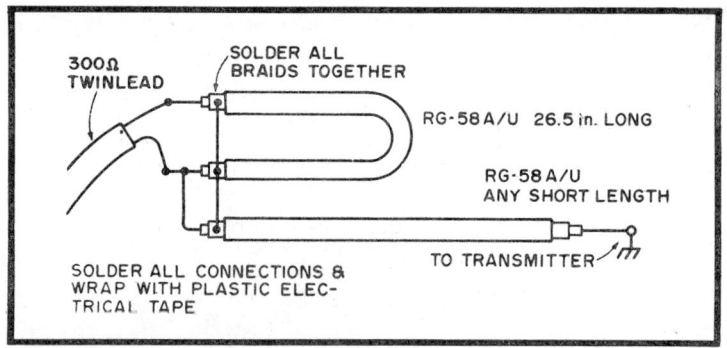

Fig. 6-31. Make a simple balun.

mid-point of each element. Match these pencil marks with the center line on the boom, so that the elements will be centered on the boom. Fasten the elements to the boom with "U" shaped brads. Dab on some epoxy glue and the beam is finished.

Perhaps your transmitter or transceiver, like many that are available today, is not designed to work with a 300Ω transmission line. Again, this is no problem. Just make and use the simple balun shown in Fig. 6-31. However, you should use the best grade of 300Ω line that you can afford.

All that's left now is to get your beam up in the air as clear and as high as possible. You can mount it vertically or horizontally. If you have a rotator, by all means, use it.

With my beam mounted vertically on top of a twenty foot high two-by-four, I consistently work through a repeater that is forty miles away, and this is with only 10 watts input. Now, tune up your transmitter and enjoy the benefits of a really "cheep" beam.

THE VERTICAL J

The Vertical "J" is *not* a radical new antenna. Its basic principle has been included in the VHF antenna portion of many antenna handbooks. Here is a case of something being so old it's new again!

Figure 6-32 shows what it looks like; the long element is ¾ wave long while the shorter one is ¼ wave. In essence, this is an end-fed half-wave, using a shorted quarter-wave section of parallel transmission line (the lower sections) as an impedance transformer. While it's possible to feed it directly with 50 ohm coax by connecting the shield to the grounded strap across the bottom and tapping the inner conductor several inches up either element, the preferred feed method is shown in Fig. 6-33—a half-wave "trombone" balun to provide 208 ohm balanced feed, which is then tapped up both elements at the proper point.

Where is this "proper point?" It will depend to a large degree on just how the antenna is put together; best practice is to determine it with the aid of a SWR bridge as will be explained later, but it's usually within 6 inches of the bottom.

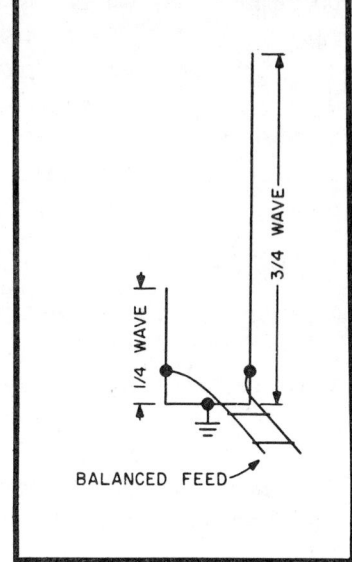

Fig. 6-32. Example of the Vertical J.

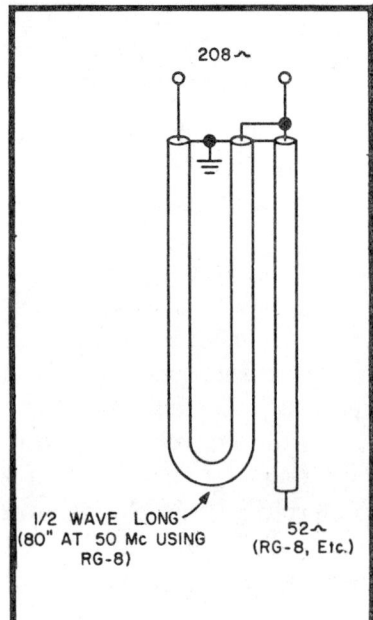

Fig. 6-33. The preferred method to feed a Vertical J.

Before looking at some more-or-less detailed construction data, let's examine the advantages and disadvantages of this antenna. On the advantage side there is omnidirectional pattern resulting from the vertical polarization; lack of cross-polarization loss when working to whip-equipped mobiles; ease of construction; and positive grounding if recommended construction practice is followed. On the disadvantage side of the ledger is the introduction of cross-polarization when working horizontal stations and lack of any antenna gain (although this antenna is usually credited with 3 db gain over a ground plane, for no tenable theoretical reason known).

Construction

Start out with a long supporting mast. Telescoping TV poles will do. Extend the top end of the mast the required ¾ wave distance above the upper guying point. This will be approximately 15 feet, requiring a 5-foot extension if a telescoping stick is used.

Shape two straps similar to Fig. 6-34 from ⅛ inch aluminum (a 1⅝ inch relay rack panel comes in handy as a source of raw material at this state), and bracket the quarter-wave element to the pole just above the guy point as shown in Fig. 6-35. Scrape all metal surfaces clean and tighten screws fully, since this is a high-current point and any resistance will cause power loss.

Form a similar set of straps from ⅛ inch Plexiglass or Lucite. To bend the plastic, soak it in boiling water until it softens and then bend rapidly, holding in place until cool. Attach this insulating bracket near the top of the quarter-wave element to maintain spacing and to support the element.

Now prepare the balun as shown in Fig. 6-33 and solder each center conductor (the 208 ohm connection points) to a radiator-hose clamp of proper size to slip over the antenna elements. Slip the clamps over the elements (this may require partial disassembly of the antenna or may not, depending on the clamps; when using the "universal" variety, no disassembly should be necessary) and tighten them just enough to hold in position but not so tight that they cannot be moved. Using strap braid or salvage shielding from a short chunk of coax, connect the common ground point of the balun assembly to the center of the aluminum bracket holding the quarter-wave element on. This provides an electrical ground at this point.

Now hoist the antenna to an approximately vertical position and feed in some rf. If a source of about 5 watts or less is available adjustments can be made with power in the line; with higher power it's best to turn the rig off. With an SWR bridge in the coax, preferably as closely as possible to the balun, slide

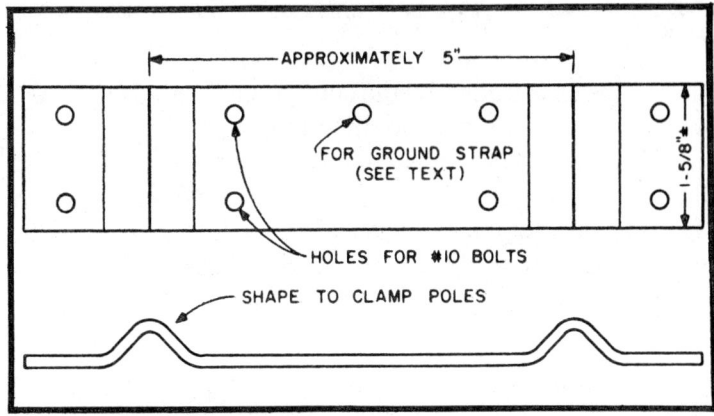

Fig. 6-34. How to shape bracket.

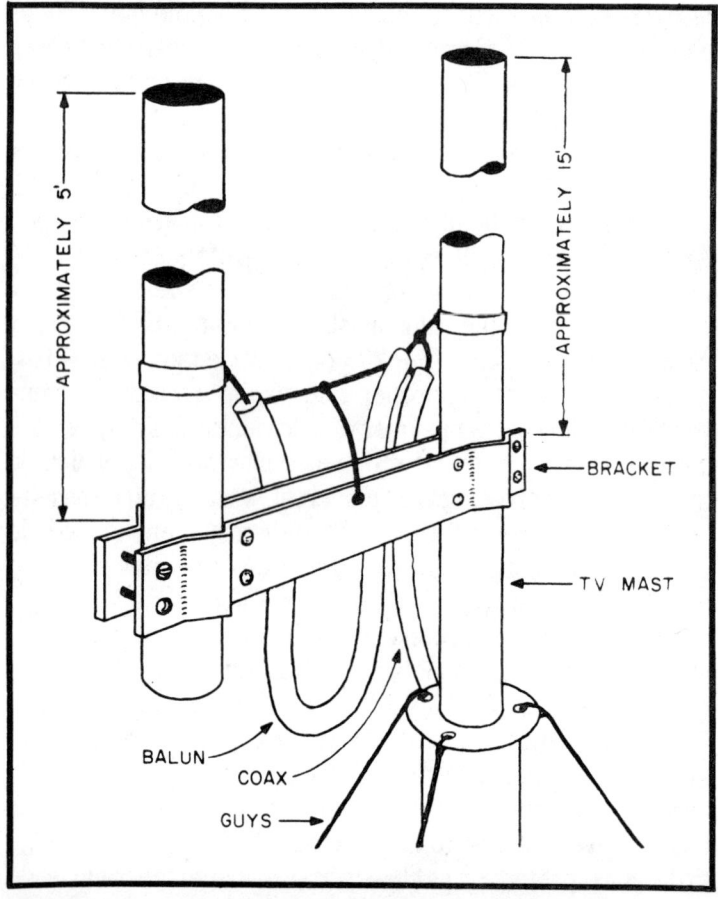

Fig. 6-35. Bracket the quarter-wave element to the pole just above the guy point.

the clamps up and down on the antenna until a reading of 1.0 is obtained at the desired operating frequency.

The only remaining step is to tighten the clamps down firmly so they won't slide, and waterproof all connections by spraying with Krylon or similar plastic. Tape the coax to the side of the mast as the antenna is raised into position.

PASSIVE REFLECTORS FOR AMATEURS

An often overlooked method in amateur circles, and even in the professional engineering society, of increasing the effective system gains at microwave frequencies is through the use of the antenna/reflector system. This system uses a parabolic antenna colocated with the transmitter-receiver and a remotely located plane faced reflector. Figure 6-36 diagramatically illustrates the arrangement.

Depending on various factors, such as operating frequency, antenna/reflector spacing, antenna diameter and reflector size, gains upward of 4 db to 5 db are easy to achieve in a practical arrangement. This is gain over and above the antenna gain available using just the parabola.

Before going into the techniques of calculating antenna/reflector combinations, there are some other advantages available through their use that bear a little discussion.

Suppose the microwave equipment is all peaked for performance. A hill, mountain or other high projection has been found that provides proper microwave path clearance to a location with which to communicate. However, power may not be available at the exact spot to be used. One solution would be to locate the microwave gear down low and run a long transmission line up the hill and mount the parabola at the top of the tower or on the roof of the building on the hill.

This very thing is often done in the common carrier, public safety, and industrial radio services except waveguide of one type or another is generally used. Even so, the ingenious microwave engineer often uses the reflector technique because it offers lower losses, and is generally less expensive. The reflector technique also allows the active equipment to be located at just about any point within a wide area. Of course, the reflector must be optimized for performance each time the active terminal is moved.

One additional advantage that receives much consideration in the western states, where high mountains and deep winter snows are not compatible with mountain top microwave repeater stations, is the fact that the reflector may be placed up high while the active equipment may be placed at a low elevation where it enjoys year around accessibility. This way, one can enjoy the availability of amateur microwave propagation testing or communication on a year-round basis.

Transmission Line is Part of Antenna System

The main drawback to direct mounting a parabolic antenna a long distance from the active equipment is the transmission line. The transmission line must be considered a part of the antenna system since the length of the line will affect the overall efficiency of the antenna. In Fig. 6-36, the overall length of the tranmission line is 205 feet for the remotely located 10-foot parabola. The gain of the 10-foot parabola at 2.3 ghz is 34.5 db. However, commonly available ⅞" coaxial transmission line has a loss of 3.3 db per 100 feet. So 205 feet of this line will have a loss of 6.8 db. This loss must be subtracted from the antenna gain (34.5 db) to obtain the antenna system gain which is 27.7 db.

Using the Antenna-Reflector

Converting the direct mounted parabola with its long transmission line to an antenna-reflector arrangement will provide a more efficient system as can readily be seen by looking at Fig. 6-36. It is necessary to use only a 6-foot parabola to achieve equal or greater gains than with the 10-foot parabola.

A 6-foot parabola has 29.9 db gain at 2.3 ghz. With only 20 feet of transmission line there will be only 0.7 db transmission line loss. Then, if a 10 × 15 reflector, ellipitcally shaped, is used and if the slant distance between the parabola and reflector is 121.7 feet, there will be 1.9 db Near Field coupling gain. So, adding all the gains and losses, an antenna system gain of 31.1 db which is 3.4 db greater than with the direct mounted 10-foot parabola is obtained. This is better than double the effective power and well worth the effort.

Calculations

The above comments and statements are just that. It now becomes necessary to provide curves and charts and the method of their use to prove the numbers just mentioned.

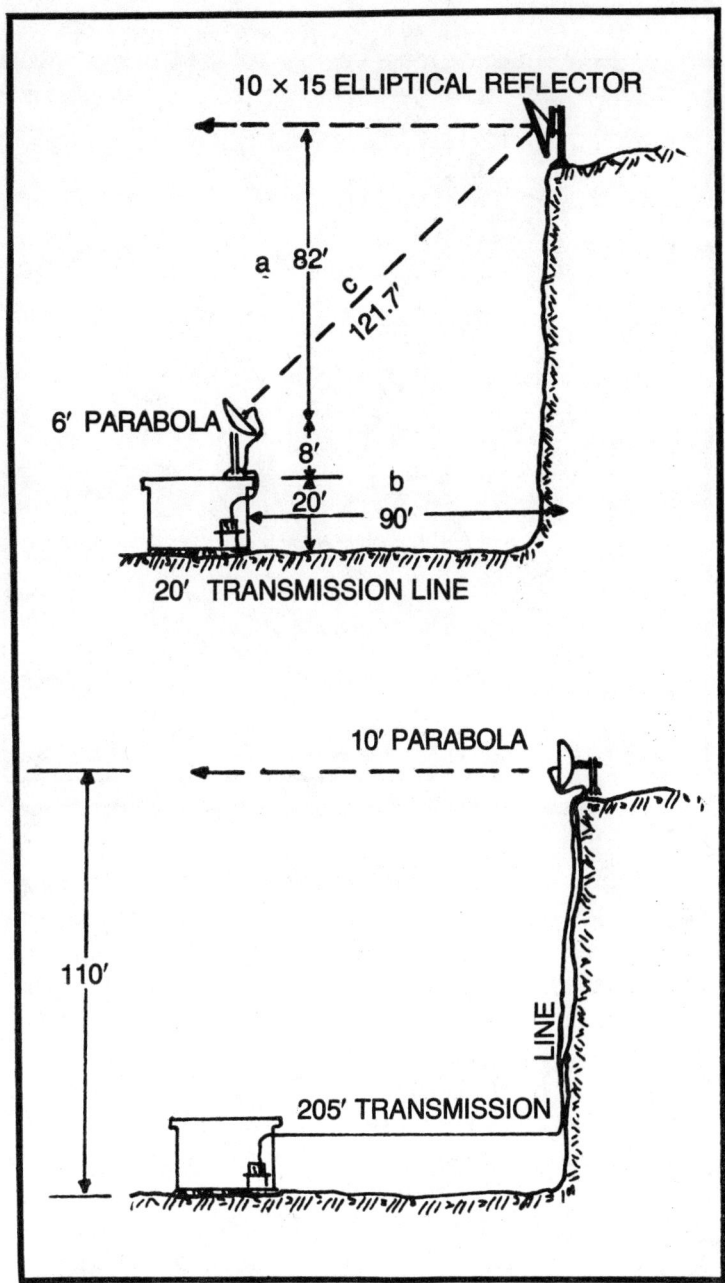

Fig. 6-36. Two methods of radiating a 2.3 ghz signal. "Periscope" system at left has effective antenna system gain of 31.1 dB and is less expensive than its direct mounted counterpart which requires a 10' parabola and provides only 27.7 dB gain. The example at left is often called a "Skewed" periscope shot.

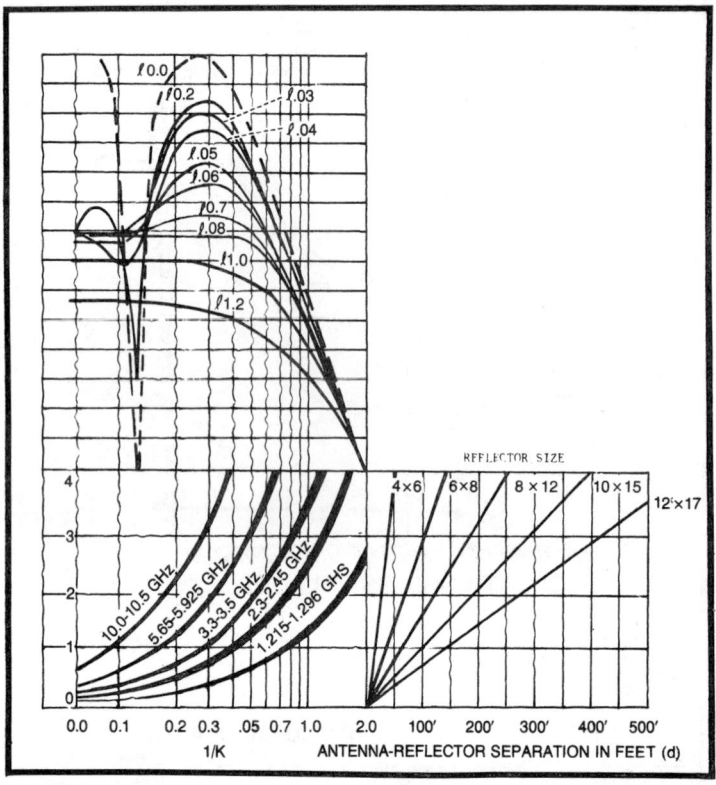

Fig. 6-37. Graphs for calculating requirements of antenna-reflector combinations.

Figures 6-37, 6-38, and 6-39 take most of the work out of calculating the requirements of antenna-reflector combinations. Figure 6-37 will be used in two basic ways. One way, using standardized size elliptically shaped reflectors and with the reflector mounted directly above the parabola (or any arrangement where the angle formed by the two beam paths is 90° ±5°, (as shown in Fig. 6-40) will use the complete chart of 6-37. Assuming a 12 × 17 reflector, spaced 190 feet from a 6-foot parabola, and an operating frequency of 2.3 ghz we proceed as follows: Refer to Fig. 6-38. Determine the value of φ. With a 6-foot parabola and 12 × 17 elliptical reflector there will be a value of 0.5. Note this number and retain it for a moment. Going to Fig. 6-37, enter the graph on the bottom right hand side at 190 feet (the distance between reflector and parabola). Read up to the reflector size (12 × 17). At this point, read left horizontally to intersect the curve corresponding to the frequency of interest, 2.3 ghz (curves for other frequencies may be computed using the formulas shown). Now, read vertically up to intersect the φ curve corresponding to the number noted earlier (0.5). Again, reading horizontally to the left, read the gain or loss in db. Here, a gain of about 0.1 is found.

Adding up the gains and losses (remember, a 6-foot parabola has 29.9 db gain), the antenna system, with 20 feet of transmission line of the type previously mentioned, will give a gain of 29.3 db. (29.9 db antenna gain +0.1 db Near Field Gain—20 feet coax at 2.3 ghz; 0.7 db).

As mentioned before, the above is to be used when the so-called "periscope" antenna arrangement is used. In offset arrangements, as in Fig. 6-36, it is necessary to perform a few calculations.

"Skewed" Shot Calculations

Using the parameters of Fig. 6-36 at 2.3 ghz there are three things to determine before performing the necessary calculations. First, the slant distance from parabola to reflector. The old ($c^2 = b^2 + a^2$) method works very well here with the right triangle. For Fig. 6-36, the distance is 121.7 feet.

With this distance determined we calculate the angle formed by the two paths (parabola to reflector and reflector to

	VALUES FOR $\ell = \frac{D}{W}$						
REFLECTOR (W)	ANTENNA SIZE (D)						
	2'	4'	5'	6'	8'	10'	12'
4 × 6	0.5	1	1.25	1.5	2	2.5	3
6 × 8	0.33	0.67	0.83	1	1.33	1.67	2
8 × 12	0.25	0.5	0.62	0.75	1	1.25	1.5
10 × 15	0.2	0.4	0.5	0.6	0.8	1	1.2
12 × 17	0.16	0.33	0.41	0.5	0.67	0.83	1

Fig. 6-38. Antenna and reflector sizes.

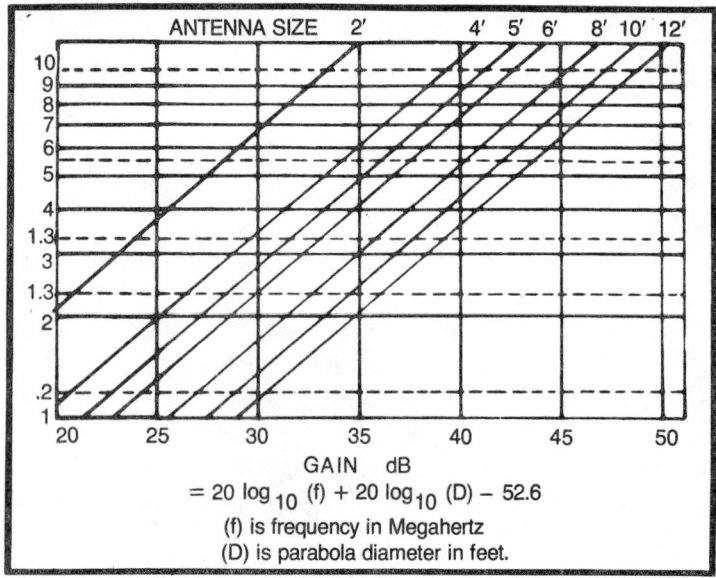

Fig. 6-39. Gain vs. antenna size.

far terminal). Using the base line of 90 feet and the slant distance of 121.7 and working with the cosine function, there is cos Θ b/c = 90/121.7 = .740. Referring to a book of trigonometric tables, .740 is represented by an angle of 42°44′. This is called the vertical included angle.

With this angle determined, we calculate the final item required before getting into the real calculation of the antenna-reflector system.

The effective area of the reflector must now be determined. The effective area will be something less than the actual area depending on the angle at which the reflector is viewed by both terminal parabolas. Since the angle of reflection is equal to the incident angle, the angles for both paths will be the same and the effective areas in both directions will be the same.

To determine the effective area, use one-half the included angle. So, 42°44′/2 × 21°22″. Then, referring again to the book of trig tables, it will be found that the cosine of 21°22′ is .9313.

Now, multiply the long axis of the 10 × 15 elliptical reflector by .9313 and this will give the effective length of the 15-foot dimension of the reflector. This is 13.97 feet.

Then, to determine the area of the ellipse, the following is applied:

$$A_e = \frac{(a)\,(b)\,(\pi)}{4}$$

where a is the 10′ dimension, b is the 13.97′ dimension and π is a constant value 3.14.

In this case we have:

$$A_e = \frac{(10)\,(13.97)\,(3.14)}{4}$$

$$= \frac{438}{4} = 109.9$$

At this point, determine the value of 1/K. Once this has been done, enter Fig. 6-37 directly at 1/K, ignoring both the

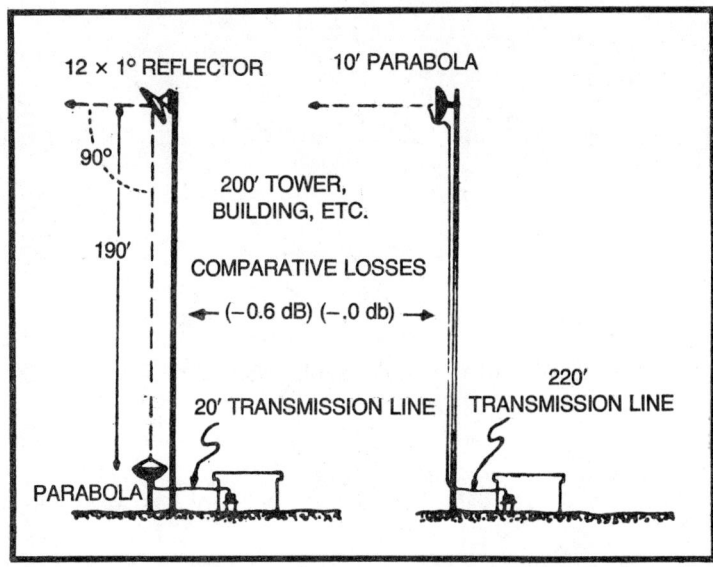

Fig. 6-40. Another example of the "Periscope" antenna system. A question often asked with regard to using small transmission lines (such as ⅞″ diameter) is, "Why not go to a larger size coax to cut down on the losses?" There is nothing to prevent this except the economic consideration. As the transmission line gets larger in diameter and its loss goes down the cost goes up drastically.

frequency curve and the Reflector Size-Reflector Separation curve. Entering at 1/K, read up to the appropriate φ curve and then read left to gain or loss in db.

$$1/K = \frac{\pi \lambda d}{4 A_e}$$

where π is the constant 3.14, λ is the wavelength in feet (985)/(Freq. in mhz), d is the distance from parabola to reflector and, A_e is the effective area of the reflector at the angle of interest. In this case:

$$1/K = \frac{(3.14)\frac{(985)}{(2300)}(121.7)}{(4)(109.9)}$$

$$= \frac{163}{438}$$

$$= .381$$

Note this value, .381.

Determining the value of φ, to intersect the correct curve, is easy and is done as follows:

$$\varphi = D \sqrt{\frac{\pi}{4 A_e}}$$

where π is the constant 3.14, A_e is the effective area of the reflector and D is the diameter of the parabola.

Therefore:

$$= 6 \sqrt{\frac{(3.14)}{(4)(109.9)}} = 6 \sqrt{.00717} = (6)(.0845)$$

$$= .508.$$

Now, entering the 1/K at .381, read up to where the φ curve is equal to .508. Then read horizontally left and read 1.9 db gain.

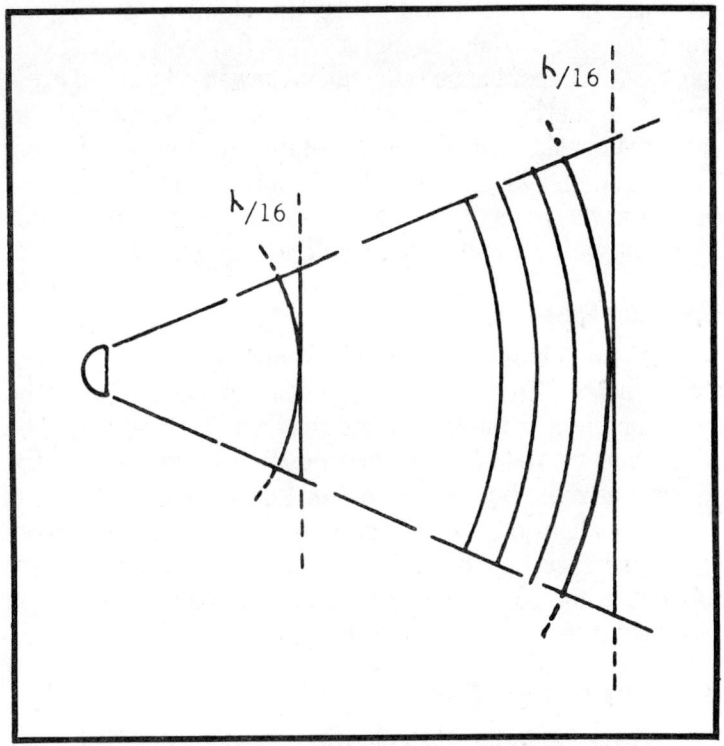

Fig. 6-41. What are near field and far field? The classical definition of the occurrence of near field is the point when a radiated wave (the microwave beam) intercepts a plane surface (the passive reflector) and the difference between any point along the radiated wave and the plane surface is 1/16 wavelength or greater. Note that either moving the passive reflector closer to the source or making the passive reflector larger can result in greating a near field situation. It will also be found that near field occurs when the computered value of 1/K is equal to 2.5 or less, i.e., 2.3, 2.3, 0.3, etc.

Add 1.9 db gain to 29.9 db gain available from a 6-foot parabola and subtract the 0.7 db transmission line loss, and a 31.1 db antenna system gain is arrived at.

Elliptical Reflectors Versus Rectangular Reflectors

There is nothing to prevent use of rectangular reflector. In fact, a rectangular reflector has about 20% more surface area than the elliptical reflector and will provide about 2 db more gain accordingly.

However, the elliptical reflector has the advantages of having lower sidelobe levels and sharper nulls between lobes, which are important considerations if troubled with co-channel or adjacent channel interference. In addition, the elliptical reflector offers 20% less surface area for wind loading purposes. Also, there is no reflection of unwanted Second Fresnel Zone energy with the ellipitcal reflector as might occur with the over-projecting corners of the rectangular reflector. The keep notch that occurs in Fig. 6-37 where 1/K is about .13 is due to the reflection of Second Fresnel Zone energy. For optimum performance, 1/K should be designed around 0.3.

Reflector Face Flatness

The reflector face should be essentially flat at microwave frequencies. The face must be flat to within ⅛ of the wavelength over the face of the reflector. If there should be deviations from flat (not to exceed ⅛ of a wavelength), the deviations must be concave rather than convex.

Deviations of about ¾" over the entire face of the reflector can be accommodated at 2.3 ghz without appreciably degrading the signal. Various construction methods may be used with the most successful being a cross ribbing technique.

Flat Reflectors and Gain

There is often some confusion, even in engineering circles, concerning how a flat surface can have "gain." Gain is usually defined as an increase (or decrease) over a predetermined level. The predetermined level serves as a reference point. In microwave antenna work the most common reference point has been established as the isotropic radiator (a point source). Sometimes a dipole will be the reference source, though. The dipole has a gain of 2.15 db over the isotropic source, so if conversions are necessary, that value should be used.

Any increase in aperture over the isotropic point source will result in more energy being radiated or redirected. Thus, the more energy, the more "gain."

Quite probably, the difficulty in realizing how a flat surface can have gain relates back to another popular misconception concerning parabolic antennas. This misconception is that it is the focusing effect of the parabolic antenna that provides "gain." Therefore, goes the faulty conclusion, since there is no focusing with a reflector that is flat, how can there be gain?

The simple answer is that it is not focusing, as a characteristic of the parabolic antenna, that provides the gain. The focusing is merely a convenient means of transistion from a large aperture (the dish) to a closely spaced small aperture (the feed device). It is projected aperture of the dish that provides the gain, not the focusing. Since it is the projected aperture that provides gain, the flat reflector, with its projected aperture, will provide gain which can be reliably calculated and measured.

Plane Reflectors Rather Than Back to Back Parabolas

Plane reflectors are extremely efficient devices. Commercial production line models have an efficiency of 99% as compared with the 55% efficiency of parabolas. So, if a passive repeater were to be made from a pair of back to back parabolas, the passive could be made from a plane reflector that would be about 6 db smaller.

As an example, a single 7 × 8 rectangular reflector will provide as much gain (with a horizontal included angle of 90° or less) as two 10' parabolas back to back.

A single reflector may be used where the angle to be turned is about 135° − 140° or less. With angles greater than that, the effective area is greatly reduced and efficiency of aperture drops off. Then, the common practice is to use a pair of reflectors closely spaced to get the "in-line" microwave beam over a hill or other in line obstruction. Depending on the aperture of the reflectors, the operating frequency and the spacing between them, the coupling loss with the double reflector will be less than 1 db.

Conclusion

Reflector technology has been overlooked in amateur microwave work. Little emphasis has been given to it. With the development of microwave nets it would be a simple matter to interconnect many stations in an area where they normally would not enjoy line of sight and proper Fresnel Zone clearance. Use of reflector technology would eliminate the need for tall towers to provide proper path clearance.

Reflectors may be used at frequencies as low as 420 − 450 mhz. They may be used even at lower frequencies but unless the spacing is close and the reflector rather large, the aperture gains are somewhat low with the reduced frequency.

However, each situation should be mathematically calculated to determine whether or not it will perform. Use of any rule of thumb is to be avoided since no rule of thumb can possibly include all the system design parameters and therefore no rule of thumb can possibly be accurate except by accident.

Chapter 7
Accessories

VHF DUMMY LOAD WATTMETER

This unit is similar to some 60 watt modes which may be found around many commercial two-way radio shops. This dummy load has provision for connection to an external relative output meter. This external output meter may become an accurate wattmeter if the following criteria are met:

1. Frequency bandwidth of ± 10% of calibration frequency.
2. RF output kept within power dissipation of dummy load.
3. Accurate initial calibration.

These criteria may be easily met in amateur vhf operation if only one band is considered for each set of calibration data. Since most vhf amateurs operate on 50 mhz, 144 mhz or 432 mhz, the ± 10% frequency limitations may be easily met. This limitation gives a 10 mhz, bandwidth at 50 mhz, 29 mhz bandwidth at 144 mhz, and 86 mhz at 432 mhz. The limitation to the power ratings of the dummy load is only common sense, for if a resistive network is overloaded, the impedance may be drastically increased, caused by damage to the load resistors.

The calibration limitation may be overcome if a standard previously calibrated unit, or commercial unit is used.

The unit may be built for less than six dollars (less meter movement) if all new parts are purchased. It consists basically of 16 220 ohm resistors in a series parallel arrangement. The metering circuit consists of a germanium diode pickup with necessary *rf* filtering. The meter movement is generally a vom, but any 50 μa meter movement should suffice. Exact physical layout is not extremely critical, but it is suggested that the layout be made similar to the unit shown in the accompanying photographs. This unit is acceptable for 60 watt output transmitters without modification. The power capability may be increased to about 200 watts if the resistive network is uspended in 1 quart of oil. If this is done, care must be taken to keep the metering circuit out of the oil. The lead from the diode to the resistive network must, of course, be partly submerged, but keep the diode itself out of the oil.

Calibration

Calibration is best accomplished by using a Bird "Thruline" or similar commercial vhf inline wattmeter. Second choice is a Bird "Termaline" or similar dummy toad-

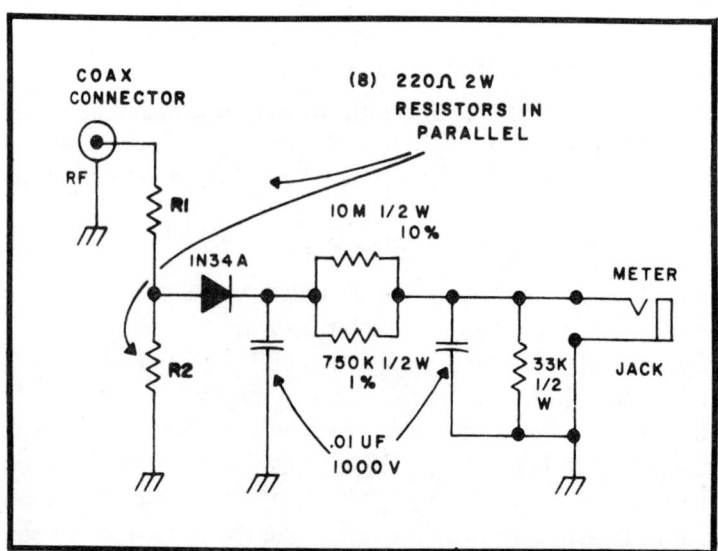

Fig. 7-1. Schematic R1 and R2 consist of eight resistors each, in parallel.

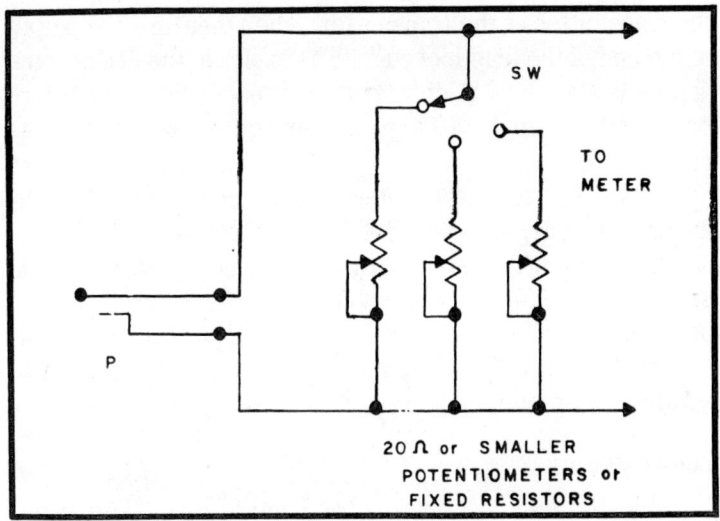

Fig. 7-2. Metering shunts.

wattmeter. In both cases, a graph should be created by plotting meter divisions on the horizontal axis, and power on the vertical axis. The meter shunt should be placed at minimum resistance and increased to give maximum reading at the desired power level or, if a vom or vtvm is being used the range switch should be placed on a high voltage setting and reduced a setting at a time unitl the desired reading is obtained. The transmitter should be adjusted for various power levels on the standard wattmeter and the voltage or current reading on the new meter recorded on the graph. In the case where the standard meter is of the dummy load-wattmeter type, it will be necessary to switch the coax from one unit to the other. Do not retune the transmitter, for each unit will present almost the same load to the transmitter (50 ohms). Take the reading and record as with an inline type of meter. The points on the graph should now be connected with a smooth curve. If multi-band use is expected, the graphs must be made for each band. Use of the wattmeter now requires only the connection to the transmitter, setting of range switch to the proper level, and reading the graph.

The uses of this dummy load-wattmeter are as varied as the amateur mind can devise. One very important use is determining the losses of 50 ohm coax. Measure the output of the transmitter at the transmitter. Then measure the output at the end of the length of coax. The losses in the line become apparent. The loss in db may be calculated by the standard power ratio formula, 10 $\log_{10}$ Power out of coax/Power into coax.

Another use is the determination of efficiency of final amplifier stages. This efficiency may be calculated by Power out (measured by dummy load-wattmeter)/Power in (measured by plate current/plate voltage meter) × 100%. A third use is determining once and for all which amateur really has the most output. This list may be expanded by the builder to suit his own tastes.

Conclusion

This dummy load-wattmeter is not a Bird "Termaline" nor should it be regarded as a substitute for any other laboratory equipment, However, with a little care in calibration, (assuming a 5% accuracy standard is used for initial calibration) the accuracy should be within 10%, and this is not bad for a wattmeter costing less than $10.

A RUGGED ROTATOR

Whether old-timers or raw beginners, there are probably very few hams who are not somewhat familiar with the old, reliable prop-pitch motor. Just after World War 2, the surplus market saw a veritable flood of them, and, along with coaxial cable, they caused quite a revolution in the construction and operation of rotary antennas. Many of the motors which were put into use right after the war have performed faithfully over the years, and even today there is really nothing which can approach them for sheer power and ease of operation.

Before the prop-pitch motor may be used as a rotator, the brake assembly and limiting dogs are usually removed and some sort of coupling or bracket is welded to the output gear. The motor is then mounted vertically inside the tower and supplied with 24 volts AC or DC. Refinements such as remote controls and direction indicators are left up to the imagination of the owner.

One of the very few faults with his arrangement is the tendency for moisture to collect inside the motor housing, especially on the brushes, with the result that the motor turns very erratically or not at all. It goes without saying that this usually happens in the winter, just when the rig is being used the most and when the working conditions at the top of the tower are at their worst.

Another point, particularly in cold climates—the oil in the gear box congeals putting quite a load on the motor. This causes very slow starts and extremely slow rotation of the antenna. Since the usual rotation speed of the output gear is only about three quarters of an rpm, any further decrease is intolerable. It is possible to open the gear box and pin one set of planetary gears, which will approximately double the output speed, but many hams apparently would rather put up with the slow speed than monkey with the gear train.

The biggest factor in the rugged rotator was the decision to mount the motor in the horizontal plane, just as it was in the aircraft. The ideal approach at this point would be to procure the mating gear which originally was on the end of the propeller blade, but this is apparently impossible to find. It seems equally difficult to find any other suitable gear, so this idea was soon forgotten. The problem is to find a way to obtain a right-angle drive, along with a bearing, which will support the heaviest possible antenna load. The most common item which comes to mind is the differential or automobile "rear-end." Certainly it is the most available. In addition to being a right-angle device, it is extremely sturdy, and if properly set up, it will provide a step-up speed ratio. The exact ratio will depend upon the original design of the associated car, but it will be somewhere around 2 to 1, or a little better. When the prop-pitch motor is coupled to the axle of the differential, the *drive shaft* turns at 1½ to 2 rpm, thus solving one of the major drawbacks of the prop-pitch motor without modifying it. There is no problem with the differential being able to support the antenna, and the strength of the internal gears is far in excess of any torque which the antenna will exert.

By mounting the motor horizontally, any moisture which collects inside the housing will run to the bottom and a small drain hole will take care of it. With the infrequent operation and very high quality of the gears, it is doubtful if any oil is actually required for lubrication. It can either be drained out completely, or a small amount left inside. With the horizontal mounting, the gears will pick up some lubrication during each revolution if a small quantity remains in the box. In any case, the problem of congealing is eliminated.

Because of the length of a car rear-end it would be too ungainly to mount in the average tower. In fact, it must be understood at once that this rotator is quite heavy, but since it is intended for use with large and heavy arrays it would be necessary to have a heavy-duty tower. It is ideal, of course, for the windmill type of tower, or for base mounting with a long drive shaft.

A quick inspection of the nearest junk yard will provide the rear-end needed. It is best to look for small cars, since even the smallest unit will be satisfactory. Make sure the brake drums are still on the unit. One brake drum is positioned on the output gear of the prop-pitch and welded to it. Make sure that this is done carefully and accurately, as it is going to provide the coupling between the prop-pitch motor and rear-end after it is modified. This is also a good time to remove the brake assembly and dogs from the motor, and drain out the oil. Also pull out the motor power leads, and label them—clockwise, counter-clockwise and common. Hook up 24 volts temporarily to find out which is which. The motor should also be fitted with filtering capacitors from each brush holder to ground. Small micas can be used, but the .002, .001 or even .01 disc ceramic is more convenient because of its small size. These steps will complete the modification of the motor itself.

The next step is to cut down the rearend. First of all, remove both axles and set them aside. Then cut off both axle housings within one or two inches of the gear housing. If a power hacksaw is available, the job may seem easier, but the shape of the unit makes it difficult to hold steady. An ordinary hand hack-saw is entirely adequate and is actually easier to use. In order for the unit to transfer power from one axle to the antenna drive shaft, the other axle must be prevented from turning. The easiest way to do this is to insert one axle into its normal gear in the rear-end, then cut it off flush with the gear housing, and weld it to the housing itself. The axle housing can be cut off near the brake shoe assembly. The axle is then inserted through the bearing in the brake assembly and on into its mating gear. The brake assembly is welded to the gear housing, the axle is cut off flush with the end of the brake-assembly bearing, and then welded to the bearing. The bearing in turn is welded to the brake-assembly.

The end result is the same, but now the backing plate can be used to help form a mounting bracket arrangement for the complete rotator.

The other side of the rear end must be modified with more care to ensure proper coupling to the motor. The actual coupling is accomplished by using the wheel mounting studs on the axle to mate with the corresponding holes in the brake drum which was welded to the prop-pitch output gear. By using this arrangement, the motor can be quickly removed for servicing by merely removing the mounting brackets on the motor itself and sliding it away from the differential.

In order to construct the coupling, both ends of the axle must be used. The spliced end should be inserted completely into its mating gear, and the axle cut off flush with the edge of the gear housing. Now, note that the remaining piece of axle must be cut off about three inches from the round plate on the end. This piece of axle (with the plate on it) must be welded to the piece which is inserted into the gear. The result will be a much shorter version of the original axle. It is best to have the two pieces welded together. Any reasonably good machine shop can do this. When this new axle is inserted in place, the axle housing must be carefully measured and cut off in such a way that it can be inserted through the brake-assembly bearing; the plate on the axle can mate properly with the studs on the brake assembly, and the housing can be welded to the differential housing. Figure 7-3 will probably make this more clear than the verbal description. After the axle has been fitted and the housings welded to each other, the brake drum can be slipped over the studs and fastened onto the axle plate with the two or three original screws. The studs are now in place, ready for insertion into the holes on the prop-pitch drum.

The entire assembly must now be mounted, and probably the simplest way is to use two lengths of heavy angle iron or aluminum to form support rails. The prop-pitch motor is laid horizontally between the rails, with the gear case housing providing its own support. Two short pieces of angle are bolted to the side rails and to the large flange of the motor gear box. This prevents any rotary movement of the motor. Two other pieces of angle are used to hold the rails together, otherwise the weight of the motor tends to force the rails apart and put the entire weight on the bolts in the side pieces of angle. See Fig. 7-5.

The differential assembly is coupled to the prop-pitch motor and carefully adjusted so that the whole affair is level both end to end, and side to side. Mounting brackets can then be made out of mild steel strap. The brackets are first fastened to the housings by standard muffler clamps, making sure the assembly is level. Then the legs of the brackets are bolted to the side rails. Unless a large unit is used, it's not likely that the bottom of the differential will be resting on the edges of the rails. To provide additional support, a piece of channel steel should be used to support the under-side of the differential,

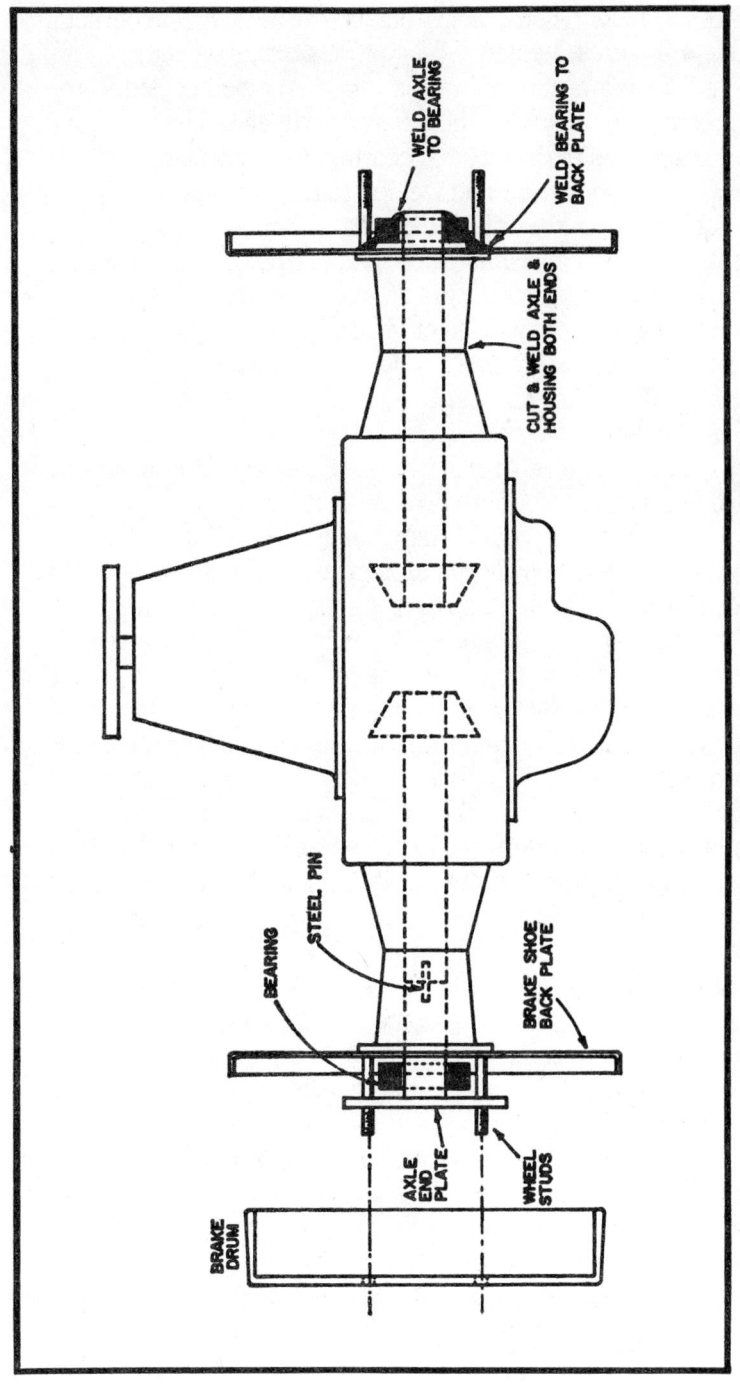

Fig. 7-3. Mechanical details of modifying the automobile differential before installing the prop-pitch motor.

with its height being adjusted by washers or sheet-metal shims. Figures 7-4 and 7-5 show this arrangement.

The top of the differential must now be coupled to the antenna drive-shaft. As the differential itself will have a heavy plate attached to the internal gearing, a mating plate or flange can be cut from stock and welded to the antenna drive-shaft. Matching holes are drilled in the flange and then the flange is bolted to the differential plate. A more flexible arrangement can be made by procuring the universal joint and possibly even the entire drive shaft of the original car. By using the universal joint, some mis-alignment between rotator and antenna can be tolerated, and the shaft can be quickly uncoupled if necessary. See Fig. 7-4.

The last step in the rotator itself is the direction indicator, and for this purpose there is really nothing that can do as good a job as a pair of selsyns. Many units are still available on the surplus market, though the easiest to use are those designed for 115 VAC. The hook-up for the selsyns is shown in Fig. 7-6. Mechanically, the antenna selsyn is mounted on a metal plate with two home-made U-bolts. The plate is mounted on the wheel studs of the brake assembly opposite the prop-pitch motor. The drive system from the antenna shaft to the selsyn consists of two V-belt pulleys of equal size, and the V-belt itself. The cheapest suitable pulleys are made for use on laundromats or dryers, and are available from any appliance service shop. The correct size V-belt can be obtained at the same time. The first pulley is drilled to accept the same bolts which couple the differential to the antenna shaft, and the second is mounted onto the shaft of the selsyn. It will probably be necessary to make up some sort of mounting plate and possibly a set-screw arrangement. This depends upon the type of pulley used as well as the type of synchro. See Fig. 7-4 for details.

A suitable terminal board should be provided to handle the three motor-control wires and the five selsyn wires. In addition, a weather-proof cover should be made up to cover at least the selsyn and terminal board. The motor and differential are well able to stand the weather just as they are.

The control method which is used will depend upon the desires of the individual and the available materials. All that is

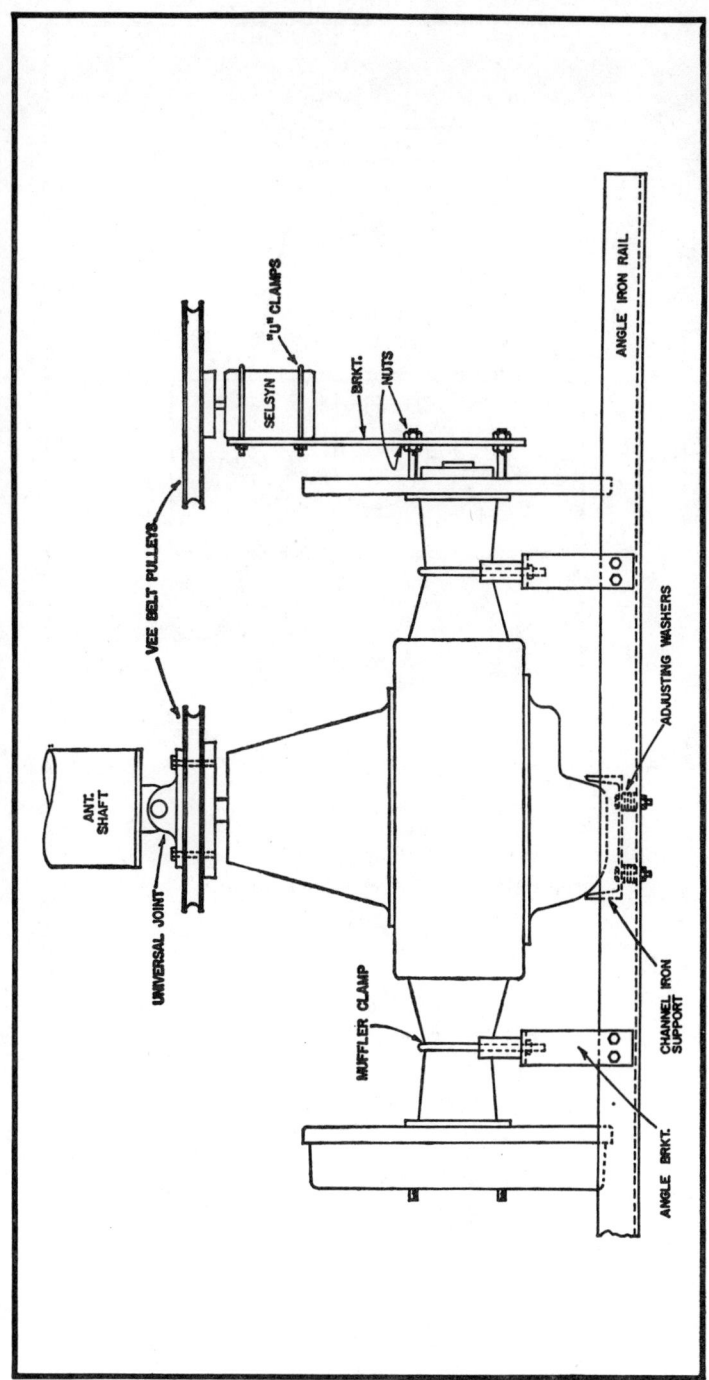

Fig. 7-4. Mounting details of the differential, selsyn drive, and antenna drive shaft. The mounting brackets can be made as shown here, but in the photo, brackets were made of scrap material and the muffler clamps mounted upside down. Either method is satisfactory.

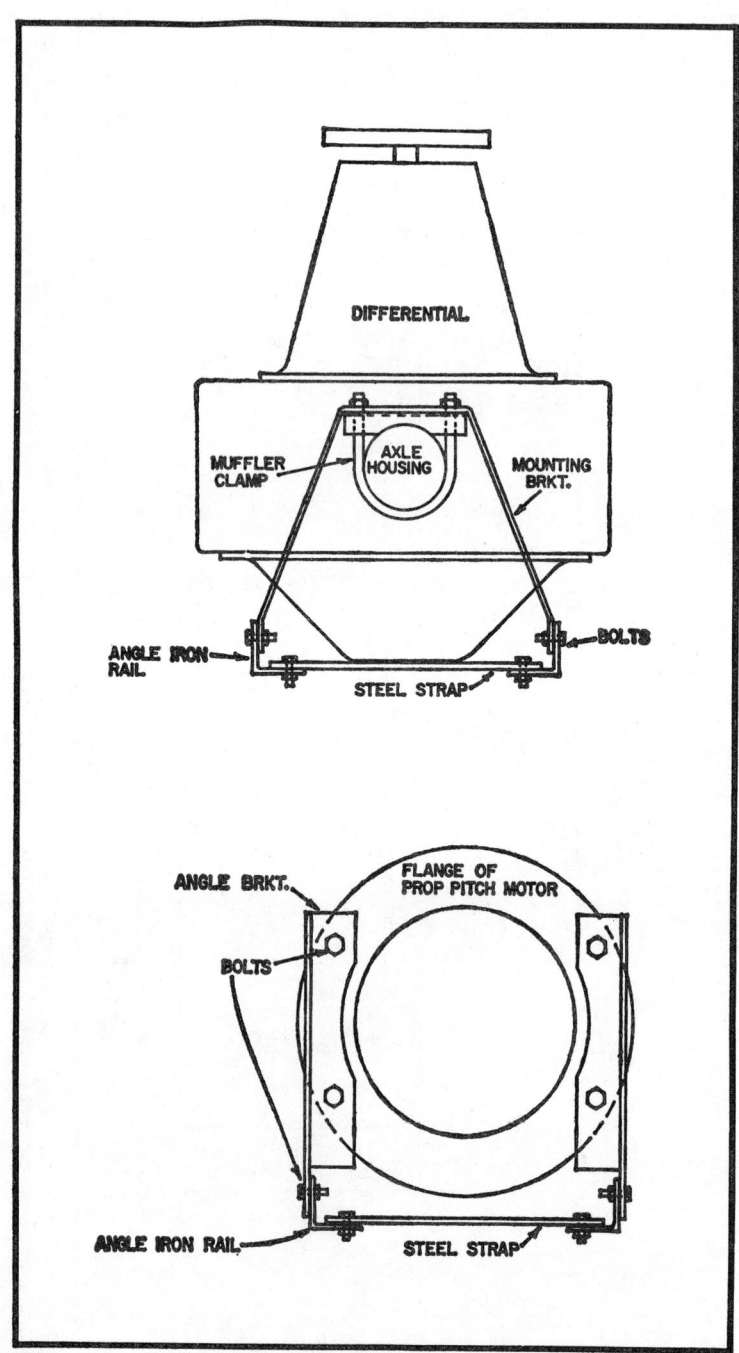

Fig. 7-5. Side view of the rail-mounted differential and prop-pitch motor.

really necessary is a means of getting 24 volts AC or DC to the motor, and switching it between the cw and ccw wires. The selsyn wires are merely connected to the remote selsyn, and the remote unit is fitted with an indicator needle and some sort of compass scale. The antenna is pointed north, the remote selsyn housing is loosened and turned until the needle points north on the scale, and the housing tightened up.

However, a more elaborate control unit is illustrated in Fig. 7-7. Since the unit is mounted on the tower, the low voltage, high current leads to the motor should be as short as possible. A pair of 4-pole double-throw relays handle the motor switching, with all poles wired in parallel to minimize the possibility of arcing or burned contacts. The 24 volt AC supply is rectified with a high-current silicon diode, filtered and the resultant DC used to feed the motor. While not necessary, the motor seems to run quieter and smoother on DC than AC. The 24-volt transformers and relays used in this unit were salvaged from pin-setting machines in a defunct bowling alley, but any similar component will be suitable. Alternatively, a pair of

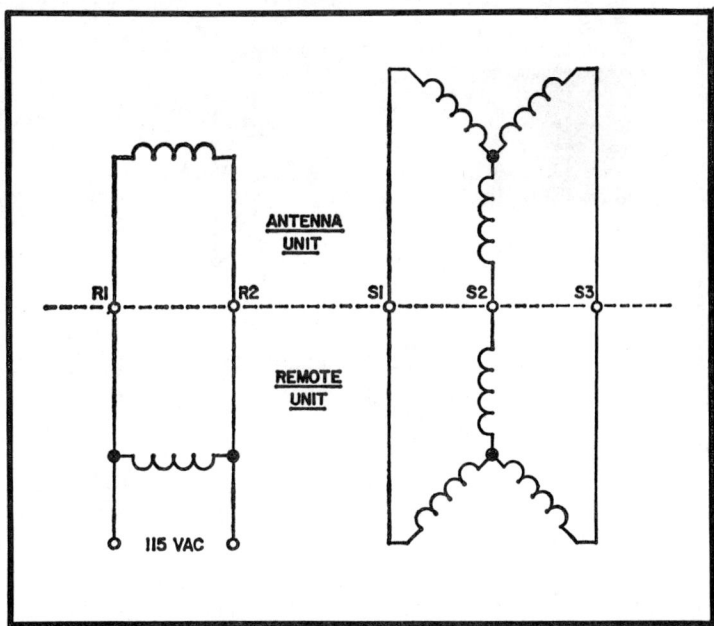

Fig. 7-6. Interconnection for the selsyn motors. A four-wire system may be used if terminals R2 and S3 are run on the same wire.

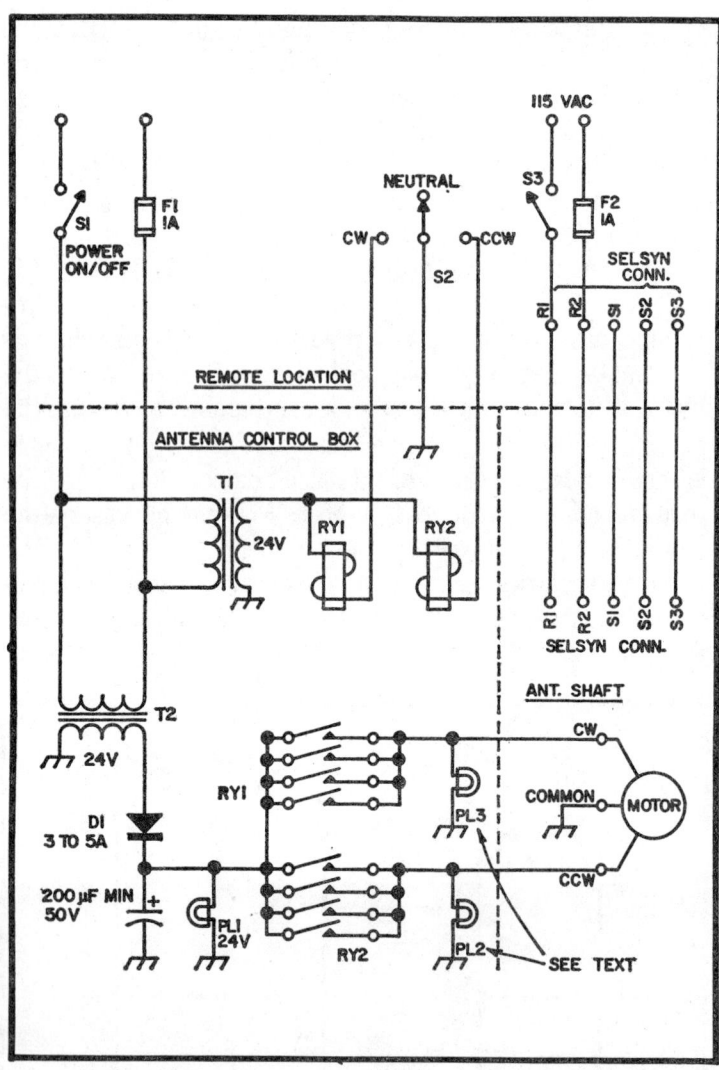

Fig. 7-7. The complete remote-control system for the really rugged rotator. If suitable 24-volt relays cannot be found, two 12-volt automobile horn relays may be used with a dropping resistor.

12-volt units can be hooked in series, and a couple of automobile horn relays used for switching.

On this particular unit, a large 24-volt pilot lamp is mounted on the side of the control box and is visible from the ground. It monitors the output DC immediately before it is applied to the direction relays. This is being modified to two lights, which will be connected directly to the motor input leads. Should the antenna fail to turn at any time, a quick look at the tower will show whether or not voltage is getting to the motor. This will immediately point out motor trouble or control unit trouble and is a handy trouble-shooting aid.

The resultant rotator is very rugged and will provide suitable for the largest and heaviest stacked arrays.

AUTOMATIC TRANSMISSION LINE TUNER

Reducing aggravations around the shack is a job that keeps the average homebrewer happily scheming and building throughout his hamming life. At the top of the "bother" list is the matchbox that must be readjusted after making a large frequency change within a band, and the box is located a great distance from the transmitter. Anyone that has put up with this problem for years, must have wondered as I, why can't the signal that deflects the SWR meter do the adjusting?

While working on the problem of getting the meter to twist the knob, it was soon discovered that the mismatch defying adjustment at the transmitter is the one that causes a reactive current to appear at the load end of the transmission line. A resistive mismatch as large as 3:1 causes little fuss and can be tuned out with ease. For this reason, the change in the resistive load presented by the same antenna at different locations in a given band is not large enough to require an adjustment, so one setting of the matchbox input loading capacitor is good for the entire band. This is not true of the output capacitor. It must be reset to bring the antenna—transmission line combination back into resonance, or a large inductive or capacitive phased current will be reflected back to the transmitter along the coaxial cable. It then follows that only one shaft requires adjustment after a frequency change is made, and if any change in the phase relationship of the current to the voltage in the coaxial cable at its load end could be detected and converted into a DC voltage, this signal could become the start of the automatic control. Such phase changes can be detected and converted into rotary motion.

The automatic transmission line tuner, ATLT, is designed as an outboard unit to work in conjunction with a simple matchbox of the fundamental type described in most handbooks. Figure 7-8 shows how it is used in a typical antenna feed system. The match-box consists of a small input coil that has a loading capacitor in series with it and ground, inductively coupled to a large parallel-tuned coil which has the balanced transmission line fastened to it by means of taps. The ATLT

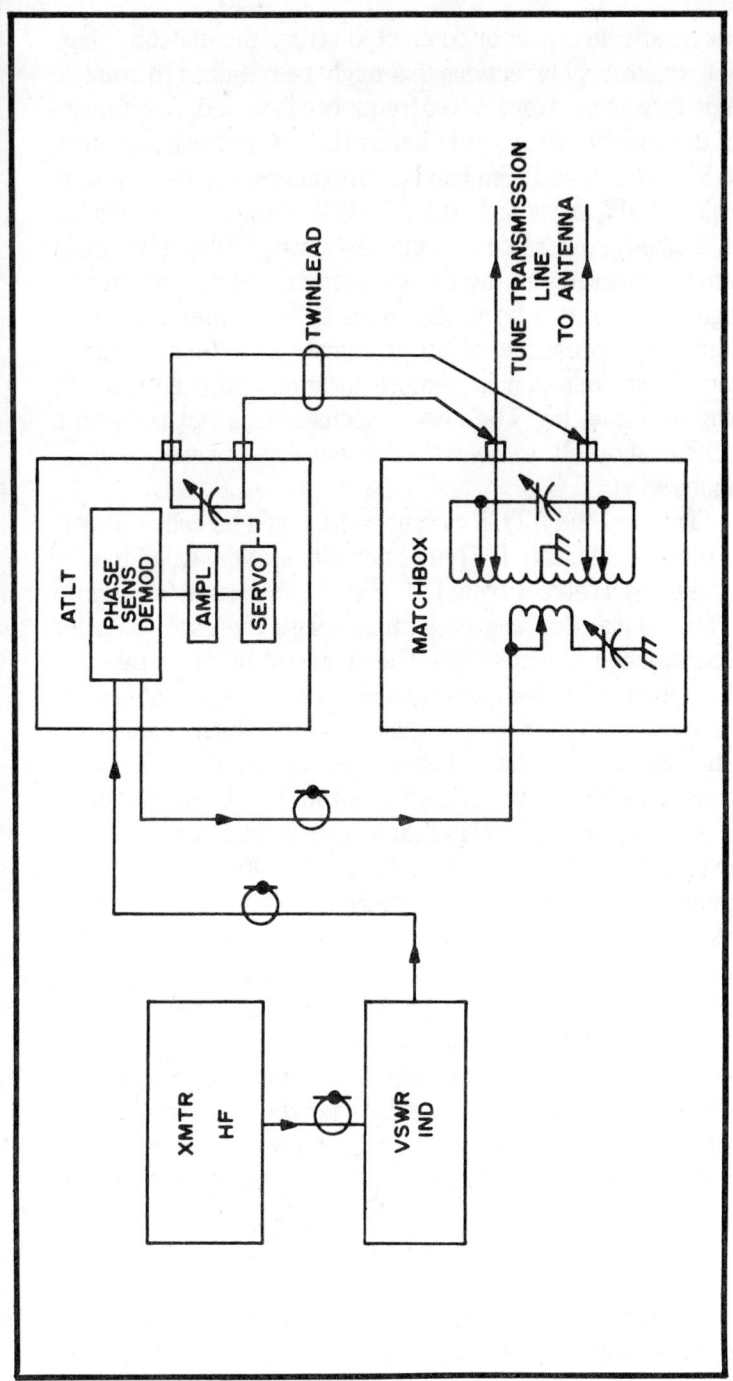

Fig. 7-8. Automatic transmission line tuner used with a simple matchbox.

uses a variable capacitor connected across the matchbox output to make any corrections that might be required to maintain resonance as the transmitter frequency is varied. It will maintain the coaxial cable vswr to better than 2:1 and will operate in the 80, 40, 20, and 15m bands. Transmitter inputs can be as small as 90W, or as high as its variable capacitor will handle. The highest power used during tests was 800W. The unit is battery powered and the design is such that no current of a magnitude that can be detected on a 50 μA meter will flow without the presence of an rf signal. This feature allows unattended operation in remote locations with battery life going into months. The servo-mechanism is not made in a machine shop. It came out of a battery-powered toy automobile.

The way the ATLT circuits work could be called clever, but never complicated. There is an element designed to pick up the coaxial cable current (L1, Fig. 7-9) and transform it into a useful signal that can be summed with the cable voltage, much reduced in amplitude. The resultant of these two signals, detected by diodes, produces a DC voltage that is positive, negative, or zero, depending upon the phase relationship of the current to the voltage. This circuit is known as a phase-sensitive demodulator and is the ATLT nerve center.

Current coil L1 is a laterally wound coil that encircles the coaxial cable inner conductor. When the alternating current through the cable increases, decreases, and then reverses direction, an expanding and contracting magnetic field is produced, inducing a potential across this coil. This voltage is 90 degrees out of phase with the cable current because of the electrical law stating that magnitude of a voltage induced into a conductor by an electromagnetic field is directly proportional to the rate of change of this field. When the current, alternating from its peak negative to its peak positive value passes through the zero point, the rate of change is at its maximum and is displaced by 90 degrees from either current peak.

Without the voltage signal present, the rectified currents out of the ends of L1, through CR1 and R2, CR2 and R1, and back to its centertap, will be equal in each circuit branch. This produces a zero potential difference across the R1-R2 divider.

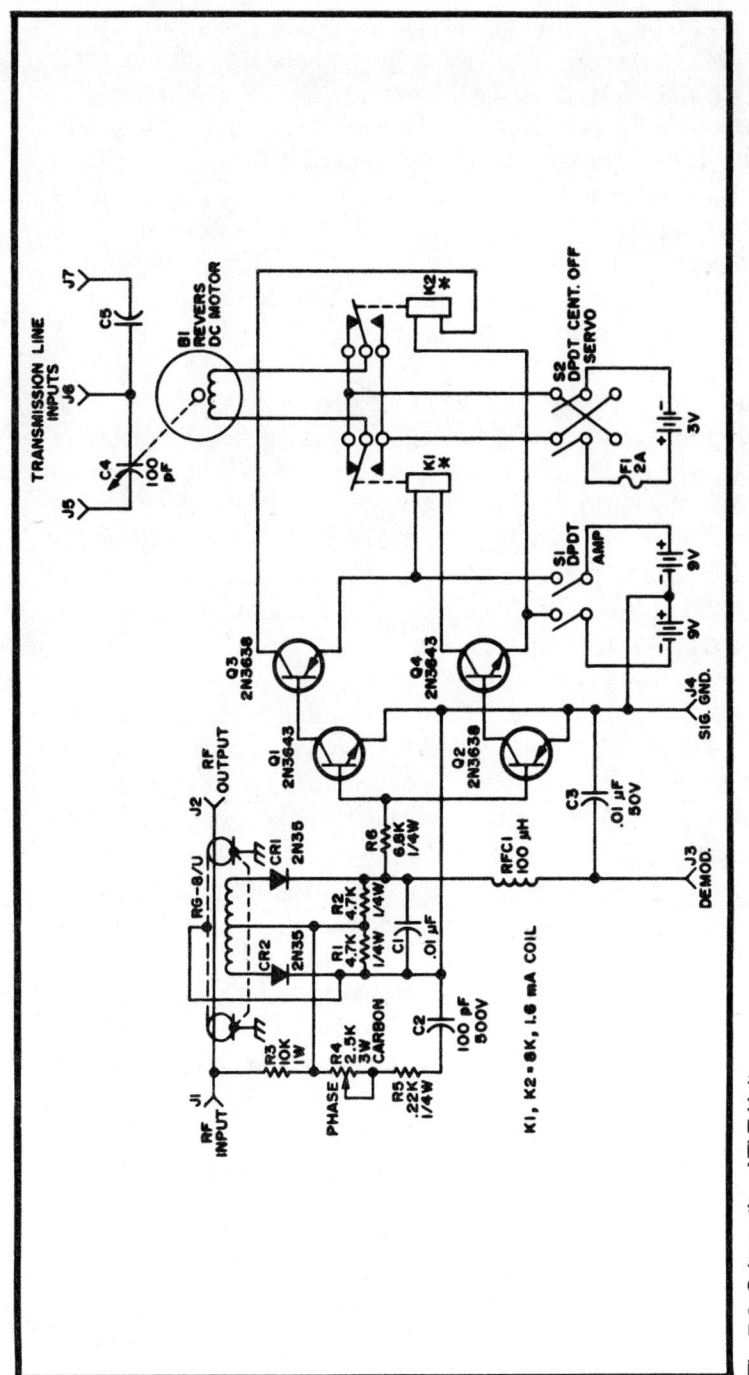

Fig. 7-9. Schematic—ATLT Unit.

The primary voltage signal path is through R3, to the center-tap of L1 and out its ends, through CR1 and CR2 to ground. C1 provides an AC ground to both ends of the R1-R2 divider, plus smooths the pulsating DC. The secondary path through R4, R5, and C2 reduces the voltage applied to L1 and allows small phase correction to be made to it. When the voltage signal is present, it will add to or subtract from the current signal, depending upon their phase relationship. If the voltage and current signals are directly in phase, then across one coil of L1 they will add while across the other they will subtract. Unequal currents will flow through R1 and R2 creating a potential difference across the divider. If the phase is changed 180 degrees, this potential will be of the same magnitude but of opposite polarity. When the signals are phased 90 degrees to each other, right in the center between 0 and 180 degrees, equal currents again flow and there is no potential difference. The voltage signal, applied to this phase-sensitive demodulator, is not shifted from its original phase, while the current produces a signal that is shifted 90 degrees. Therefore, when the coaxial cable voltage and current are in phase (unity power factor) there will be no demodulator output measurable across J3 and J4. Any small change in this phase relationship will result in an output that will be either positive or negative, depending upon whether the angular difference is leading or lagging.

The demodulator output is applied between the base and the emitter of Q1 and Q2. Q1, being an NPN transistor, will be switched on when the output is positive, and Q2, a PNP, will turn on when it is negative. Both transistors are operating class C and require about 0.5V of signal before either will conduct. Q1 and Q2 are direct coupled to Q3 and Q4. Either one or the other can be switched on, actuating K1 or K2, depending upon the state of the demodulator output. K1 and K2 are high-resistance, low-coil current relays with their contacts wired so that a high-current positive or negative, 3V DC output is made available to operate the servo motor, which positions variable capacitor C4. C4 is connected across the matchbox output and will automatically be repositioned every time the power factor in the coaxial cable changes from unity, bringing the antenna system back into resonance.

The reversing switch (S2) is used to eliminate the construction problems that would be brought about by trying to keep all phases and outputs aligned so that the servo motor will turn in the correct direction when called on to make an adjustment of C4. A complicated switch of this type is not needed if C4 does not have stops and will rotate 360 degrees. This is because C4 can be turned in either direction to its correct location.

A homemade fixed capacitor (C5) is used during 15m operation to prevent "hunting." Hunting is the oscillating motion of the servo-mechanism about null that takes place when variable capacitor C4 coasts through the null point after the motor power drops to zero, causing the demodulator to produce an output that drives the motor in the opposite direction, where it coasts past null again repeating the whole process. Series connected C5 makes it necessary for C4 to move a reasonable number of degrees to make an effective adjustment.

The ATLT is assembled on a 2 in. chassis of about 5 by 7 in. area. The front and rear panels are across the narrow ends of the chassis and are 5 in. high with rounded corners. The cover is a wrap-around type that snaps in place by holes in its lower side picking up the protruding heads of screws mounted in the chassis. Easy cover removal is necessary unless some sort of dial is provided to indicate the rotor position of capacitor C5. The front panel controls are: *phase* potentiometer R4; *amp* and *servo* switches S1 and S2; *demod* and *sig gnd* jacks J3 and J4.

Connections to the fixed and variable capacitors (C5 and C4) are brought through the rear panel using high-voltage ceramic feedthrough insulator posts. A shield, run down the center of the chassis, provides a place to mount the battery holders. The one for the 9V batteries is made of thin aluminum, while the 1.5V cells are held in the battery box removed from the toy auto which also provided the necessary motor gearbox assembly.

The current pickup coil details can be seen on Fig. 7-10. A piece of RG-8/U coaxial cable is cut so that it is long enough to be soldered between box connectors J1 and J2, located either side of the chassis at the back. The outer insulating jacket is slit lengthwise and peeled off in one undamaged piece, because it will be used for a form for L1. The braid is also pushed off undamaged so it can be later replaced. The outer jacket insulating material is wrapped around the insulated center conductor and trimmed so that it will fit snugly without overlapping. Its length is trimmed to 2 in. and a hole is punched in its end either side of the slit for the coil leads to pass through. Seventeen bifilar turns of 28-gage enamel wire are wrapped around the jacket material lengthwise. (Seventeen bifilar turns are equal to thirty-four single turns.) The start of coil B and the end of coil A are twisted together. These leads will make the centertap when they are soldered. The other leads, along with the centertap, are slipped into an insulating tube, and the completed L1 is thinly taped around the insulated coax center conductor. The braid is expanded and slipped back over L1 and the center conductor, fishing the L1 leads out through a hole worked into it about ¾ in. from one end. The whole thing now looks like a sort piece of RG-8/U without its outer jacket, and a small lump where L1 is located under the braid. The tubing, with the L1 leads in it, will be coming through the braid at one end of this lump, and when the current pickup assembly is soldered to the box connectors, this end will be near J2. The braid ends are wrapped and soldered to one turn of bare wire, and these wires are connected to two ground lugs at each connector shell.

Constructing a servomechanism turns out to be a lot less frightening than first thoughts make it seem. Timer assemblies make excellent servomechanisms for this application, but require power sources not readily portable. Luckily, battery-powered motor-gearbox assemblies abound in any toy store. There are more snappy battery toys made today than one who has been away from such things has a right to realize.

There is no reason for not being selective about picking out a toy that operates rather slowly, and has an easy to-get-at drive assembly.

The electronics is built on a perforated Vector board and arranged to look neat. The transistor types can be changed to practically any silicon NPN and PNP type and still work fine. The balanced 2N35 diodes are rather old-fashioned and can be replaced by other types if both have equal conduction. The

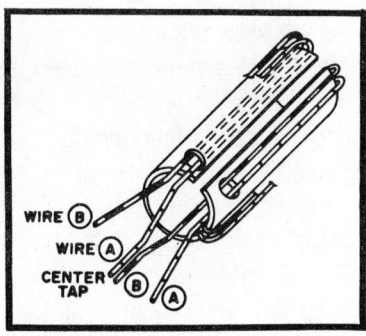

Fig. 7-10. L1 construction is on 2 in. long RG-8/U casing; 17 turns (lengthwise) of 28-gage enamel wire.

relays have 8 kΩ coils and will operate with 15V across them. The circuit board is mounted on spacers under the chassis and wired to the switches, jacks, L1, and signal ground (which is a point on the coax braid over the center of L1); R3 is connected to the center conductor of J1.

C5, the homemade fixed capacitor, is made from a U-shaped square (1½ in.) of aluminum. The top of the "U" end is trimmed ¾ in. deep so only the lower ½ in. of it is left. This portion is drilled and bolted to the 15m (bottom) feedthrough post. The center plate is a 1½ × 2 in. right angle shape with a ½ in. flange. The lower portion of the flange is trimmed like the "U" so it will clear the 15m feedthrough post. The tab that is left is connected to the center post that also feeds the stator of C5. The plates of the capacitor are bent until they are spaced ¼ in. from each other.

Before operating the completed ATLT, a few precautions should be taken. The plate spacing of C5 is about .070 in. This makes it necessary that the length of the tuned transmission line be selected so that a very high voltage will not wind up at the matchbox output. Remember, a quarterwave line inverts the load connected to it, and a half-wave line will repeat it. Be careful that rf does not get into the unit from the test meter, or indications will be erratic. Check that the transistors do not have a current flow before an rf signal is present. Keep in mind that the ATLT will detect and correct *only* reactive power conditions in the coaxial cable, so the vswr indicator can be indicating a high vswr of a resistive nature that will not be seen by the ATLT. The resistive mismatch must be tuned out with the matchbox input capacitor, but only one time for each band.

After the above checks are made, connect the ATLT in series with the coaxial cable running to the matchbox input, along with a vswr indicator ahead of it. Connect C5 across the matchbox output using a short piece of TV twinlead, and set the rotor halfway open. Load the matchbox with a lamp so that its output capacitor will tune rather sharply (coil loaded to a medium Q). Set all ATLT switches off, and apply a 40 or 80m signal to the setup. Adjust the matchbox for optimum output.

Rotate *phase* control R4 to the center of its travel. Measure the demodulator output between J3 and J4 using a 20 kΩ/V meter, and readjust the *phase* control for a null indication. Move the matchbox output capacitor slightly one way and then the other from the set position and the meter will indicate a potential, a null when returned to set, and an opposite polarity potential when on the other side of set. Place the *amp* switch S1 on. K1 and K2 will click on and off as the capacitor is rotated through null.

Place the *servo* switch S2 on and the servomechanism will relocate variable capacitor C4 to compensate for these position changes. If C4 drives against one of its stops, set the *servo* switch to its other position and the motor will drive in the correct direction to produce a demodulator null. When the transmitter frequency is changed, C4 will be driven open or closed as the frequency is increased or decreased, and the vswr indicator will indicate a low ratio.

C4 is not a large capacitor and with some matchboxes it might not be capable of adjusting throughout the entire spectrum of the lower bands, but it most certainly will cover whole CW or phone portions. Try the setup on the other bands, using C5 in series with C4 on 15m. If trouble is encountered, the ATLT may be tested like a vswr indicator by connecting a dummy load to coaxial box connector J2. Under this condition the demodulator output should be less than 0.5V.

After these tests are completed, the ATLT is ready to be connected to the antenna feed system in Fig. 7-8. It wil operate just like it did during bench tests, correcting for all transmitter frequency changes large enough to cause objectionable standing waves along the coaxial cable.

This method of antenna tuning is as reliable as a vswr indicator. It can tune a center-fed 33 ft antenna with wonderful results on all bands.

C5 could have been a roller inductor at the base of a short vertical antenna, continuously adjusting it over the whole band, or the servo-mechanism could be rotating the matchbox capacitor instead of an outboard unit. What may not be obvious is that this idea can be expanded to include the automatic tuning of tank circuits by placing L1 in one of the circuit leads and shifting the voltage signal 90 degrees before it is summed with the current signal. By using varactors and motors, an entire transmitter could be automatically tuned.

REMOTELY TUNED, DUAL-BAND ANTENNA COUPLER

Simplified and inexpensive motor-operated remote control of tuning and provision for remote reflected power monitoring are features of this coupler. The ideas presented can be applied to a variety of situations where remote control and band-switching of an antenna coupler unit is desired while utilizing only a single pair control cable for all remote control functions.

Instead of running resonant feedlines or the end of a voltage-feed antenna into the shack, it would be better to have the coupler used remotely located and controlled. In many instances, this would avoid most of the problems of rf in the shack and possibly permit a more optimum placement of the radiating portion of the antenna system. In apartment situations, it may also prevent TVI/BCI difficulties due to being able to use a coaxial transmission line to the remotely located coupler.

Many ideas have been discussed for the remote control of antenna couplers. One can get involved with expensive motors, special relays, elaborate control circuits and the need for multi-conductor control cables. The coupler described, however, uses inexpensive components that make it only slightly more expensive than a regular antenna coupler. The coupler is built for dual band operation, but the ideas used can be incorporated into more elaborate designs as required to fit a specific need.

Basic Scheme

The functional units comprising the coupler are shown in Fig. 7-11. The coupler contains a matching network with motor controlled tuning. The motor tuning is also arranged to provide a switching function. The reflected power sensor is simply half an in-line SWR meter. An SWR meter located at the transmitter could also be used to indicate the effect of tuning the coupler but a reflected power indication directly between the coupler and transmission line is much more accurate. It is also easier to use than a field-strength indicator

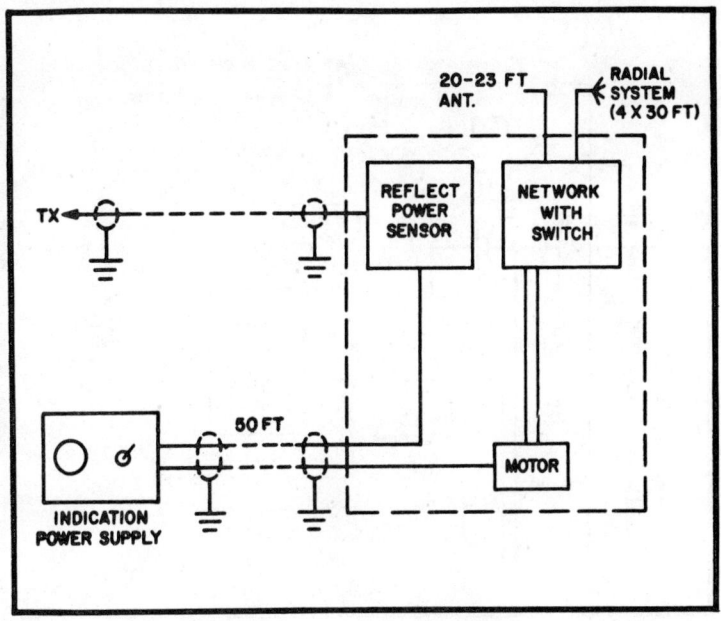

Fig. 7-11. Basic components of the dual-band remote antenna coupler system.

when tuning. The indicator voltage from the reflected power sensing unit and the DC control voltage for the motor are both transferred over a shielded 2 conductor cable which is completely independent of the transmission line. The motor itself is a DC type and the direction of rotation is controlled simply by changing the polarity of the motor voltage supply.

Coupler Unit

The coupler unit circuit shown in Fig. 7-12 was designed for use with a 20 foot rod antenna on 80 and 40 meters. The basic circuit shown can be used with many coupler arrangements, but make certain that the coupler will tune manually with a given antenna before any attempt is made to control it remotely.

As shown in Fig. 7-12, the transmission line is link coupled to the loading coil. On 80 meters, the variable capacitor is used to allow tuning of the series resonant antenna/loading coil circuit across the band.

The reflected power sensor is formed as part of the transmission line. The pickup link is made somewhat longer

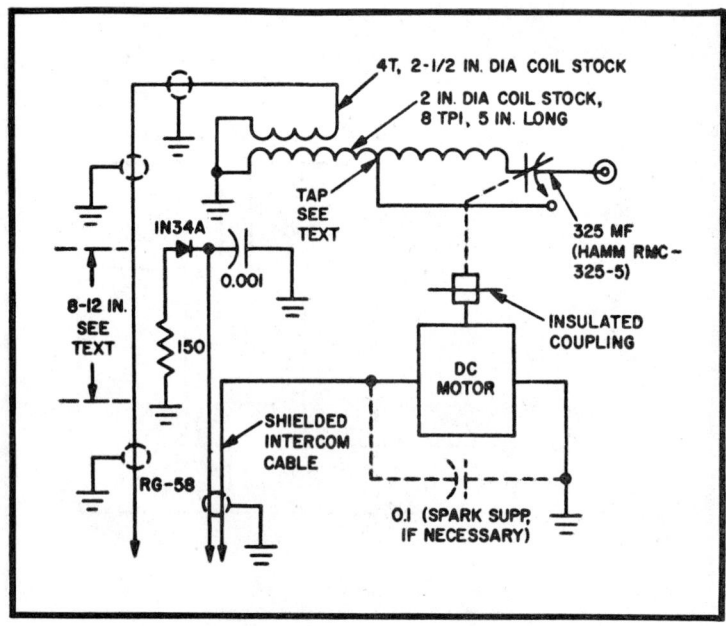

Fig. 7-12. Diagram of the remote antenna coupler.

than usual in order to develop a reasonable current over the control cable back to the indicator unit in the shack. The unit is not designed to read the actual SWR, although this can be done if the system is calibrated with a dummy load resistor simulating different SWRs. Assuming that the SWR was checked initially using a calibrated SWR meter, the reflected power indicator is only used to set the motor driven variable capacitor for minimum reflected power at any given operating frequency. An SWR meter is also used in the transmission line by the transmitter; there is a possibility of having a means available to monitor the condition of the entire transmission line/antenna system. For any reference setting of the forward power indicator on the SWR meter by the transmitter, the same reference reading from the reflected power sensor in the coupler unit should be obtained as long as all components remain in good condition.

Coupler Unit Construction

There is nothing critical about the construction as long as the variable capacitor is kept insulated. A small metal enclosure is used and the capacitor is placed on standoff insulators. The motor shaft is connected to the capacitor by an insulated coupling which mates the ¼" capacitor shaft and the ⅛" motor shaft. A small piece of dowel with appropriate sized holes drilled in each end and epoxy cement will work just as well. The motor used was a "junk box" item which works on 12V DC. Internal gearing and the friction provided by the capacitor results in a very slow tuning rate. Surplus and hobbyist outlets are sources for suitable motors. Almost any intermittent duty 12-24V DC type that has been geared down to 50 rpm or less will work.

The rotor plate nearest the back frame of the capacitor is bent slightly, as shown in Fig. 7-13, to form a sliding switch contact with one end of a chassis feed-through insulator mounted on the capacitor back-frame. A miniature chassis type feed-through is used. Any small piece of plastic may be used as a stop to prevent further rotation of the rotor plates once the switch is engaged by placing it across the stator plates on the same side as the switch (see Fig. 7-13). Epoxy cement should be used to secure the top.

The reflected power sensor is constructed by carefully slitting the jacket away for about 8-12" at the end of the coaxial cable to be connected to the link. A length of #20 enamelled wire is then manipulated under the shield; the ends are con-

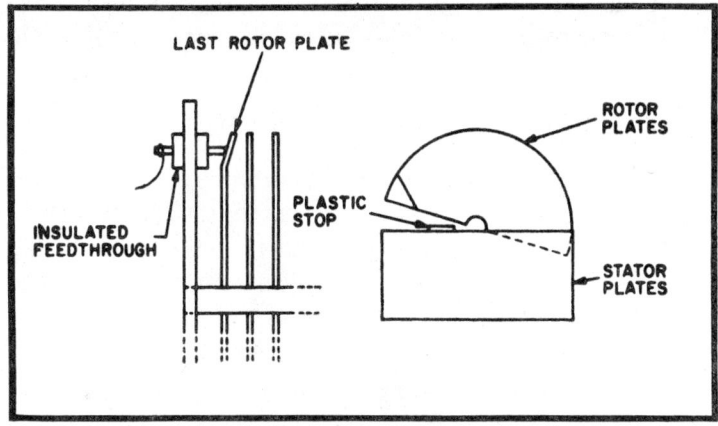

Fig. 7-13. Details of capacitor switch top view (A) and back view (B). Feed-through is placed in capacitor back frame such that rotor plate makes contact with plates almost fully disengaged from stator plates.

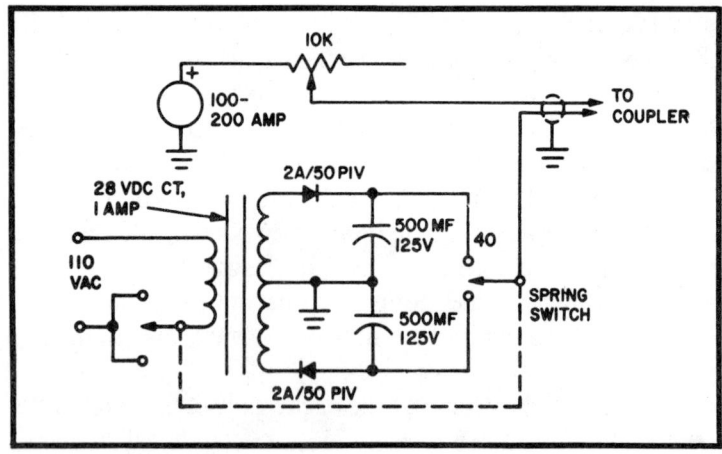

Fig. 7-14. Diagram of indicator/power supply unit. Batteries may also be used to power most small DC motors.

nected as shown in Fig. 7-12. The jacket is then replaced and the length of cable coiled together, if necessary, for compactness to fit the enclosure. The few components for the sensor are assembled on a terminal strip.

Indicator Unit

The indicator units contain only a microammeter and potentiometer as a reflected power indictor and a dual polarity source for the tuning motor. Batteries may be used for the latter function, although a small AC supply is shown in Fig. 7-14. Some overvoltage from the supply must be available in order to account for the drop in the control cable.

Adjustment and Operation

Make certain that the coupler operates manually before an attempt is made to remotely control the unit. In the case of the unit described, the numbers of turns in the main coil was established by trimming the coil on 80 meters and checking the frequency range of operation possible and the variable capacitor. On 40 meters, the capacitor switch was engaged and the top on the main coil established. These tests were done with a conventional SWR meter connected in the transmission line immediately before the coupler which was mounted in its operating position. Simultaneously, operation of the reflected power sensor was checked to see that the minimum output corresponded to the minimum reflected power indication on the SWR meter. Operation of the motor should be checked with a length of cable equal to that actually used for the control cable. This is necessary to insure that the voltage drop is not enough to affect motor torque.

Operation of the unit is not complicated, but it does require a little practice since its simple design does not provide for automatic motor cutoff or signalling of motor position. By observing the speed of motor rotation, one can learn to anticipate when a complete revolution has been made. One position of the control switch on the indicator unit should be marked "40" (or something similar) to indicate one extreme position of the capacitor.

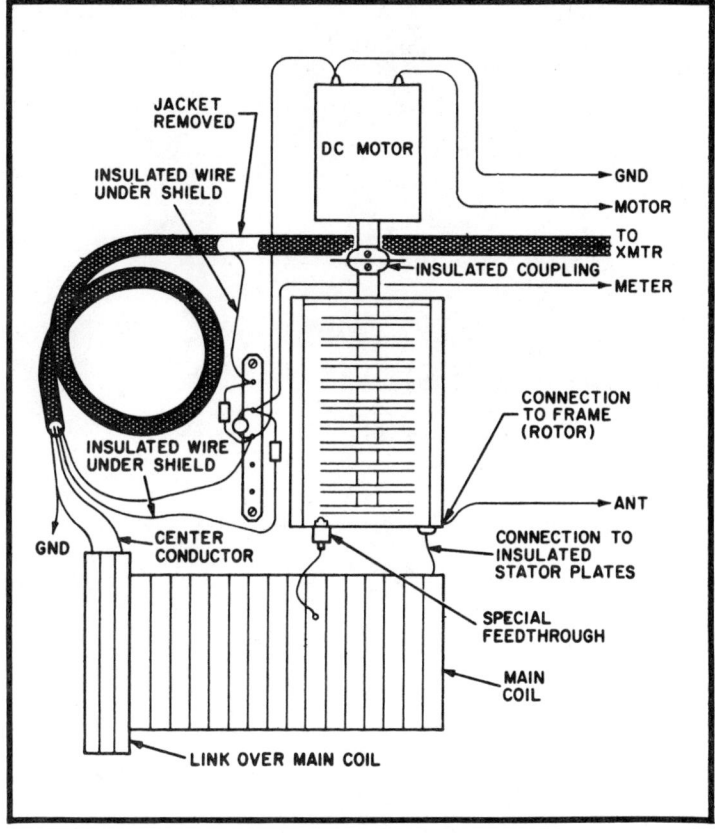

Fig. 7-15. Sketch of parts layout for dual band coupler.

Summary

The coupler shown is intended for use with low power equipment—up to about 100/150 watts. For higher power levels, an increase in the coil wire size, the insulation path between the motor and capacitor shafts and the spacing of the capacitor plates will be necessary. A larger capacitor will probably also require a heavier motor. Nonetheless, the basic construction can be modified as necessary for any power level.

TRANSMITTER TUNING OF MOBILE ANTENNAS

The usual approach to low-frequency mobile antenna construction is illustrated in Fig. 7-16A. A short whip antenna is used which is either base or center loaded so that the antenna has an effective electrical length of ¼λ and can be directly connected to a non-resonant coaxial transmission line. In order to reduce ohmic losses in the loading inductor, its "Q" is made as high as possible. The high "Q" results in greater radiation efficiency due to the reduced I^2R loss but it also results in a very restricted bandwidth for the antenna—10 to 20 kHz being typical for many 80 meter mobile whips. Whenever one wishes to change the transmitter frequency to any great degree, it is necessary to readjust either the loading inductor on the antenna or the length of the whip antenna itself.

It should be recognized that even with high-Q loading inductors, the radiation efficiency of an 8 ft whip on a low-

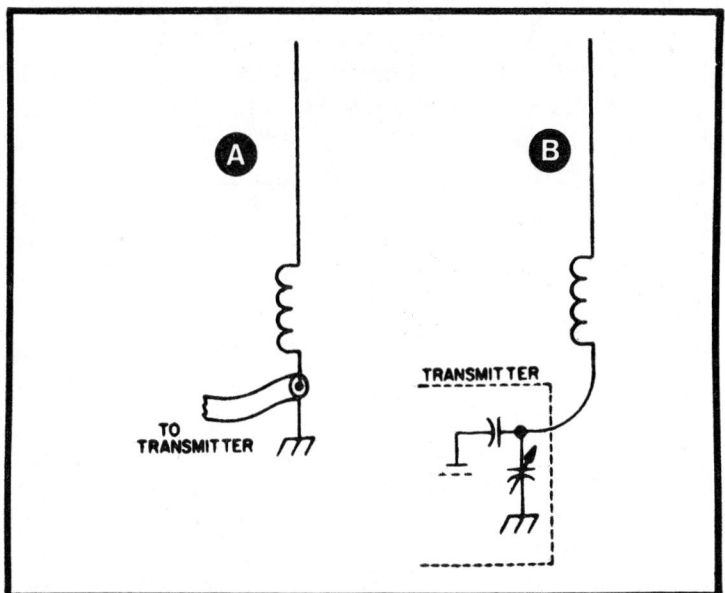

Fig. 7-16. Usual base or center loaded whip antenna which utilizes a non-resonant transmission line (A). Concept of using the high Q loading coil directly as part of the transmitter tank circuit.

frequency band is a matter of a few percent. No manner of loading inductor is going to make an 8 ft whip radiate like a 60 ft whip unless the losses in the loading inductor can be reduced to zero, a condition only possible if the *resistance* of the loading inductor can be reduced to *absolute* zero. The restricted bandwidth of low-frequency whip antennas is due to the fact that it is desired to have them self-resonant and operate into a nonresonant transmission line. This concept has no real basis as far as improving the radiation efficiency of an antenna in a low-frequency mobile installation and simply imposes a severe bandwidth restriction upon the operation of the mobile installation.

Non-Resonant Loaded Whips

If one considers the loading inductor used either at the base or center of a whip as necessary to increase the effective electrical length of the antenna to the point where the antenna system can resonate at a given frequency, there is no theoretical reason why this resonant circuit cannot simultaneously act as both the resonant output circuit for a transmitter and as the radiating medium or antenna for the transmitter. This concept is illustrated in Fig. 7-16B. The line section between the whip and transmitter also becomes part of the radiating antenna.

This idea is not really new and, indeed, in basic concept goes back to the earliest days of radio. Some readers will immediately relive some of the nightmares of harmonic radiation and interference that were present using AM transmitters with class C output stages when looking at Fig. 7-16B. However, several factors have changed which make the scheme much more practical now, mainly the use of linear output stages and the high Q of most loading inductors. The scheme is certainly not recommended for fixed station usage in a location where TVI is a problem already, since the output is not filtered and harmonic reduction is mainly a function of that provided by the single tuned circuit which is involved. However, for mobile use, the scheme does have particular appeal. Mobile operation is often conducted remotely from housing areas and the approximate 30—40 dB second harmonic attenuation provided by most good loading coils tuned as shown suffices with low power transmitters to avoid any interference problem. This is especially true for mobile operation on 80 meters.

Practical Considerations

When the scheme of Fig. 7-16B is used, only the variable capacitor in the transmitter need be adjusted for wide frequency excursions once the loading inductor and capacitor values have been balanced to provide proper transmitter loading. This adjustment is described in more detail later, but in most cases, the adjustment range required is within the existing range of the component values of the variable plate tuning capacitor within the transmitter and of the whip loading inductor. No component modifications need be made in most cases.

One area that does require some minor modification, however, is the line between the whip antenna base and the transmitter. Since the line becomes part of the antenna circuit, it will radiate and, also, it will carry the greatest portion of the antenna current. The radiation that takes place from this line where it runs in the automobile is, of course, shielded by the automobile body and lost. However, as was mentioned be-

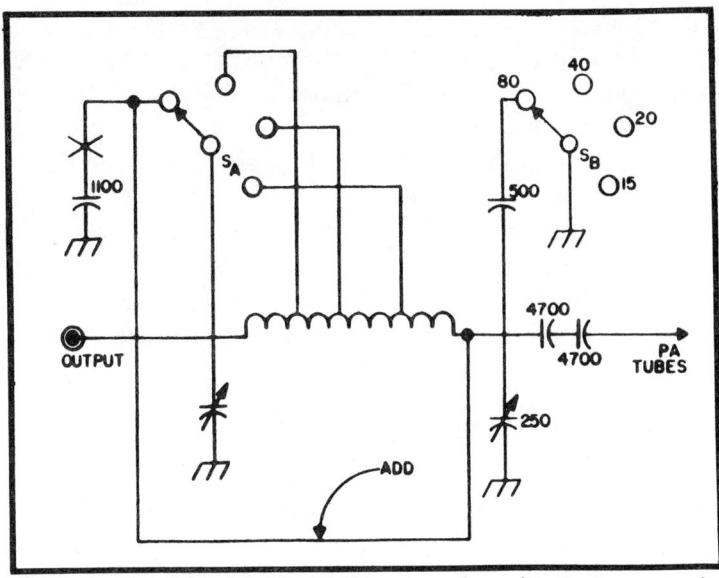

Fig. 7-17. Simple tank circuit switching connections changes necessary to bypass pi-network coil on 80 meters only. Circuit shown is typical of a wide variety of transceivers.

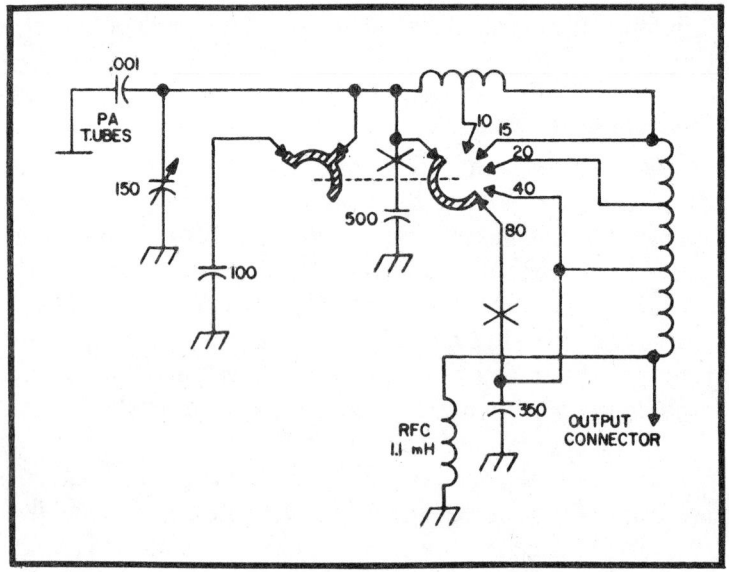

Fig. 7-18. Pi-network circuits employing continuous shorting type switches may be slightly more difficult to modify.

fore, the radiation efficiency of a loaded low-frequency whip is only several percent anyway and the additional radiation lost is not significant in most practical situations. It certainly is a small price to pay for the ability to tune a mobile rig freely across major portions of a band. Nonetheless, the line should be kept as short as possible.

Because of the heavy current that flows in the line, it should be made from heavy wire—and not just the inner conductor of a small coaxial cable. Heavy battery cable of the type with a thick covering, in order to provide the necessary voltage insulation, or the inner conductor and dielectric of a really heavy coaxial cable (shield removed) such as RG 14/U or a larger cable should be used.

Output Circuit Modifications

If a multiband transceiver is used for mobile operation, it may be found advantageous to use the method described for mobile antenna coupling on 80 meters, or both 80 and 40 meters, and a conventional loaded mobile antenna feed by a coaxial transmission line on the higher frequency bands. This situation occurs since the radiating portion of the feedline which is enclosed within the automobile increases in terms of wavelength with higher frequency and the losses encountered with the antenna coupling method of Fig. 7-16B exceed those of the method illustrated in Fig. 7-16A.

Therefore, it is necessary to provide some modification to the output circuit of a transceiver or transmitter such that it can be used with the antenna coupling arrangement of Fig. 7-16B on one or two low-frequency bands and with a conventional unmodified coupling scheme on the higher frequency bands. Figures 7-17 to 7-19 show details of various modification methods to typical pi-network output circuits. The only other switching involved is that concerning the transmission line which can be done manually or with relays.

Adjustment

Initial adjustment should be done at a low power level by reducing the drive to the output stage of a transceiver. The output loading capacitor (which is in parallel with the plate tuning capacitor once the transceiver is modified) is intially set

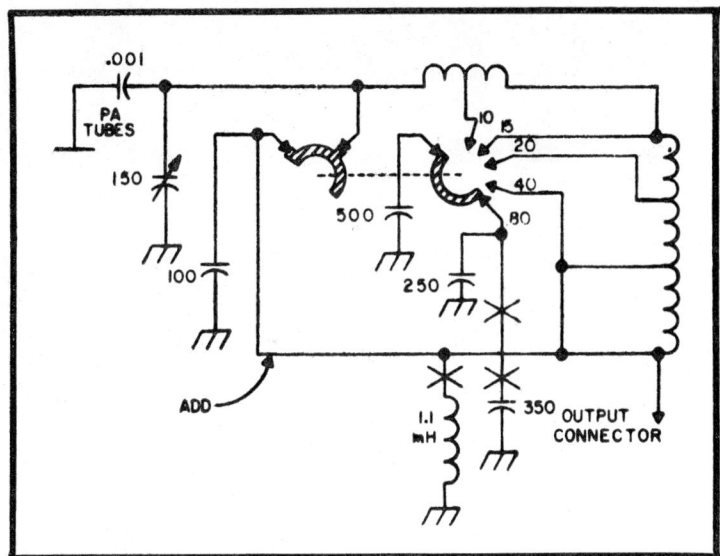

Fig. 7-19. Although an extra component must be used, this modification to the output circuit provides a far better switching arrangement than that shown in Fig. 7-18.

at minimum although it can be used later if more capacitance is required. The drive is increased and the transceiver checked for the usual meter indications of resonance and proper loading. If resonance is indicated but the loading is not correct, one can try increasing the value of the plate tuning capacitor by changing its setting and reducing the value of the antenna loading inductor (or vice versa) until adequate loading is achieved.

One must use a wavemeter or some other device to check the radiated frequency since it is possible to have the system falsely tuned in some cases. When the system is tuned correctly, one should also check the harmonic radiation and interference levels as compared to the usual setup. A check with a receiver tuned to the various harmonic frequencies will quickly indicate if the harmonics are at least 30-40 dB down from the fundamental frequency.

WIDE-RANGE ANTENNA TUNER

Most modern ham rigs are designed with single-ended outputs for direct connection to coax-fed antennas, but there are still many cases where a good wide-range antenna tuner can be very valuable. The most obvious are:

1. To couple the transmitter to any one of the wide variety of antennas which require high impedance feedlines, such as V-beams, rhombics, Lazy H's, Zepps, folded dipoles, etc. This is especially important when the QTH will not allow the erection of several antennas, and the use of a wire antenna and tuner will permit multiband operation.
2. In moderate to severe cases of TVI, where the high harmonic attenuation of the tuner will be of considerable assistance.
3. To permit optimum coupling between the antenna feedline and the *receiver*, an advantage which seems to get little attention.

The tuner can be built in a variety of ways, using circuits which have proved to be reliable over the years. However, to be as useful as possible, the tuner should be capable of either series or parallel tuning, and should be continuously useful over the entire 3-30 MHz range. All controls should be operated from the front panel, and no bandswitching, coil changing, or coil tapping should be required. The tuner described satisfies these requirements.

The general circuit arrangement is known as a Z-match and has been well known for many years. Two coils are used in conjunction with a split-stator capacitor to cover two frequency ranges at the same time. This allows a smooth transition from series to parallel tuning and no bandswitch is required. An SWR bridge is incorporated, and is in the input line at all times, making tuneup a simple matter.

The tuner is built on a 10 × 17 × 3 in. chassis, with an 8½ × 19 in. front panel. Depending upon the power range of the station transmitter, smaller components could be used and thus reduce the size of the chassis. However, the unit shown will safely handle more than the legal limit on all bands and modes.

A close study of the schematics will reveal most of the construction features. The input coaxial line is first routed through the small shielded enclosure used for the rf components of the SWR bridge. This will be described later. From this enclosure, the line is connected to the variable coupling capacitor C1, which must be insulated from the chassis and panel. A small vernier dial drives this capacitor through an insulated coupling and provides smooth control. (To aid in setting the tuner to the proper tuning point during rapid band changing, calibrated vernier dials are used on both capacitors along with a tuning chart on the front panel.)

The main tuning capacitor, C2A and C2B, is mounted parallel to C1, but *not* insulated from the chassis. A heavy-duty, two-pole, two-position ceramic switch (S1) is mounted on the front panel directly under the two large feedthrough insulators. The two coils are mounted at right angles to each other between C2 and S1 and are supported by their own leads. Commercial air-wound coil stock was used in this model, which made this method of mounting quite feasible. However, homebrew coils should be wound on ceramic forms and firmly mounted.

The SWR bridge is built in two sections, one containing the rf pickup, diodes, and matching resistors, and enclosed in the metal shell at the rear of the chassis, underneath. A small piece of terminal board holds the compnents associated with the meter, and the meter itself is mounted on the front panel, flanked by the sensitivity control R4 and the FORWARD-REFLECTED switch (S2).

The construction of the rf section of the bridge can be seen from Fig. 7-21. The inner line conductor is a 4 in. length of ¼ in. copper tubing. The outer line conductor is a flat piece of copper flashing mounted under the tubing and secured to small mounting bolts at each end of the enclosure. Small square pieces of Plexiglas are used to separate the two conductors, with the tubing being inserted in holes drilled in the center of the squares. Small holes are drilled in the outer edges of the squares to hold the two pieces of #14 copper wire

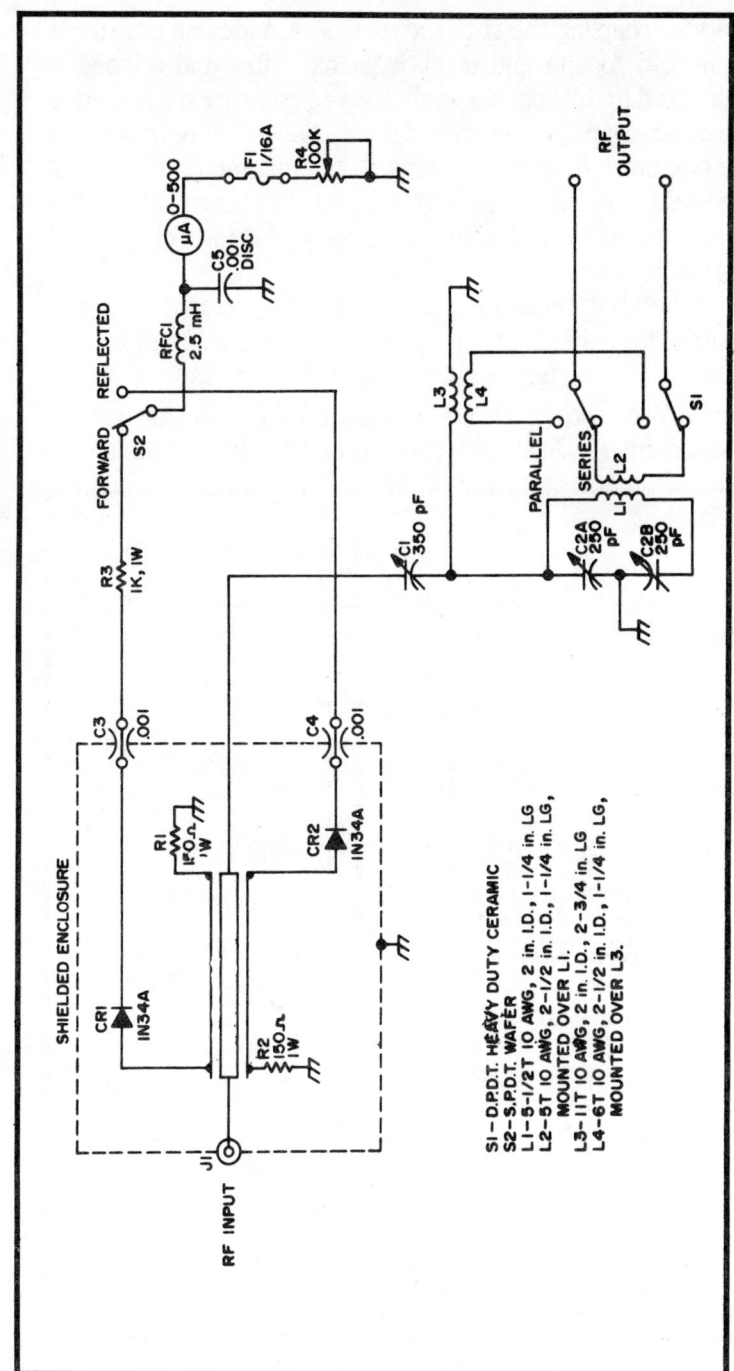

Fig. 7-20. Overall schematic of the tuner.

used for coupling the rf to the diodes. A matching resistor is connected to one end of each pickup wire, and a diode is connected to the opposite end. The resistors are connected to the outer copper conductor, but the diodes are connected to the two small feedthrough capacitors mounted on the wall of the enclosure. Small shielded wires run from these capacitors to the terminal board, then via S2 and the rf filter (RFC1 and C5) to the meter.

Operation of the unit is similar to most tuners: a matter of getting the unit set up on each band, and then marking the capacitor dial settings on the panel chart for easy reference. Figure 7-22 shows the setup required. With low power fed into the tuner, adjust R4 for full-scale deflection on the meter

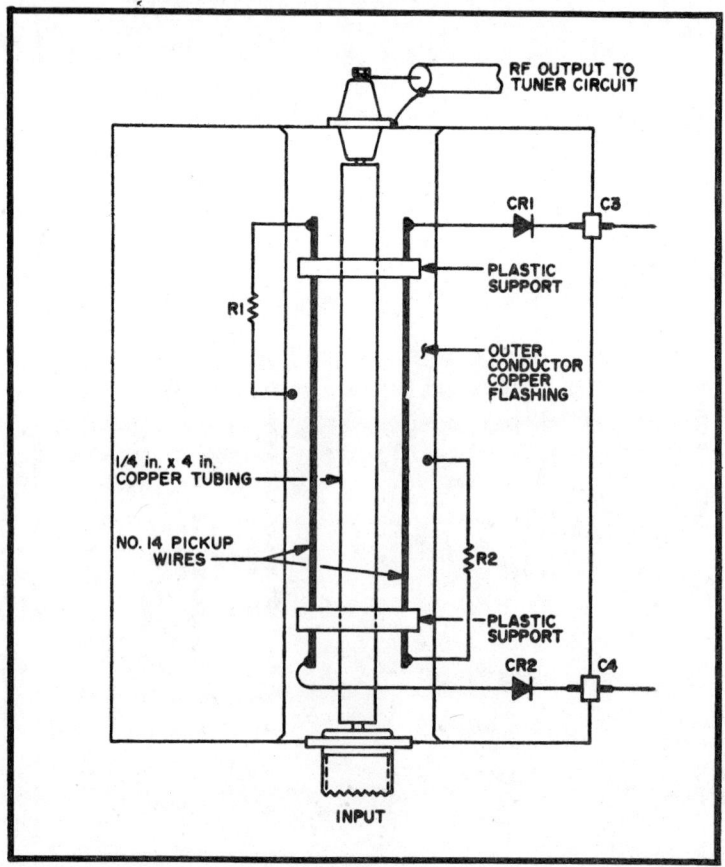

Fig. 7-21. Construction of the SWR bridge.

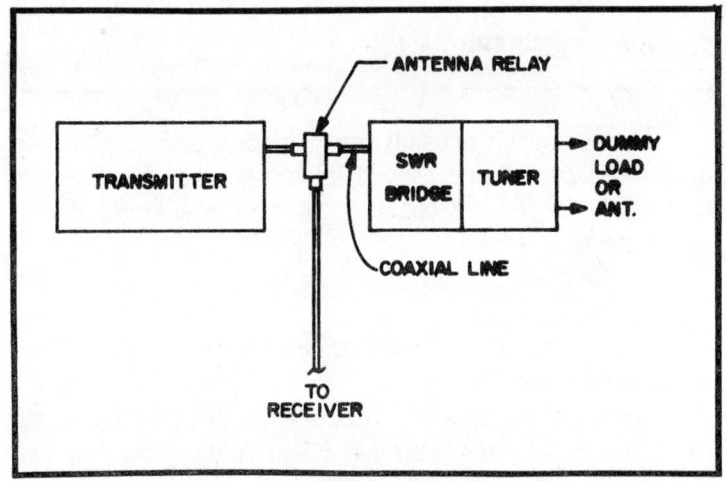

Fig. 7-22. Proper setup for using the tuner. The coax line to the transmitter may be any reasonable length, and a separate SWR bridge may be used if available.

with S2 in the FORWARD position. Switch S2 is REFLECTED and adjust C1 and C2 to obtain minimum reading on the meter. There is considerable interaction between the capacitors, so make sure they are both set properly and that the SWR is as low as possible. There is also a considerable reaction on the transmitter's plate tuning, so check this often and make sure it is in resonance at all times.

This can be a slow proceeding the first time. Only a small amount of power should be fed into the tuner, and a dummy load should be connected to the output until the settings have been determined. Then a short final adjustment with the antenna connected will suffice. Because there are no bandswitches or coil taps, the setup is very easy to follow and should result in an SWR of almost 1:1.

AN EASY, WIDE-BAND BALUN

The desirability of using a network or transformer to feed a balanced coaxial line is well known and has been widely discussed. Single-band baluns have been well covered with this in mind.

A little research can produce a simple, cheap and easily constructed balun.

It covers the 40, 20, and 15 meter bands, using 20 meters as the design center.

A number of turns of RG-8/U cable are coiled up using the diameter of the desired finished balun. The resonant frequency is then checked with a grid-dip meter. A little trimming is necessary to obtain resonance in the 20-meter band. The resultant coil consists of ten feet of cable wound in

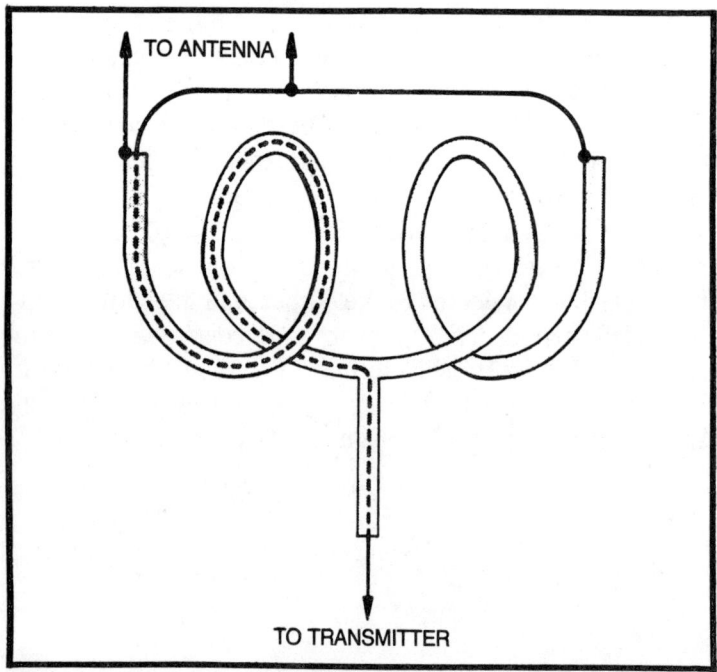

Fig. 7-23. The 1:1 balun constructed from a length of coaxial cable. When designed for the center of the desired frequency range, it will cover a 3:1 frequency operating range. The unit described here for 40, 20, and 15 meters consisted of two 66-inch lengths of RG-8/U.

five turns. The odd number of turns are purposely worked out so that the center-tap feed point will come on the opposite side of the coil from the load point for mechanical reasons. The five-turn coil is then cut in half, and the inner conductor and the shield are connected according to the diagram and then recoiled. In actual practice, the original coil was scrapped and two new 66-inch lengths of cable were cut. This allowed for three inches to be skinned back on each end to make the connections and still maintain the original length. The joints were carefully soldered and taped to keep out the moisture. The coil was then bound with lacing cord, and it was ready for installation using the shortest possible leads to the antenna.

Measurements on the experimental 1:1 balun using a 50-ohm dummy load gave SWR readings of 1.34:1 on 40, 1.15:1 on 20, and 1.43:1 on 15 meters. This was considered reasonable so the finished product was installed at the antenna. It is still necessary to tune the antenna when using a balun transformer as it works better when looking into a nonreactive load. Its purpose is to take the rf off the shield of the coaxial feed line when feeding balanced antennas and make the antenna the only radiating device in the system.

AN EASY GAMMA MATCH CAPACITOR

The yagi antenna fed with a coaxial feedline is the most widely used antenna on the HF and VHF bands. It is mechanically simple and quite strong. But the driven element of a yagi has a very low input impedance (on the order of 15 ohms) and presents a problem when matching the driven element to a coaxial feed-line. Many antennas use the familiar gamma match to raise the input impedance to a value compatible with the coaxial feedline.

Most gamma matches use an air variable capacitor to tune out the reactance introduced by the gamma rod. This introduces both mounting and weatherproofing complications. Often the gamma matching system is more difficult to construct than the rest of the yagi. Fixed capacitors, such as mica transmitting capacitors, have been tried as a means to eliminate the need for a weatherproof enclosure, but the added inconvenience of tuning the antenna with an air variable capacitor, then substituting a fixed capacitor of the approximate value, has proven to be more trouble than it is worth.

While searching for a simpler method, coax cable itself was tried as a substitute. It has a capacitance of from 20 pF to 30 pF per foot depending on the type used. It can be trimmed to the proper value needed, needs no difficult mounting assembly, is just as weatherproof as the feedline itself, and is considerably cheaper than an air variable.

Figure 7-24 shows the mechanical details of the capacitor. The details of the gamma rod are up to the reader. The first step is to determine the approximate capacitance needed for a gamma system on the design frequency of the antenna. The following table shows approximate guide values for a standard three or four element yagi.

20 meters	100 pF
15 meters	75 pF
10 meters	50 pF
6 meters	30 pF

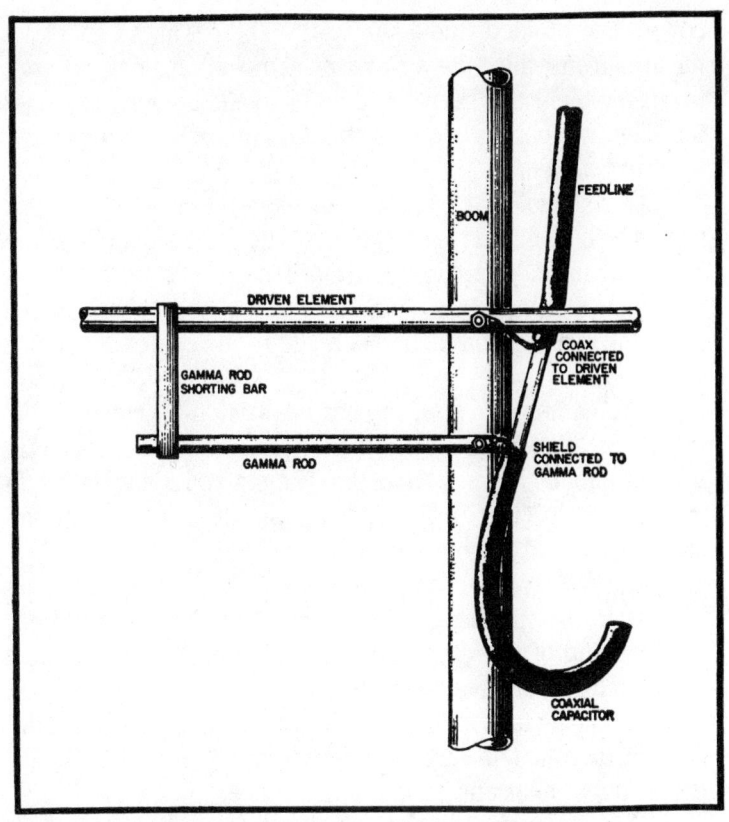

Fig. 7-24. Mechanical layout of the coax-cable gamma capacitor. After it is adjusted to the proper length, it is sealed with plastic tape.

Then determine the capacitance per foot of the coax cable to be used for the feedline.

RG- 8/U	29.5 pF per foot
RG-11/U	20.5 pF per foot
RG-58/U	28.5 pF per foot
RG-59/U	21.5 pF per foot

From this, the length of coax needed to provide the desired capacitance can be determined. For instance, a ten meter beam would require about twenty inches of RG-8/U for the gamma capacitor. Measure back this distance from the end of the cable and remove three or four inches of the outer vinyl cover. The braided shield should then be cut in the middle of the area from which the cover was removed, making sure not to cut the dielectric between the shield and the inner conductor. The two lengths of shield should be unbraided and twisted to form two leads.

The shield lead from the transmitter end of the coax should be connected to the center of the driven element in the normal manner. The shield lead from the short section used as the gamma capacitor is connected to the end of the gamma matching rod. No connection is made to the center conductor of the coax.

The gamma rod shorting bar is adjusted for lowest SWR at the operating frequency, and then the free end of the coax is trimmed about an inch, and the gamma rod shorting bar is adjusted again. This procedure of alternatively adjusting the shorting bar and trimming the coax is continued until the SWR is reduced to 1:1 at the operating frequency.

After the matching adjustments have been completed, the free end of the coax and the area from which the outer jacket and shield have been removed are sealed with a good grade of plastic tape to keep out moisture. The free end of the coax is then coiled up and taped to the boom of the yagi, presenting a neat and simple appearance.

This system has proven to be easy to construct, rugged, and quite effective.

MATCHING STUBS

This article will show how to match any feedline to an antenna and get an SWR of less than 2 to 1. There is no restriction on antenna type or feeder length. All that is required is an SWR meter; however, knowing the approximate antenna impedance will do.

When a line is hooked up to an antenna and the impedance of the line is different from that of the antenna, then there is a reactive component present on the line. The SWR is high and losses increase. A great deal of power may be lost without putting much of a signal on the air. To rectify this, put in an equal amount of reactance, but of the opposite kind. This will reduce the SWR and power losses.

A transmission line which is a quarter-wavelength long behaves like a resonant circuit. If it is shorted at one end, it appears as a parallel resonant circuit with high resistance; if

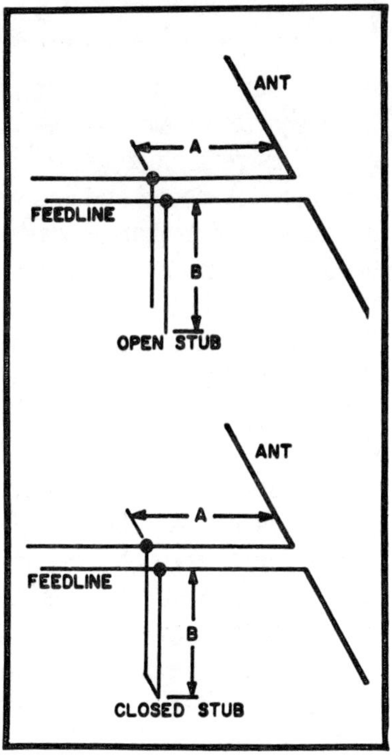

Fig. 7-25. How reactance is inserted. A shows the distance from the antenna for inserting the stub and B is the length of the stub.

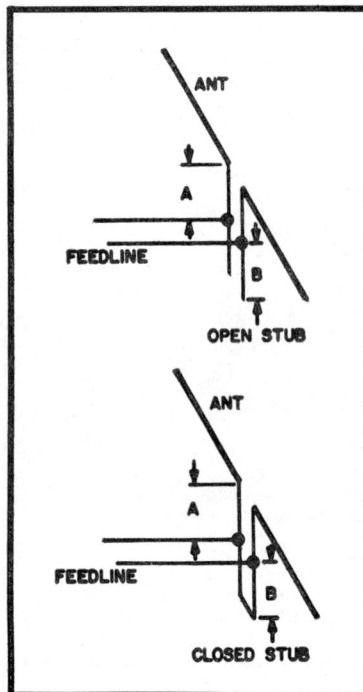

Fig. 7-26. This arrangement gives the same performance as Fig. 7-25 but is different in appearance.

open-ended, a series resonant circuit with low resistance. Lines shorter than a quarter-wavelength will exhibit reactance as well as resistance. An open end line will have capacitive reactance. A closed end line will have inductive reactance. A line less than a quarter-wavelength long may therefore be used to match antenna and line impedances for low SWR. Figure 7-25 shows how the reactance is inserted. A is the distance from the antenna at which the stub is inserted. B is the length of the stub itself. Figure 7-26 shows another arrangement which performs the same way, but looks a little different.

If either the standing wave ratio or the impedance of the antenna is known, then lengths A and B can be computed easily. There are only two requirements. One, the stub and the feedline must have the same impedance, and two, the antenna must be resonant at the intended frequency of operation. These are easy requirements to fulfill.

The first decision to make is whether to use an open-end stub or a closed stub. This will depend on the ratio of antenna to line impedance. When the antenna impedance is less than the line impedance, a capacitive or open end stub is used. If the antenna impedance is greater than the characteristic line impedance, an inductive or closed end stub is needed. When using 150, 300, 450, or 600 ohm twinlead or ladderline and the antenna is current-fed, it will probably need an open stub. If the antenna is voltage-fed, then a closed stub will probably be needed. There are exceptions.

Figure 7-27 shows the current and voltage distribution along a half-wavelength of antenna. If the feedline intersects the antenna at a current loop (maximum) and a voltage node (minimum) then the antenna is current-fed. The old standby, the half-wave dipole, is current-fed. If the antenna is fed at a voltage loop and current node, then it is voltage-fed. Note that these terms do not correspond to the terms end feed or center feed.

Having decided what type of stub to use, the next step is to measure the standing wave ratio. Hook the feedline directly to the antenna and tune up. CAUTION: do not load to maximum. When manufacturers say that their transceiver will deliver 400 watts to a load with an SWR of 2 to 1 or less, they mean it. These finals will not dissipate the reflected power. The feedline may not take the extremes caused by mismatch either. And while measuring, remember that some power is being radiated despite a monstrous SWR. If, the impedance of

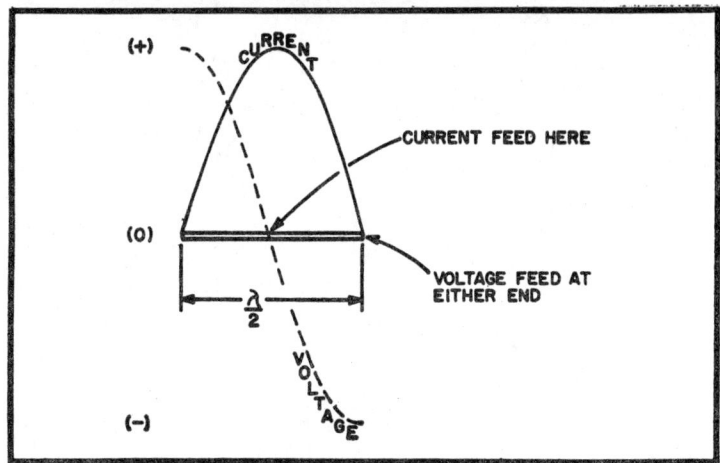

Fig. 7-27. Current and voltage distribution.

441

the antenna is known, then forget about measuring SWR unless the antenna is not very high or is very near anything that might affect its impedance value. Divide the antenna impedance by the line impedance, or vice versa if the line has the larger value. Use this or the SWR reading (they should be about the same).

Due to the different dielectrics used, radio waves travel along transmission lines at different speeds, always less than the speed of radio waves in free space. Assuming the wavelength required, the frequency, and the velocity factor are known, the exact lengths needed can be computed. Use the following equation:

$$\text{length in feet} = \frac{985}{\text{frequency}} \times \text{velocity}$$

factor X length in wavelengths. Use it twice; once for A, and again for B.

Now, merely break into the line at the appropriate point and connect the stub. If using the arrangement in Fig. 7-26, just connect a stub equal in length to A plus B and hook the feedline onto it at a distance from the antenna equal to A.

Here is a sample problem. Assume that a two half-waves-in-phase collinear for ten meters has been erected. This looks like a dipole except each side is a half-wave-length long. This makes it voltage-fed. Therefore, use a closed stub. Hook up the 300 ohm twinlead and measure the SWR and get a reading of about 15 to 1. Next, take a look at chart 5. From this comes a value of .21 wavelengths for A and .04 for B. Using the velocity factor of .82 for twinlead and an operating frequency of 28.1, comes the following equations:

$$\text{for A } \frac{985}{28.1} \times .82 \times .21 = \text{approximately } 6.05 \text{ feet}$$

$$\text{for B } \frac{985}{28.1} \times .82 \times .04 = \text{approximately } 1.15 \text{ feet}$$

Which means that a distance of six feet and one-half inch from the antenna, a stub is inserted one foot two inches long and shorted at the end.

It's hard to say just how broadband this type of thing will be. On fifteen meters, by designing around a frequency of 21.050 MHz, the test antenna gets a standing wave ratio of 1.1 to 1 at 21.0 MHz, and 1.5 to 1 at 21.350 MHz on a bisquare.

There is no reason why the stubs cannot be made out of coax. Use a T-connector or splice it and seal the joint with tape.

This method should eliminate a lot of unnecessary work in tuning the feed system of any antenna that uses stub matching. No more hit-or-miss adjustments will ever have to be made.

A DURABLE GAMMA-MATCH

The subject is a **sturdy** gamma match with a one piece gamma rod. It can be made inexpensively from scrap metal and discarded parts. The gamma rod, labelled **A** in Fig. 7-28, is a piece of 50 ohm foam-flex aluminum shielded coax. No more than 3 feet is required for 20, 15, or 10 meter beams, and a piece this size can usually be picked up gratis from most any twoway radio store. The SO 239 coax connector, **B**, is joined to the center of the driven element by an aluminum strap, **C**. At an arbitrary length from **B** a second aluminum strap **D** was placed which acts as the shorting bar. The distance, **d**, between the driven element of the beam and the gamma rod should be at least 3 inches or else the gamma rod will be inconveniently long.

The following steps were necessary in adjusting the gamma match:

1. First, the ends of the aluminum shielding next to the coax connector end of the gamma rod were bent outward slightly, and a coat of good sealer was applied for weather protection. The inner conductor of the coax was not connected to the SO-239 but was left floating, and a 100 $\mu\mu$F variable capacitor wired between the aluminum shielding and the coax connector as shown in the inset of Fig. 7-28.

2. The beam was taken up the tower to a height near that to be used in the final installation, as settings made near the ground will not be valid at normal operating heights. A signal was then fed to the beam from the transmitter, and the shorting bar and variable capacitor varied simultaneously until the setting with minimum SWR was determined. A Heathkit Reflected Power Meter and SWR Bridge was used for this operation. The shorting bar was fixed at this point.

3. The variable capacitor was brought down very carefully without disturbing the setting so that the capacitance could be measured with a capacity meter.

4. At this juncture knowledge of the capacitance per foot of the foamflex was necessary in order to use the proper length to equal the capacitance from the variable capacitor.

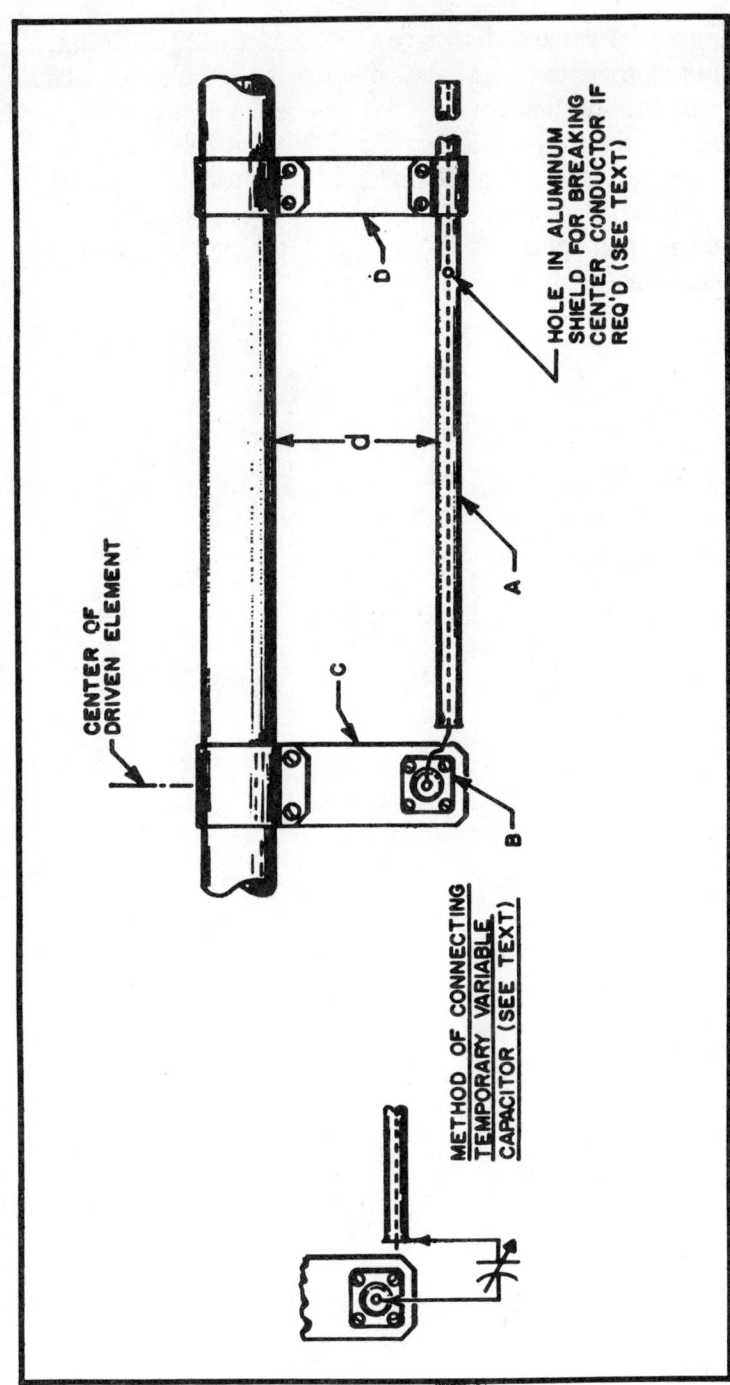

Fig. 7-28. The essential features of the gamma-match.

Our material was known to have a capacitance of 29 $\mu\mu$F/foot, and an appropriate distance was measured. If the distance falls beyond the shorting bar the coax can be cut at this point as shown by the diagonal line in Fig. 7-28. If it falls inside the shorting bar, a hole must be drilled into the foamflex breaking the inner conductor as shown by the circle-dot on Fig 7-28. The hole was subsequently filled with sealing compound to insulate and weatherproof.

A HIGH POWER LOW PASS FILTER

The construction of a low pass filter capable of handling maximum legal power levels is usually complicated by the following factors.

1. Most construction articles describe units for 250 watts or less.
2. Filters for higher power levels require special capacitors which are not readily available (not to mention cost).
3. The physical size of such capacitors increase the over-all size of the unit, if variable, require equipment for alignment and usually will not lend themselves to following the original layout.

The filter described herein requires no capacitors; double-sided copperclad board is used as the capacitive elements. If the dimensions are followed NO alignment is necessary, and the overall size is *small* 5×5×24.5 cm.

The materials needed for construction are, double clad copper board 1.5mm (1/16"), #10 solid copper wire and 2 SO-239 connectors.

The circuit for this filter is not new, but the use of copper clad board for the capacitors is. This filter is designed for use in 52 ohm lines, but any standard filter may be built by applying the capacitance value of the board per square inch and calculating the box size accordingly. The capacitance of 1.5mm (1/16") double clad board was measured at 14 pf per 6.5 sq. cm. Phenolic or epoxy measured essentially the same. And 2mm (3/32 in.) board measured 8 pf per 6.5 sq. cm.

A line drawing of the low pass filter is shown in Fig. 7-29. There are four shielded compartments. The inside walls of each section form one plate of the capacitor with the outside of the box forming the other plate.

The box ends, dividers and foil track are all at ground potential. Figure 7-30 shows the electrical circuit of the filter. The copperclad board parts are all soft soldered in place. Figure 7-31 is the dimensional drawing of the board which makes up two sides of the box. The .3cm (⅛") cm wide

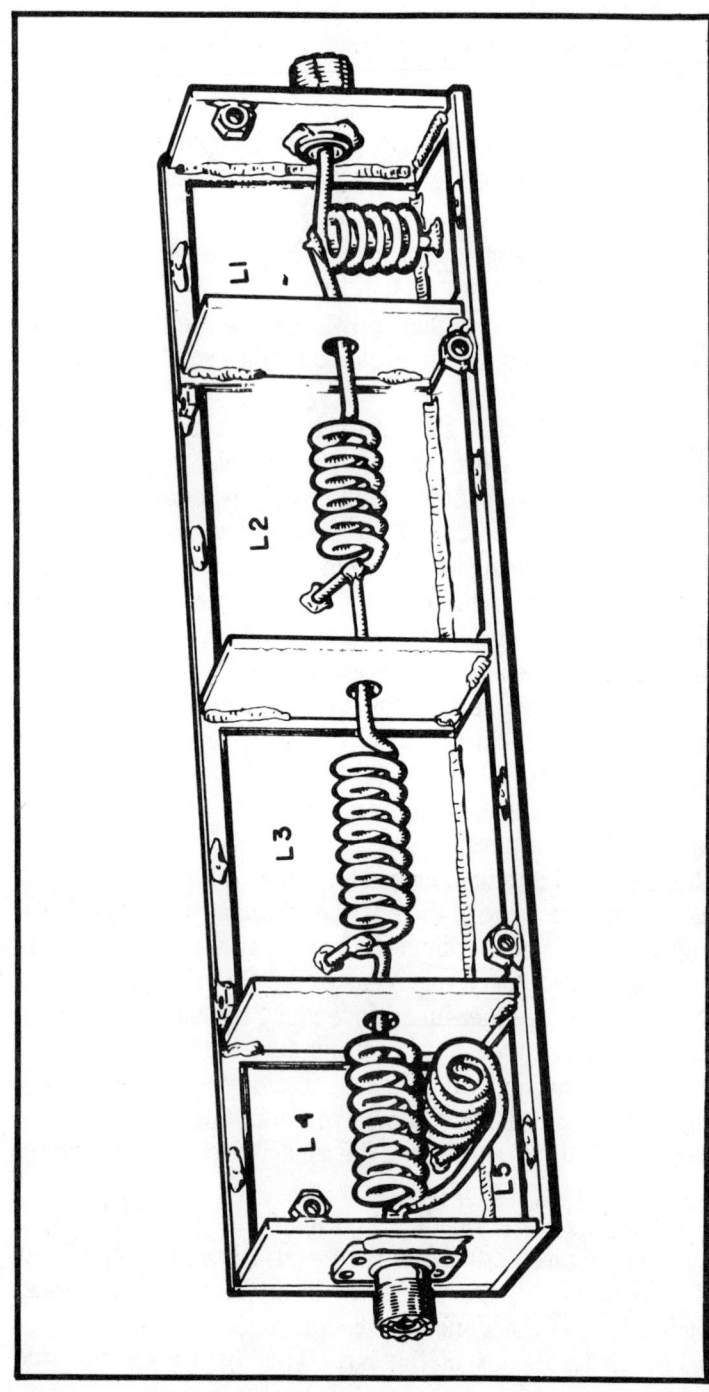

Fig. 7-29. Pictorial view of the low pass filter. It is built entirely from double sided copper clad stock, with etched out sections of the board serving as capacitors.

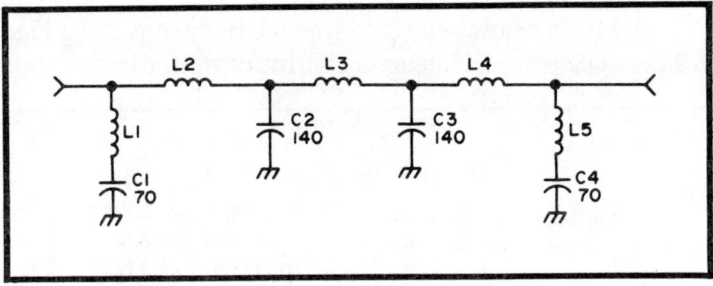

Fig. 7-30. Schematic of the filter. All coils are #10 copper 1.2 cm (½") inside diameter. L1 and L5 are 5T, 2cm (¾") long; L2 and L4 are 6T, 2.4cm (1") long; L3 is 8½T, 3.8cm (1½") long.

insulating tracks may be etched or cut using a hobby or carpet knife and the foil peeled away. The board is then cut in half, the cut ends filed to a 45 degree angle and the two halves soldered together being sure to solder both the inside and outside surfaces. The method of bonding the inner ground track and the outer surface together is via a number of holes drilled through the board with pins or wires passed through and soldered to each copper surface. The shields can also be made of copper clad with (.6cm) holes bored through their centers for coil connections.

The cover may simply be light weight aluminum bent at a 90 degree angle and holes drilled to line up with the mounting nuts soldered to the inner ground track. Wind the five coils

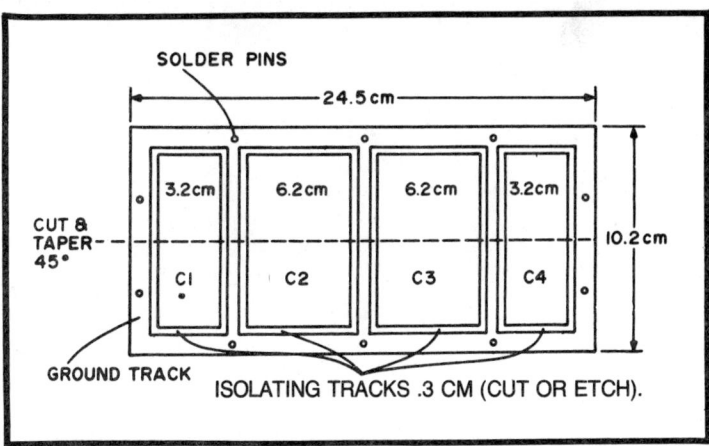

Fig. 7-31. The capacitors are formed by etching isolation tracks around sections of foil.

449

from the table below and solder them into the box using Fig. 7-29 as a guide to positioning. Install the cover and hook-er up.

COIL TABLE

	No. of Turns	Length
L1 & L5	5 turns	3/4 in. 1.9cm
L2 & L4	6 turns	15/16 in. 2.4cm
L3	8½ turns	1½ in. 3.8cm

All coils #10 solid copper wire 1.1cm (7/16) inside diameter.

The filter constructed worked fine when 1200 watts were run through it into a 50Ω dummy load. No increase in swr was noticeable. The frequency cut-off is at 300 MHz, with the attenuation falling sharply to 40 MHz.

Appendix

Converting Wavelength to Frequency.

HERTZ	KILOHERTZ	MEGAHERTZ
300,000,000 meters 984,250,000 feet 11,811,000,000 inches 3×10^{10} centimeters	300,000/meters 984,250/feet 11,811,000/inches 3×10^7 centimeters	300/meters 984/feet 11,811/inches 30,000/centimeters

Converting Frequency To Wavelength.

METERS	FEET	INCHES	CENTIMETERS
300,000,000/hertz 300,000/kilohertz 300/megahertz	984,250,000/hertz 984,250/kilohertz 984/megahertz	11,811,000,000/hertz 11,811,000/kilohertz 11,811/megahertz	3×10^{10} hertz 3×10^7 kilohertz 300,000/megahertz

One-Half Wavelength At Common Radio Frequencies.

FREQ (MHz)	$\lambda/2$ (ft)	FREQ (MHz)	$\lambda/2$ (ft)
0.5	984	20	24.5
1	492	50	9.84
2	245	100	4.92
5	98.4	200	2.45
10	49.2	300	1.54

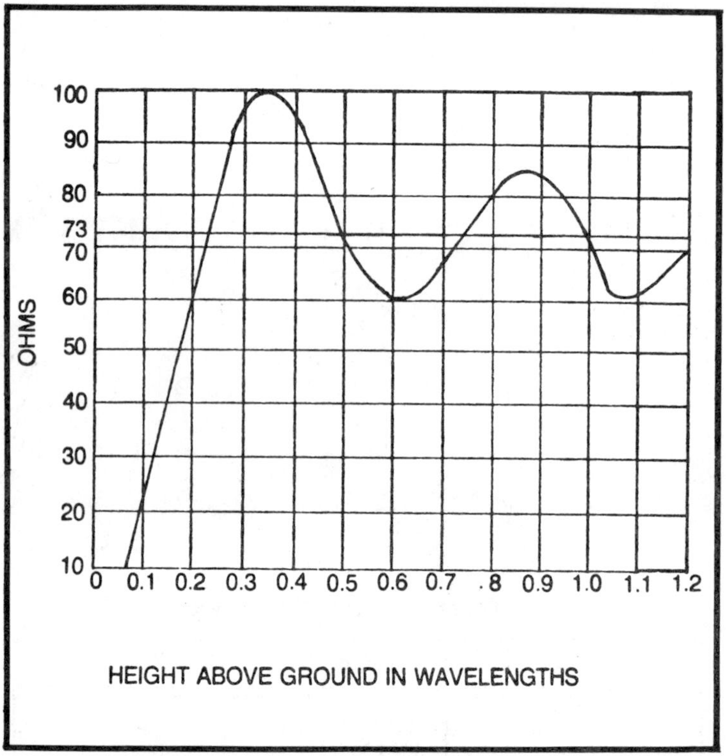

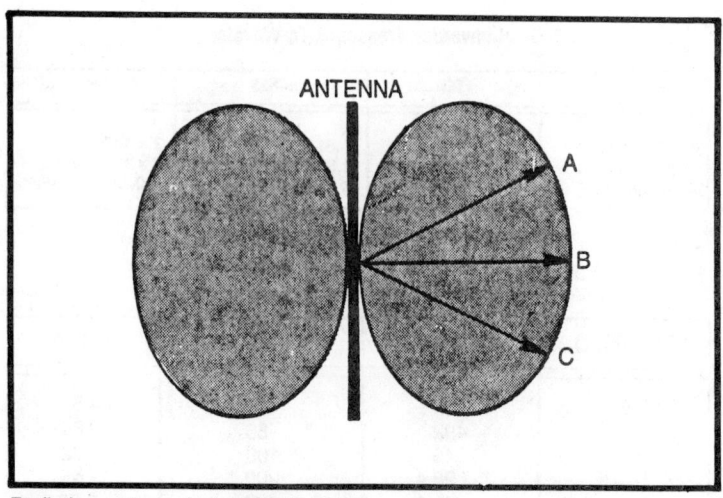

Radiation resistance at center of half-wave antenna.

Radiation pattern of ideal dipole antenna.

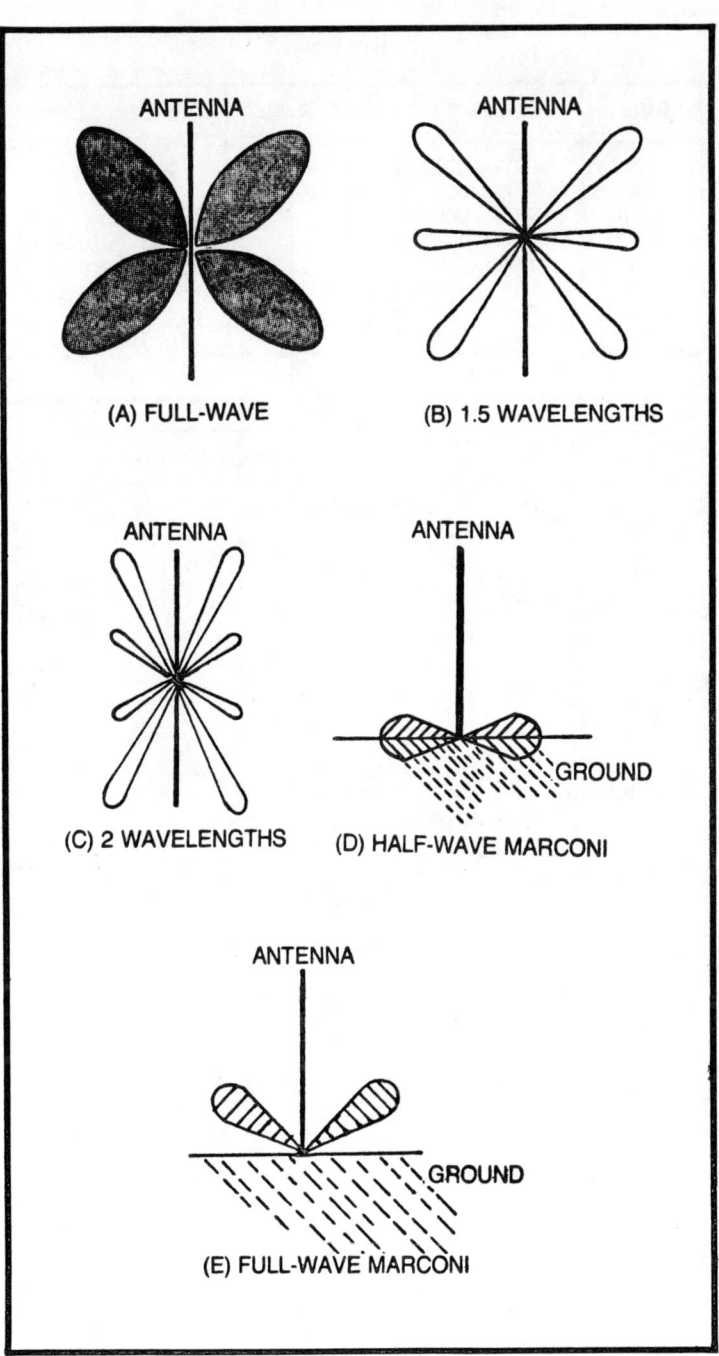

Additional radiation patterns.

Ham Bands.

BAND	FREQUENCY LIMITS	BAND	FREQUENCY LIMITS
160m	1800–2000 kHz	1.25m	220–225 MHz
80m	3500–4000 kHz	70 cm	420–450 MHz
40m	7000–7300 kHz	–	1215–1300 MHz
20m	14,000–14,350 kHz	–	2300–2450 MHz
15m	21.00–21.45 MHz	–	3300–3500 MHz
10m	28.0–29.7 MHz	–	5650–5925 MHz
6m	50.0–54.0 MHz	–	10,000–10,500 MHz
2m	144–148 MHz	–	24,000–24,050 MHz

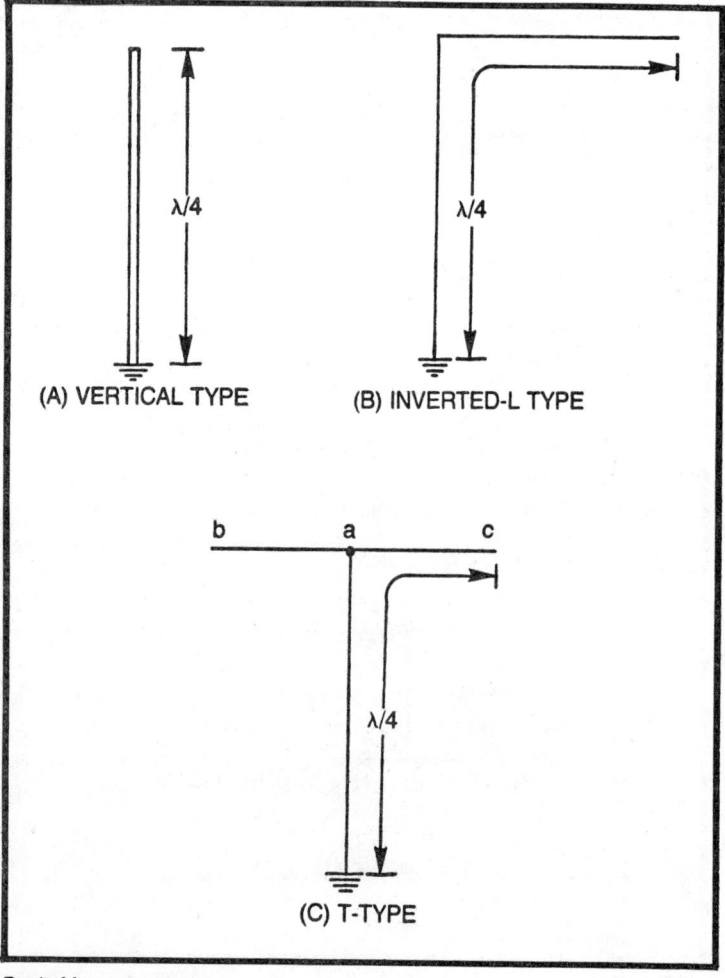

Basic Marconi antennas.

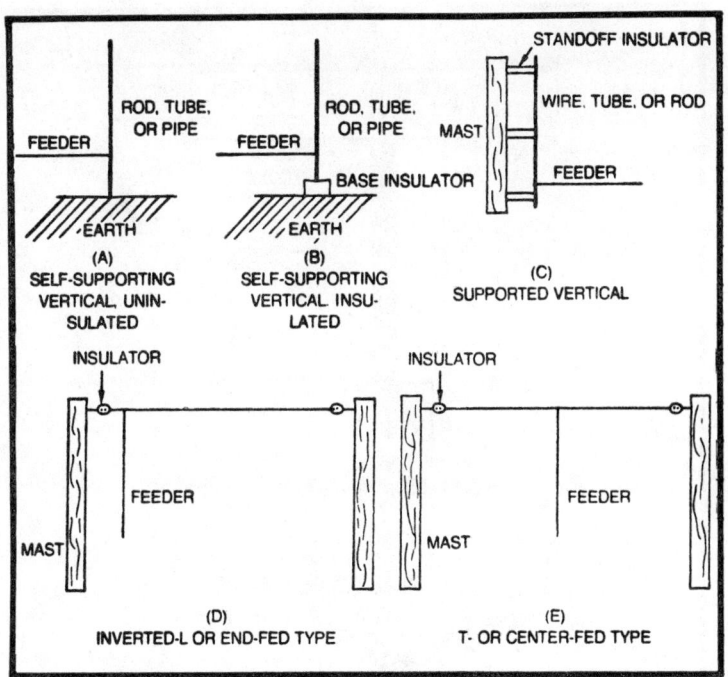

Marconi constructional details.

Marconi Antenna Dimensions.

METERS	MEGAHERTZ	λ/4 ANTENNA
160	1.8 1.9 2.0	130'0" 123'2" 117'0"
80	3.5 3.75 4.0	67'10" 62'5" 58'6"
40	7.0 7.15 7.3	33'6" 32'9" 32'0"
20	14.0 14.175 14.35	16'8" 16'6" 16'4"
15	21.0 21.225 21.45	11'2" 11'0" 10'11"
10	28.0 28.85 29.7	8' 4.25" 8'1.25" 7' 10.5"

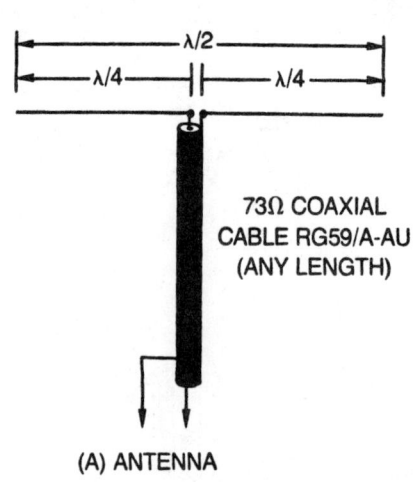

(A) ANTENNA

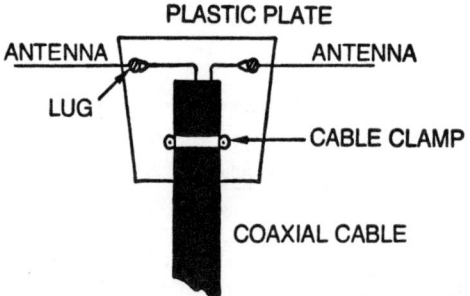

(B) FEEDER CONNECTION

METERS	MIDBAND LENGTH	
	$\lambda/2$	$\lambda/4$
160	246' 4"	123' 2"
80	124' 10"	62' 5"
40	65' 6"	32' 9"
20	33' 0"	16' 6"
15	22' 0"	11' 0"
10	16' 2.5"	8' 1.25"

(C) DIMENSIONS

Simple dipole.

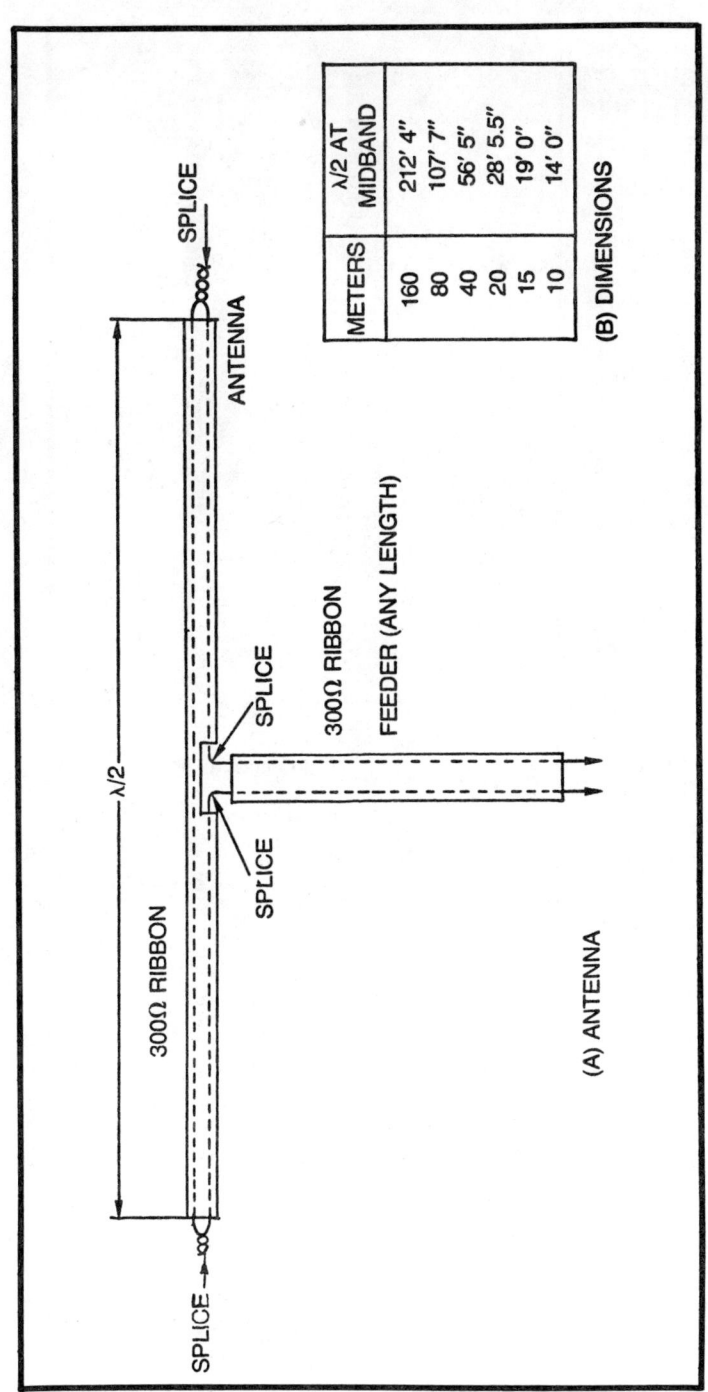

Folded dipole.

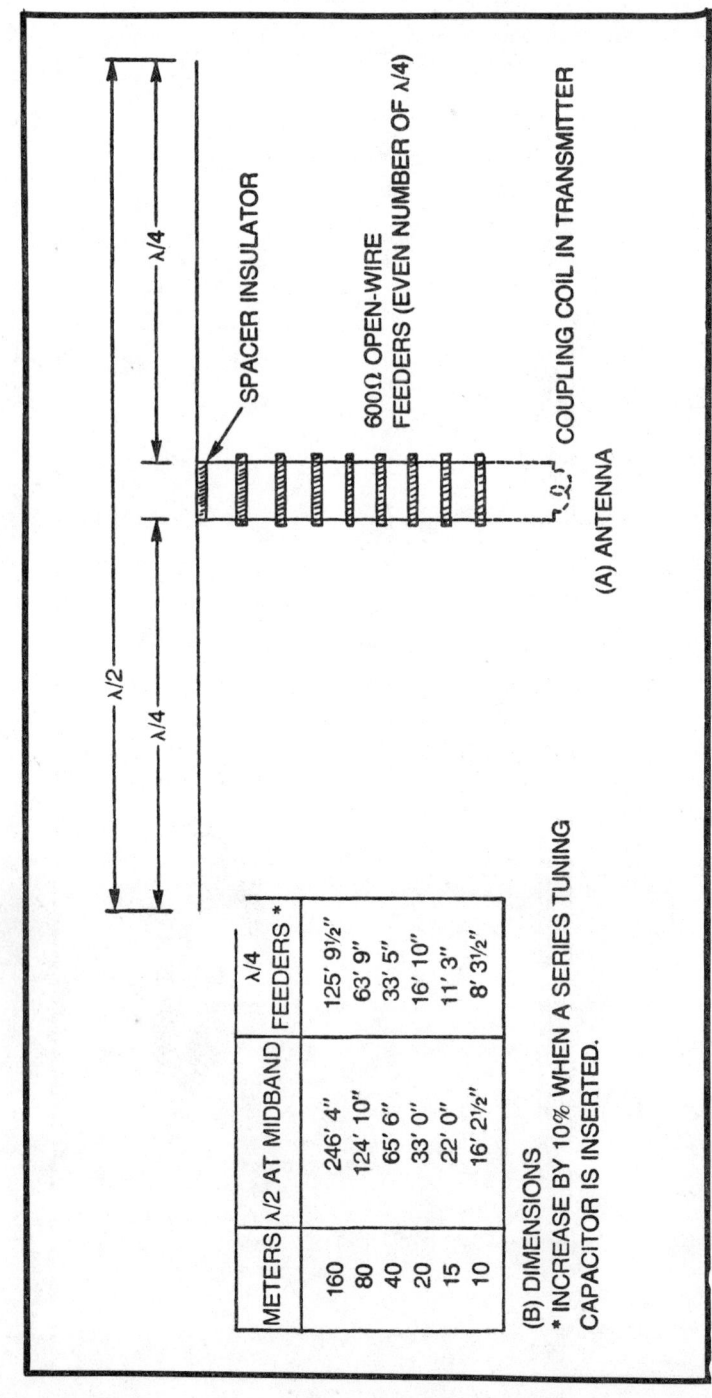

Center-fed Hertz with open-wire feeders.

Index

A

A/D ratio	103
All-band rotatable dipole antenna	156
Alpha element taper	274
A new quad antenna	331
Antenna	
equivalent circuit	14
gain	10, 16
height	51
parameters	281
reflector system	379
size	51
testing	139
tuner	216
Apartment dweller's antenna	202
Aperture	279

B

Back-to-back parabolic reflectors	389
Balanced dipole antenna	158
(without balun)	145
Baluns	162, 225, 374, 434
Base-tuned center-loaded antenna	205
BCI	416
Beginner's 10-meter beam antenna	254
Bird termaline	394
Bobtail antenna	181
Bridge	432
SWR	32, 228, 432
Broadband amplifier	240
Bruce array	62

C

Capacitor-FET antenna	240
Capacity hat	111
Capture area	20
City dweller's multiband antenna	212
Cloverleaf effect	371
Coaxial	
cable	334
dipole	333
Collinear-gain antenna	347
Copper sheeting	221
Corona loss	15
Cost versus performance	53
Coupler, antenna	416
Critical angle of radiation	28
Cross section	279
Cycle	11

D

dB	
gain	88
loss	88
de Broglie, Louis	30
Decibel (dB)	49, 86
Diamond array antenna	271
Dielectric constant	46
Differential, automobile	396
Diffraction	30

Dipole antenna	63, 223	plane antenna	92, 99, 362
Directive		resistance	94
array	17	Grounded antenna	171
gain	16		
Directivity	16		

H

DJ4BQ double-dipole antenna 149
Double inverted vee antenna 143
Double-Zepp antenna 164
Double log periodic antenna 56, 57
Dual band ¾-wavelength dipole 165
Dummy load, VHF 391

Half-wave length feedline 43
Haywire test bridge 140
Heat 15
Helix antenna 278
Horizontal dipole antenna 263
Horizontal linear dipole 212

E

I

Effective height 21
Eighty-meter phased array
  antenna 261
18″ all-band antenna 238
Einstein, Albert 25
Electrostatic field 10, 23
Electromagnetic speed 12
Element spacing 284
Elliptical reflectors 387
Ether 24

Impedance 15, 36
Indoor antenna 221
Inexpensive quad antenna 296
Intercept area 279
Interference
  adjacent channel 387
  Co-channel 387
Intrinsic impedance of space 13
Inverted vee long periodic
  antenna 69
Ion 28
Isotropic antenna 16

F

J

F/B ratio 258
Feedline 82
Feed point impedances 286
FET source follower 238
Field strength 13
Filter, low-pass 447
⅝-wave vertical antenna 100
Fixed beam antenna 65
Fixed-direction beam antenna 249
Flat reflectors 388
F-layer 102
Folded dipole 212
  antenna 153, 373
Fortran 48
Four-element
  parasitic loop antenna 278
  quad antenna 308
  10-meter beam antenna 257
Forty-meter inverted vee
  beam antenna 260
Full size 7 MHz quad antenna 322
450 MHz antenna 340

J antenna 356, 375

K

Krylon 378

L

L network 220
Lissajous display 265
Loading coil 18, 112
Lobes 387
Logarithmic element taper 274
Log periodic
  antenna (L-P) 55, 276
  VHF antennas 275
Long, circular quad antenna 313
Long John Yagi antenna 61
Long wire antenna 216, 292
Loop circumferences 236
Losses 15
Loss resistance 16

G

M

Gam antenna 362
Gamma match 244, 436, 444
G5RV antenna 148
Ground
  effect 102

Magnetic field 23
Match, gamma 436, 444

Matching	244, 249, 264
Matching stubs	439
Maxwell's equations	11
Maxwell, James Clerk	11
Megahertz	11
Michelson, Albert A.	24
Microwave	379
Miniature antenna	196
Minimum conductor diameters	236
Miniquad antenna	301
Mini vee-beam	215
Mobile antenna	174
Morely, Edward	25
Mounting beam elements	256
Multiband indoor antenna	222
Multiband log periodic antenna	274
Multiband vertical antenna	129
Multiple loop beam antenna	278

N

Nonradiative energy losses	233
Nulls	387

O

Open-circuit voltage	21
½-wavelength dipole antenna	191
160-meter antenna	171
Oscilloscope	265

P

Parabolic antenna	379
Passive reflectors	379
Pattern switch	248
Phase	24, 76
Phasing stub	177
Photon	25, 29
Planck, Max	25
Plane reflectors	389
PL239 Plug	334
Polarization	279, 368
Power	
density	12, 18
gain	16
Prop-pitch motor	395
PVC plastic pipe	341

Q

Quantum theory	25, 28
Quick band-change mobile antenna	183

R

Radiation	
efficiency	15
pattern	17, 265, 368
resistance	16, 229, 232
Random-length antenna	171
Reactance	36
transfer	206
Reciprocity	10
Rectangular reflectors	387
Repeater antenna	347
Resonance	14, 115
indicator	219
Rhombic	66
antenna	216
Roof mounted VHF whip antenna	354
Rotary antenna	65
Rotator	395
control	403

S

Saw tooth long periodic antenna	57
Servomechanism	412
Selectivity	368
Selsyn drive	401
Shape	17
Sinusoidal	11
Skewed shot calculations	383
Skip distance	28
Small loop antenna	227
S-meter readings	64
Spherical divergence	13
Spider-leg configuration	180
Square versus triangular loop	279
Super loop antenna	165
Switching manifold	264
SWR (VSWR)	43, 67

T

Take-apart beam antenna	251
TCI	66
Terminal	
reactance	105
resistance	104
Three-band, two-element quad antenna	297
Three-element quad antenna	305
3½ wavelength dipole antenna	194
¾-wavelength dipole antenna	164
Triangular loop beam antenna	278
Transfer function	21
Transformation	11
Transmission line	380

Trap antenna	212
Trapezoidal zig-zag log periodic antenna	57
Trapless multiband antenna	148
Traps	18
Traveling wave	11
Tri-band dipole antenna	223
Tuneable 432 MHz antenna	343
Tuner	216
antenna	429
transmission line	406
Tuning	
capacitances	236
mobile antenna	423
TVI	416
Twinlead	
dipole	135
phased array antenna	267, 269
Two-element	
beam	247
horizontal phased array antenna	266
Two-meter	
beam antenna	251, 372
halo antenna	337
mobile ship antenna	364
quarter-wave antenna	333
2-½ wavelength dipole antenna	193

U

Ultraviolet	28
Umbrella antenna	124

V

Variable impedance toroid balun	162
Velocity factor	35
Vertical	
antenna	90
dipole log periodic antenna	71
monopole long periodic antenna	63, 84
Velocity factor	334
Vertical antenna	263
VSWR bridge	43

W

Wavemeter	428
Whip	111
Widespaced beam antenna	241
Wind loads	280, 288, 290
Windom antenna	150, 178, 213, 345
Window antenna	179
Wire	77, 286
WWV	240
W8JK antenna	76
W4AEO	62

Y

Yagi antenna	18, 66
YV5DLT	62

Z

Zig-zag configuration	181
ZL special antenna	76